Primitive Culture

Researches into the Development of
Mythology, Philosophy, Religion,
Language, Art and Custom

in Two Volumes

—— Volume II ——

Edward Burnett Tylor

Dover Publications, Inc.
Mineola, New York

Bibliographical Note

This Dover edition, first published in 2016, is an unabridged republication of the second (1873) edition of the work originally published by John Murray, London, in 1871.

Library of Congress Cataloging-in-Publication Data

Names: Tylor, Edward B. (Edward Burnett), 1832-1917, author.
Title: Primitive culture / Edward Burnett Tylor.
Description: Mineola, New York : Dover Publications, 2016. | Series: Dover thrift editions
Identifiers: LCCN 2015046363| ISBN 9780486807508 (v. 1) | ISBN 0486807509 (v. 1) | ISBN 9780486807515 (v. 2) | ISBN 0486807517 (v. 2)
Subjects: LCSH: Ethnology. | Mythology. | Language and languages. | Animism. | BISAC: SOCIAL SCIENCE / Anthropology / Cultural. | SOCIAL SCIENCE / Anthropology / General. | SOCIAL SCIENCE / Anthropology / Physical. | SCIENCE / Philosophy & Social Aspects. | HISTORY / World.

Classification: LCC GN315 .T948 2016 | DDC 305.8—dc23 LC record available at https://lccn.loc.gov/2015046363

Manufactured in the United States by RR Donnelley
80751701   2016
www.doverpublications.com

CONTENTS

OF

THE SECOND VOLUME.

CHAPTER XII.

ANIMISM (continued).

Doctrine of Soul's Existence after Death ; its main divisions, Transmigration and Future Life—Transmigration of Souls: re-birth in Human and Animal Bodies, transference to Plants and Objects—Resurrection of Body scarcely held in savage religion—Future Life : a general though not universal doctrine of low races—Continued existence, rather than Immortality ; second death of Soul—Ghost of Dead remains on earth, especially if corpse unburied ; its attachment to bodily remains—Feasts of the Dead . 1

CHAPTER XIII.

ANIMISM (continued).

Journey of the Soul to the Land of the Dead—Visits by the Living to the Regions of Departed Souls—Connexion of such legends with myths of Sunset : the Land of the Dead thus imagined as in the West—Realization of current religious ideas, whether of savage or civilized theology, in narratives of visits to the Regions of Souls—Localization of the Future Life—Distant earthly region: Earthly Paradise, Isles of the Blest—Subterranean Hades or Sheol—Sun, Moon, Stars—Heaven—Historical course of belief as to such localization—Nature of Future Life—Continuance-theory, apparently original, belongs especially to the lower races — Transitional theories — Retribution-theory, apparently derived, belongs especially to the higher races—Doctrine of Moral Retribution as developed in the higher culture—Survey of Doctrine of Future State, from savage to civilized stages—Its practical effect on the sentiment and conduct of Mankind . . 44

CHAPTER XIV.

ANIMISM (continued).

Animism, expanding from the Doctrine of Souls to the wider Doctrine of Spirits, becomes a complete Philosophy of Natural Religion—Definition of Spirits similar to and apparently modelled on that of Souls—Transition-stage: classes of Souls passing into good and evil Demons—Manes-Worship—Doctrine of Embodiment of Spirits in human, animal, vegetable, and inert bodies—Demoniacal Possession and Obsession as causes of Disease and Oracle-inspiration—Fetishism—Disease-spirits embodied—Ghost attached to remains of Corpse—Fetish produced by a Spirit embodied in, attached to, or operating through, an Object—Analogues of Fetish-doctrine in Modern Science—Stock-and-Stone-Worship—Idolatry—Survival of Animistic Phraseology in modern Language—Decline of Animistic theory of Nature 108

CHAPTER XV.

ANIMISM (continued)

Spirits regarded as personal causes of Phenomena of the World—Pervading Spirits as good and evil Demons affecting man—Spirits manifest in Dreams and Visions: Nightmares; Incubi and Succubi; Vampires; Visionary Demons—Demons of darkness repelled by fire—Demons otherwise manifest: seen by animals; detected by footprints—Spirits conceived and treated as material—Guardian and Familiar Spirits—Nature-Spirits; historical course of the doctrine—Spirits of Volcanos, Whirlpools, Rocks—Water-Worship: Spirits of Wells, Streams, Lakes, etc.—Tree-Worship: Spirits embodied in or inhabiting Trees; Spirits of Groves and Forests—Animal-Worship: Animals worshipped, directly, or as incarnations or representatives of Deities; Totem-Worship; Serpent-Worship—Species-Deities; their relation to Archetypal Ideas 184

CHAPTER XVI.

ANIMISM (continued).

Higher Deities of Polytheism—Human characteristics applied to Deity—Lords of Spiritual Hierarchy—Polytheism: its course of development in lower and higher Culture—Principles of its investigation; classification of Deities according to central conceptions of their significance and function—Heaven-god—Rain-god—Thunder-god—Wind-god—Earth god—Water-god—Sea-god—Fire-god—Sun-god—Moon-god 247

CHAPTER XVII.

ANIMISM (*continued*).

Polytheism comprises a class of great Deities, ruling the course of Nature and the life of Man—Childbirth-god—Agriculture-god—War-god—God of the Dead—First Man as Divine Ancestor—Dualism; its rudimentary and unethical nature among low races; its development through the course of culture—Good and Evil Deity—Doctrine of Divine Supremacy, distinct from, while tending towards, the doctrine of Monotheism—Idea of Supreme Deity evolved in various forms among the lower races; its place as completion of the Polytheistic system and outcome of the Animistic philosophy; its continuance and development among higher nations—General survey of Animism as a Philosophy of Religion—Recapitulation of the theory advanced as to its development through successive stages of culture; its primary phases best represented among the lower races, while survivals of these among the higher races mark the transition from savage through barbaric to civilized faiths—Transition of Animism in the History of Religion; its earlier and later stages as a Philosophy of the Universe; its later stages as the principle of a Moral Institution 304

CHAPTER XVIII.

RITES AND CEREMONIES.

Religious Rites: their purpose practical or symbolic—Prayer: its continuity from low to high levels of Culture; its lower phases Unethical; its higher phases Ethical—Sacrifice: its original Gift-theory passes into the Homage-theory and the Abnegation-theory—Manner of reception of Sacrifice by Deity—Material Transfer to elements, fetish-animals, priests; consumption of substance by deity or idol; offering of blood; transmission by fire; incense—Essential Transfer: consumption of essence, savour, etc.—Spiritual Transfer: consumption or transmission of soul of offering—Motive of sacrificer—Transition from Gift-theory to Homage-theory: insignificant and formal offerings; sacrificial banquets—Abnegation-theory; sacrifice of children, etc.—Sacrifice of Substitutes; part given for whole; inferior life for superior; effigies—Modern survival of Sacrifice in folklore and religion—Fasting, as a means of producing ecstatic vision; its course from lower to higher Culture—Drugs used to produce ecstacy—Swoons and fits induced for religious purposes—Orientation: its relation to Sun-myth and Sun-worship; rules of East and West as to burial of dead, position of worship, and structure of temple—Lustration by Water and Fire : its transition from material to symbolic purification ; its connexion with special events of life ; its appearance among the lower races—Lustration of new-born children ; of women ; of those polluted by bloodshed or the dead—Lustration continued at higher levels of Culture—Conclusion . . 362

CHAPTER XIX.

CONCLUSION.

Practical results of the Study of Primitive Culture—Its bearing least upon Positive Science, greatest upon Intellectual, Moral, Social, and Political Philosophy—Language—Mythology—Ethics and Law—Religion—Action of the Science of Culture, as a means of furthering progress and removing hindrance, effective in the course of Civilization 448

PRIMITIVE CULTURE.

CHAPTER XII.

ANIMISM—*continued.*

Doctrine of Soul's Existence after Death; its main divisions, Transmigration and Future Life—Transmigration of Souls: re-birth in Human and Animal Bodies, transference to Plants and Objects—Resurrection of Body scarcely held in savage religion—Future Life: a general though not universal doctrine of low races—Continued existence, rather than Immortality; second death of Soul—Ghost of Dead remains on earth, especially if corpse unburied; its attachment to bodily remains—Feasts of the Dead.

HAVING thus traced upward from the lower levels of culture the opinions of mankind as to the souls, spirits, ghosts, or phantoms, considered to belong to men, to the lower animals, to plants, and to things, we are now prepared to investigate one of the great religious doctrines of mankind, the belief in the soul's continued existence in a Life after Death. Here let us once more call to mind the consideration which cannot be too strongly put forward, that the doctrine of a Future Life as held by the lower races is the all but necessary outcome of savage Animism. The evidence that the lower races believe the figures of the dead seen in dreams and visions to be their surviving souls, not only goes far to account for the comparative universality of their belief in the continued existence of the soul after the death of the body, but it gives the key to many of their speculations on the nature of this existence, speculations rational enough from the savage point of view, though apt to seem far-fetched absurdities to moderns in their much changed intellectual condition. The belief in a Future Life falls into two main divisions. Closely connected and even largely overlapping one another, both world-wide in their distribution, both ranging back in time to periods of unknown antiquity, both deeply rooted in the lowest strata of human life which lie open to our observation, these two doctrines have in the modern world passed into wonderfully different conditions. The one is the theory of the Transmigration of Souls, which has indeed risen from its lower stages to establish itself among the huge religious communities of Asia, great in history, enormous even in present mass, yet arrested and as it seems henceforth unprogressive in development; but the more highly educated world has rejected the ancient belief, and it now only survives in Europe in dwindling remnants. Far different has been the history of the other doctrine, that of the independent existence of the personal soul after the death of the body, in a Future Life. Passing onward through change after change in the condition of the human race, modified and renewed in its long ethnic course, this great belief may be traced from its crude and primitive manifestations among savage races to its establishment in the heart of modern religion, where the faith in a future existence forms at once an inducement to goodness, a sustaining hope through suffering and across the fear of death, and an answer to the perplexed problem of the allotment of happiness and misery in this present world, by the expectation of another world to set this right.

In investigating the doctrine of Transmigration, it will be well first to trace its position among the lower races, and afterwards to follow its developments, so far as they extend in the higher civilization. The temporary migration of souls into material substances, from human bodies down to morsels of wood and stone, is a most important part of the lower psychology. But it does not relate to the continued existence of the soul after death, and may be more conveniently treated of elsewhere, in connexion with such subjects as dæmoniacal possession and fetish-worship. We are here concerned with the more permanent tenancy of souls for successive lives in successive bodies.

Permanent transition, new birth, or re-incarnation of human souls in other human bodies, is especially considered to take place by the soul of a deceased person animating the body of an infant. North American Indians of the Algonquin districts, when little children died, would bury them by the wayside, that their souls might enter into mothers passing by, and so be born again.[1] In North-West America, among the Tacullis, we hear of direct transfusion of soul by the medicine-man, who, putting his hands on the breast of the dying or dead, then holds them over the head of a relative and blows through them; the next child born to this recipient of the departed soul is animated by it, and takes the rank and name of the deceased.[2] The Nutka Indians not without ingenuity accounted for the existence of a distant tribe speaking the same language as themselves, by declaring them to be the spirits of their dead.[3] In Greenland, where the wretched custom of abandoning and even plundering widows and orphans was tending to bring the whole race to extinction, a helpless widow would seek to persuade some father that the soul of a dead child of his had passed into a living child of hers, or *vice versâ*, thus gaining for herself a new relative and protector.[4] It is mostly ancestral or kindred souls that are thought to enter into children, and this kind of transmigration is therefore from the savage point of view a highly philosophical theory, accounting as it does so well for the general resemblance between parents and children, and even for the more special

[1] Brebeuf in 'Rel. des Jés. dans la Nouvelle France,' 1635, p. 130; Charlevoix, 'Nouvelle France,' vol. vi. p. 75. See Brinton, p. 253.
[2] Waitz, vol. iii. p. 195, see pp. 198, 213.
[3] Mayne, 'British Columbia,' p. 181.
[4] Cranz, 'Grönland,' pp. 248, 258, see p. 212. See also Turner, 'Polynesia,' p. 353; Meiners, vol. ii. p. 793.

phenomena of atavism. In North-West America, among the Koloshes, the mother sees in a dream the deceased relative whose transmitted soul will give his likeness to the child;[1] and in Vancouver's Island in 1860 a lad was much regarded by the Indians because he had a mark like the scar of a gun-shot wound on his hip, it being believed that a chief dead some four generations before, who had such a mark, had returned.[2] In Old Calabar, if a mother loses a child, and another is born soon after, she thinks the departed one to have come back.[3] The Wanika consider that the soul of a dead ancestor animates a child, and this is why it resembles its father or mother;[4] in Guinea a child bearing a strong resemblance, physical or mental, to a dead relative, is supposed to have inherited his soul;[5] and the Yorubas, greeting a new-born infant with the salutation, "Thou art come!" look for signs to show what ancestral soul has returned among them.[6] Among the Khonds of Orissa, births are celebrated by a feast on the seventh day, and the priest, divining by dropping rice-grains in a cup of water, and judging from observations made on the person of the infant, determines which of his progenitors has reappeared, and the child generally at least among the northern tribes receives the name of that ancestor.[7] In Europe the Lapps repeat an instructive animistic idea just noticed in America; the future mother was told in a dream what name to give her child, this message being usually given by the very spirit of the deceased ancestor, who was about to be incarnate in her.[8] Among the lower races generally the

[1] Bastian, 'Psychologie,' p. 28.
[2] Bastian, 'Zur vergl. Psychologie,' in Lazarus and Steinthal's 'Zeitschrift,' vol. v. p. 160, etc., also Papuas and other races.
[3] Burton, 'W. & W. fr. W. Afr.' p. 376.
[4] Krapf, 'E. Afr.' p. 201.
[5] J. L. Wilson, 'W. Afr.' p. 210; see also R. Clarke, 'Sierra Leone,' p. 159; Burton, Dahome, vol. ii. p. 158.
[6] Bastian, l. c.
[7] Macpherson, p. 72; also Tickell in 'Journ. As. Soc, Bengal,' vol. ix. pp. 793, etc.; Dalton in 'Tr. Eth. Soc.' vol. vi. p. 22 (similar rite of Mundas and Oraons).
[8] Klemm, 'Culturgeschichte,' vol. iii. p. 77.

renewal of old family names by giving them to new-born children may always be suspected of involving some such thought. The following is a curious pair of instances from the two halves of the globe. The New Zealand priest would repeat to the infant a long list of names of its ancestors, fixing upon that name which the child by sneezing or crying when it was uttered, was considered to select for itself; while the Cheremiss Tatar would shake the baby till it cried, and then repeat names to it, till it chose itself one by leaving off crying.[1]

The belief in the new human birth of the departed soul, which has even led West African negroes to commit suicide when in distant slavery, that they may revive in their own land, in fact amounts among several of the lower races to a distinct doctrine of an earthly resurrection. One of the most remarkable forms which this belief assumes is when dark-skinned races, wanting some reasonable theory to account for the appearance among them of human creatures of a new strange sort, the white men, and struck with their pallid deathly hue combined with powers that seem those of superhuman spiritual beings, have determined that the manes of their dead must have come back in this wondrous shape. The aborigines of Australia have expressed this theory in the simple formula, " Blackfellow tumble down, jump up Whitefellow." Thus a native who was hanged years ago at Melbourne expressed in his last moments the hopeful belief that he would jump up Whitefellow, and have lots of sixpences. The doctrine has been current among them since early days of European intercourse, and in accordance with it they habitually regarded the Englishmen as their own deceased kindred, come back to their country from an attachment to it in a former life. Real or imagined likeness completed the delusion, as when

[1] A. S. Thomson, 'New Zealand,' i. 118 ; see Shortland, 'Traditions,' p. 145; Turner, 'Polynesia,' p. 353 ; Bastian, 'Mensch,' vol. ii. p. 279 ; see also p. 276 (Samoieds). Compare Charlevoix, 'Nouvelle France,' vol. v. p. 426 ; Steller, 'Kamtschatka,' p. 353 ; Kracheninnikow, p. 117. See Plath, 'Rel der alten Chinesen,' ii. p. 98.

Sir George Grey was hugged and wept over by an old woman who found in him a son she had lost, or when a convict, recognized as a deceased relative, was endowed anew with the land he had possessed during his former life. A similar theory may be traced northward by the Torres Islands to New Caledonia, where the natives thought the white men to be the spirits of the dead who bring sickness, and assigned this as their reason for wishing to kill white men.[1] In Africa, again, the belief is found among the Western negroes that they will rise again white, and the Bari of the White Nile, believing in the resurrection of the dead on earth, considered the first white people they saw as departed spirits thus come back.[2]

Next, the lower psychology, drawing no definite line of demarcation between souls of men and of beasts, can at least admit without difficulty the transmission of human souls into the bodies of the lower animals. A series of examples from among the native tribes of America, will serve well to show the various ways in which such ideas are worked out. The Ahts of Vancouver's Island consider the living man's soul able to enter into other bodies of men and animals, going in and out like the inhabitant of a house. In old times, they say, men existed in the forms of birds, beasts, and fishes, or these had the spirits of the Indians in their bodies; some think that after death they will pass again into the bodies of the animals they occupied in this former state.[3] In another district of North-West

[1] Grey, 'Australia,' vol. i. p. 301, vol. ii. p. 363, [native's accusation against some foreign sailors who had assaulted him, "*djanga* Taal-wurt kyle-gut bomb-gur,"—"one of the dead struck Taal-wurt under the ear," etc. The word *djanga* = the dead, the spirits of deceased persons (see Grey, 'Vocab.' of S. W. Australia), had come to be the usual term for a European.] Lang, 'Queensland,' pp. 34, 336; Bonwick, 'Tasmanians,' p. 183; Scherzer, 'Voy. of Novara,' vol. iii. p. 34; Bastian, 'Psychologie,' p. 222, 'Mensch,' vol. iii. pp. 362—3, and in Lazarus and Steinthal's 'Zeitschrift,' l. c.; Turne., 'Polynesia,' p. 424.

[2] Römer, 'Guinea,' p. 85; Brun-Rollet, 'Nil Blanc,' etc. p. 234.

[3] Sproat, 'Savage Life,' ch. xviii., xix., xxi. Souls of the dead appear in dreams, either in human or animal forms, p. 174. See also Brinton, p. 145.

America, we find Indians believing the spirits of their dead to enter into bears, and travellers have heard of a tribe begging the life of a wrinkle-faced old she grizzly bear as the recipient of the soul of some particular grandam, whom they fancied the creature to resemble.[1] So, among the Esquimaux, a traveller noticed a widow who was living for conscience' sake upon birds, and would not touch walrus-meat, which the angekok had forbidden her for a time, because her late husband had entered into a walrus.[2] Among other North American tribes, we hear of the Powhatans refraining from doing harm to certain small wood-birds which received the souls of their chiefs;[3] of Huron souls turning into turtle-doves after the burial of their bones at the Feast of the Dead;[4] of that pathetic funeral rite of the Iroquois, the setting free a bird on the evening of burial, to carry away the soul.[5] In Mexico, the Tlascalans thought that after death the souls of nobles would animate beautiful singing birds, while plebeians passed into weasels and beetles and such like vile creatures.[6] So, in Brazil, the Içannas say that the souls of the brave will become beautiful birds feeding on pleasant fruits, but cowards will be turned into reptiles.[7] Among the Abipones we hear of certain little ducks which fly in flocks at night, uttering a mournful hiss, and which fancy associates with the souls of the dead;[8] while in Popayan it is said that doves were not killed, as inspired by departed souls.[9] Lastly, transmigration into brutes is also a received doctrine in South America, as when a missionary heard a Chiriquane woman of Buenos

---

[1] Schoolcraft, 'Indian Tribes,' part iii. p. 113.
[2] Hayes, 'Arctic Boat Journey,' p. 198.
[3] Brinton, 'Myths of New World,' p. 102.
[4] Brebeuf in 'Rel. des Jés.' 1636, p. 104.
[5] Morgan, 'Iroquois,' p. 174.
[6] Clavigero, 'Messico,' vol. ii. p. 5.
[7] Martius, 'Ethnog. Amer.' vol. i. p. 602; Markham in 'Tr. Eth. Soc.' vol. iii. p. 195.
[8] Dobrizhoffer, 'Abipones,' vol. ii. pp. 74, 270.
[9] Coreal in Brinton, l. c. See also J. G. Müller, p. 139 (Natchez), 223 (Caribs), 402 (Peru).

Ayres say of a fox, " May not that be the spirit of my dead daughter ?"[1]

In Africa, again, mention is made of the Maravi thinking that the souls of bad men became jackals, and good men snakes.[2] The Zulus, while admitting that a man may turn into a wasp or lizard, work out in the fullest way the idea of the dead becoming snakes, a creature whose change of skin has so often been associated with the thought of resurrection and immortality. It is especially certain green or brown harmless snakes, which come gently and fearlessly into houses, which are considered to be "amatongo" or ancestors, and therefore are treated respectfully, and have offerings of food given them. In two ways, the dead man who has become a snake can still be recognized; if the creature is one-eyed, or has a scar or some other mark, it is recognized as the "itongo" of a man who was thus marked in life; but if he had no mark, the "itongo" appears in human shape in dreams, thus revealing the personality of the snake.[3] In Guinea, monkeys found near a graveyard are supposed to be animated by the spirits of the dead, and in certain localities monkeys, crocodiles, and snakes, being thought men in metempsychosis, are held sacred.[4] It is to be borne in mind that notions of this kind may form in barbaric psychology but a portion of the wide doctrine of the soul's future existence. For a conspicuous instance of this, let us take the system of the Gold-Coast negroes. They believe that the "kla" or "kra," the vital soul, becomes at death a "sisa" or ghost, which can remain in the house with the body, plague the living, and cause sickness, till it departs or is driven by the sorcerer to the bank of the River Wolta, where the ghosts build themselves houses and dwell. But they can and do come back from

[1] Brinton, p. 254; see also Martius, vol. i. p. 446.
[2] Waitz, vol. ii. p. 419 (Maravi).
[3] Callaway, 'Rel. of Amazulu,' p. 196, etc.; Arbousset and Daumas, p. 237.
[4] J. L. Wilson, 'W. Afr.' pp. 210, 218. See also Brun-Rollet, pp. 200, 234; Meiners, vol. i. p. 211.

this Land of Souls. They can be born again as souls in new human bodies, and a soul who was poor before will now be rich. Many will not come back as men, but will become animals. To an African mother who has lost her child, it is a consolation to say, "He will come again."[1]

In higher levels of culture, the theory of re-embodiment of the soul appears in strong and varied development. Though seemingly not received by the early Aryans, the doctrine of migration was adopted and adapted by Hindu philosophy, and forms an integral part of that great system common to Brahmanism and Buddhism, wherein successive births or existences are believed to carry on the consequences of past and prepare the antecedents of future life. To the Hindu the body is but the temporary receptacle of the soul, which, "bound in the chains of deeds" and "eating the fruits of past actions," promotes or degrades itself along a series of embodiments in plant, beast, man, deity. Thus all creatures differ rather in degree than kind, all are akin to man, an elephant or ape or worm may once have been human, and may become human again, a pariah or barbarian is at once low-caste among men and high-caste among brutes. Through such bodies migrate the sinful souls which desire has drawn down from primal purity into gross material being; the world where they do penance for the guilt incurred in past existences is a huge reformatory, and life is the long grievous process of developing evil into good. The rules are set forth in the book of Manu how souls endowed with the quality of goodness acquire divine nature, while souls governed by passion take up the human state, and souls sunk in darkness are degraded to brutes. Thus the range of migration stretches downward from gods and saints, through holy ascetics, Brahmans, nymphs, kings, counsellors, to actors, drunkards, birds, dancers, cheats, elephants, horses, Sudras, barbarians, wild·beasts, snakes, worms, insects, and inert things. Obscure as the relation mostly is between the crime and its punishment in a new

[1] Steinhauser in 'Mag. der Evang. Miss.' Basel, 1856, No. 2, p. 135.

life, there may be discerned through the code of penal transmigration an attempt at appropriateness of penalty, and an intention to punish the sinner wherein he sinned. For faults committed in a previous existence men are afflicted with deformities, the stealer of food shall be dyspeptic, the scandal-monger shall have foul breath, the horse-stealer shall go lame, and in consequence of their deeds men shall be born idiots, blind, deaf and dumb, misshaped, and thus despised of good men. After expiation of their wickedness in the hells of torment, the murderer of a Brahman may pass into a wild beast or pariah; he who adulterously dishonours his guru or spiritual father shall be a hundred times re-born as grass, a bush, a creeper, a carrion bird, a beast of prey; the cruel shall become bloodthirsty beasts; stealers of grain and meat shall turn into rats and vultures; the thief who took dyed garments, kitchen-herbs, or perfumes, shall become accordingly a red partridge, a peacock, or a musk-rat. In short, "in whatever disposition of mind a man accomplishes such and such an act, he shall reap the fruit in a body endowed with such and such a quality."[1] The recognition of plants as possible receptacles of the transmigrating spirit well illustrates the conception of souls of plants. The idea is one known to lower races in the district of the world which has been more or less under Hindu influence. Thus we hear among the Dayaks of Borneo of the human soul entering the trunks of trees, where it may be seen damp and blood-like, but no longer personal and sentient;[2] and the Santals of Bengal are said to fancy that uncharitable men and childless women are eaten eternally by worms and snakes, while the good enter into fruit-bearing trees.[3] But it is an open question whether these and the Hindu ideas are originally independent of each other, and if not, did the Hindus adopt the

[1] Manu, xi. xii. Ward, 'Hindoos,' vol. i. p. 164, vol. ii. pp. 215, 347-52.
[2] St. John, 'Far East,' vol. i. p. 181.
[3] Hunter, 'Rural Bengal,' p. 210. See also Shaw in 'As. Res.' vol. iv. p. 46 (Rajmahal tribes).

ideas of the indigenes, or *vice versâ*? A curious commentary on the Hindu working out of the conception of plant-souls is to be found in a passage in a 17th century work, which describes certain Brahmans of the Coromandel Coast as eating fruits, but being careful not to pull the plants up by the roots, lest they should dislodge a soul; but few, it is remarked, are so scrupulous as this, and the consideration has occurred to them that souls in roots and herbs are in most vile and abject bodies, so that if dislodged they may become better off by entering into the bodies of men or beasts.[1] Moreover, the Brahmanic doctrine of souls transmigrating into inert things has in like manner a bearing on the savage theory of object-souls.[2]

Buddhism, like the Brahmanism from which it seceded, habitually recognized transmigration between superhuman and human beings and the lower animals, and in an exceptional way recognized a degradation even into a plant or a thing. How the Buddhist mind elaborated the doctrine of metempsychosis, may be seen in the endless legends of Gautama himself undergoing his 550 births, suffering pain and misery through countless ages to gain the power of freeing sentient beings from the misery inherent in all existence. Four times he became Maha Brahma, twenty times the dewa Sekra, and many times or few he passed through such stages as a hermit, a king, a rich man, a slave, a potter, a gambler, a curer of snake bites, an ape, an elephant, a bull, a serpent, a snipe, a fish, a frog, the dewa or genius of a tree. At last, when he became the supreme Buddha, his mind, like a vessel overflowing with honey, overflowed with the ambrosia of truth, and he proclaimed his triumph over life :—

[1] Abraham Roger, 'La Porte Ouverte,' Amst. 1670, p. 107.
[2] Manu, xii. 9 : "çarîrajaih karmmadoshaih yâti sthâvaratâm narah"—"for crimes done in the body, the man goes to the inert (motionless) state ;" xii. 42, "sthâvarâh krimakîtâçcha matsyâh sarpâh sakachhapâh paçavaçcha mrigaschaiva jaghanyâ tâmasî gatih"—"inert (motionless) things, worms and insects, fish, serpents, tortoises and beasts and deer also are the last dark form."

" Painful are repeated births.
O house-builder! I have seen thee,
Thou canst not build again a house for me.
Thy rafters are broken
Thy roof-timbers are shattered,
My mind is detached,
I have attained to the extinction of desire."

Whether the Buddhists receive the full Hindu doctrine of the migration of the individual soul from birth to birth, or whether they refine away into metaphysical subtleties the notion of continued personality, they do consistently and systematically hold that a man's life in former existences is the cause of his now being what he is, while at this moment he is accumulating merit or demerit whose result will determine his fate in future lives. Memory, it is true, fails generally to recal these past births, but memory, as we know, stops short of the beginning even of this present life. When King Bimsara's feet were burned and rubbed with salt by command of his cruel son that he might not walk, why was this torture inflicted on a man so holy? Because in a previous birth he had walked near a dagoba with his slippers on, and had trodden on a priest's carpet without washing his feet. A man may be prosperous for a time on account of the merit he has received in former births, but if he does not continue to keep the precepts, his next birth will be in one of the hells, he will then be born in this world as a beast, afterwards as a preta or sprite; a proud man may be born again ugly with large lips, or as a demon or a worm. The Buddhist theory of "karma" or "action," which controls the destiny of all sentient beings, not by judicial reward and punishment, but by the inflexible result of cause into effect, wherein the present is ever determined by the past in an unbroken line of causation, is indeed one of the world's most remarkable developments of ethical speculation.[1]

[1] Köppen, 'Religion des Buddha,' vol. i. pp. 35, 289, etc., 318 ; Barthélemy Saint-Hilaire, 'Le Bouddha et sa Religion,' p. 122 ; Hardy, 'Manual of Buddhism,' pp. 98, etc., 180, 318, 445, etc.

Within the classic world, the ancient Egyptians are described as maintaining a doctrine of migration, whether by successive embodiments in a "cycle of necessity" through creatures of earth, sea, and air, and back again to man, or by the simpler judicial penalty which sent back the wicked dead to earth as unclean beasts. The pictures and hieroglyphic sentences of the Book of the Dead are still preserved, and though the ambiguity of its formulas and the difficulty of distinguishing material from mystical meaning in its doctrine make it of little use as a check upon the classic accounts, yet it shows at least that notions of metamorphosis of the soul did hold a large place in the Egyptian religion.[1] In Greek philosophy, great teachers stood forth to proclaim it. Plato had mythic knowledge to convey of souls entering such new incarnations as their glimpse of real existence had made them fit for, from the body of a philosopher or a lover down to the body of a tyrant and usurper; of souls transmigrating into beasts and rising again to man according to the lives they led; of birds that were light-minded souls; of oysters suffering in banishment the penalty of utter ignorance. Pythagoras is made to illustrate in his own person his doctrine of metempsychosis, by recognizing where it hung in Here's temple the shield he had carried in a former birth, when he was that Euphorbos whom Menelaus slew at the siege of Troy. Afterwards he was Hermotimos, the Klazomenian prophet whose funeral rites were so prematurely celebrated while his soul was out, and after that, as Lucian tells the story, his prophetic soul passed into the body of a cock. Mikyllos asks this cock to tell him about Troy—were things there really as Homer said? But the cock replies, "How should Homer have known, O Mikyllos? When the Trojan war was going on, he was a camel in Baktria!"[2]

[1] Herod. ii. 123, see Rawlinson's Tr.; Plutarch. De Iside 31, 72; Wilkinson, 'Ancient Eg.' vol. ii. ch. xvi.; Bunsen, 'Egypt's Place in Univ. Hist.' vols. iv. and v.
[2] Plat. Phædo, Timæus, Phædrus, Repub. Pindar. Olymp. ii. antistr. 4;

In the later Jewish philosophy, the Kabbalists took up the doctrine of migration, the *gilgul* or "rolling on" of souls, and maintained it by that characteristic method of Biblical interpretation which it is good to hold up from time to time for a warning to the mystical interpreters of our own day. The soul of Adam passed into David, and shall pass into the Messiah, for are not these initials in the very name of Ad(a)m, and does not Ezekiel say that "my servant David shall be their prince for ever." Cain's soul passed into Jethro, and Abel's into Moses, and therefore it was that Jethro gave Moses his daughter to wife. Souls migrate into beasts and birds and vermin, for is not Jehovah "the lord of the spirits of all flesh?" and he who has done one sin beyond his good works shall pass into a brute. He who gives a Jew unclean meat to eat, his soul shall enter into a leaf, blown to and fro by the wind; "for ye shall be as an oak whose leaf fadeth;" and he who speaks ill words, his soul shall pass into a dumb stone, as did Nabal's, "and he became a stone."[1] Within the range of Christian influence, the Manichæans appear as the most remarkable exponents of the metempsychosis. We hear of their ideas of sinners' souls transmigrating into beasts, the viler according to their crimes; that he who kills a fowl or rat will become a fowl or rat himself; that souls can pass into plants rooted in the ground, which thus have not only life but sense; that the souls of reapers pass into beans and barley, to be cut down in their turn, and thus the elect were careful to explain to the bread when they ate it, that it was not they who reaped the corn it was made of; that the souls of the auditors, that is, the spiritually low commonalty who lived a married life, would pass into melons and cucumbers, to finish their purification by being eaten by the elect. But these details come to us from the accounts of bitter theological adversaries, and

---

Ovid. Metam. xv. 160; Lucian. Somn. 17, etc. Philostr. Vit. Apollon. Tyran. See also Meyer's Conversations-Lexicon, art. 'Seelenwanderung.' For rebirth in old Scandinavia, see Helgaqvidha, iii., in 'Edda.'

[1] Eisenmenger, part ii. p. 23, etc.

the question is, how much of them did the Manichæans really and soberly believe? Allowing for exaggeration and constructive imputation, there is reason to consider the account at least founded on fact. It seems clear that the Manichæan sect, when they fused together Zarathustrism, Buddhism, and Christianity, into a transcendental ascetic faith, adopted the Hindu theory of penance and purification of souls by migration into animals and plants, probably elaborating it meanwhile into fresh and fanciful details.[1] In later times, the doctrine of metempsychosis has been again and again noticed in a district of South-western Asia. William of Ruysbroek speaks of the notion of souls passing from body to body as general among the mediæval Nestorians, even a somewhat intelligent priest consulting him as to the souls of brutes, whether they could find refuge elsewhere so as not to be compelled to labour after death. Rabbi Benjamin of Tudela records in the 12th century of the Druses of Mount Hermon: "They say that the soul of a virtuous man is transferred to the body of a new-born child, whereas that of the vicious transmigrates into a dog, or some other animal." Such ideas indeed, seem not yet extinct in the modern Druse nation. Among the Nassairi, also, transmigration is believed in as a penance and purification: we hear of migration of unbelievers into camels, asses, dogs, or sheep, of disobedient Nassairi into Jews, Sunnis, or Christians, of the faithful into new bodies of their own people, a few such changes of "shirt" (*i.e.* body), bringing them to enter paradise or become stars.[2] An instance of the belief within the limits of modern Christian Europe may be found among the Bulgarians, whose superstition is that Turks who have never eaten pork in life will become wild boars after death. A party assembled to feast on a boar has been known to throw

---

[1] Beausobre, 'Hist. de Manichée,' etc., vol. i. pp. 245—6, vol. ii. pp. 496—9. See Augustin. Contra Faust.; De Hæres.; De Quantitate Animæ.

[2] Gul. de Rubruquis in 'Rec. des Voy. Soc. de Géographie de Paris,' vol. iv. p. 356. Benjamin of Tudela, ed. and tr. by Asher, Hebrew 22, Eng. p. 62. Niebuhr, 'Reisebeschr. nach Arabien,' etc. vol. ii. pp. 438—443; Meiners, vol. ii. p. 796.

it all away, for the meat jumped off the spit into the fire, and a piece of cotton was found in the ears, which the wise man decided to be a piece of the ci-devant Turk's turban.[1] Such cases, however, are exceptional. Metempsychosis never became one of the great doctrines of Christendom, though not unknown in mediæval scholasticism, and though maintained by an eccentric theologian here and there into our own times. It would be strange were it not so. It is in the very nature of the development of religion that speculations of the earlier culture should dwindle to survivals, yet be again and again revived. Doctrines transmigrate, if souls do not; and metempsychosis, wandering along the course of ages, came at last to animate the souls of Fourier and Soame Jenyns.[2]

Thus we have traced the ancient theory of metempsychosis in stage after stage of the world's civilization, scattered among the native races of America and Africa, established in old Egypt, elaborated by the Hindu mind into its great system of ethical philosophy, reviving and failing through classic and mediæval Europe, and lingering at last in the modern world as an intellectual crotchet, of little account but to the ethnographer who notes it down as an item of

[1] St. Clair and Brophy, 'Bulgaria,' p. 57. Compare the tenets of the Russian sect of Dukhobortzi, in Haxthausen, 'Russian Empire,' vol. i. p. 288, etc.

[2] Since the first publication of the above remark, M. Louis Figuier has supplied a perfect modern instance by his book, entitled 'Le Lendemain de la Mort,' translated into English as 'The Day after Death: Our Future Life according to Science.' His attempt to revive the ancient belief, and to connect it with the evolution-theory of modern naturalists, is carried out with more than Buddhist elaborateness. Body is the habitat of soul, which goes out when a man dies, as one forsakes a burning house. In the course of development, a soul may migrate through bodies stage after stage, zoophyte and oyster, grasshopper and eagle, crocodile and dog, till it arrives at man, thence ascending to become one of the superhuman beings or angels who dwell in the planetary ether, and thence to a still higher state, the secret of whose nature M. Figuier does not endeavour to penetrate, "because our means of investigation fail at this point." The ultimate destiny of the more glorified being is the Sun; the pure spirits who form its mass of burning gases, pour out germs and life to start the course of planetary existence. (Note to 2nd edition.)

evidence for his continuity of culture. What, we may well ask, was the original cause and motive of the doctrine of transmigration? Something may be said in answer, though not at all enough for full explanation. The theory that ancestral souls return, thus imparting their own likeness of mind and body to their descendants and kindred, has been already mentioned and commended as in itself a very reasonable and philosophical hypothesis, accounting for the phenomenon of family likeness going on from generation to generation. But why should it have been imagined that men's souls could inhabit the bodies of beasts and birds? As has been already pointed out, savages not unreasonably consider the lower animals to have souls like their own, and this state of mind makes the idea of a man's soul transmigrating into a beast's body at least seem possible. But it does not actually suggest the idea. The view stated in a previous chapter as to the origin of the conception of soul in general, may perhaps help us here. As it seems that the first conception of souls may have been that of the souls of men, this being afterwards extended by analogy to the souls of animals, plants, etc., so it may seem that the original idea of transmigration was the straightforward and reasonable one of human souls being re-born in new human bodies, where they are recognized by family likenesses in successive generations. This notion may have been afterwards extended to take in re-birth in bodies of animals, etc. There are some well-marked savage ideas which will fit with such a course of thought. The half-human features and actions and characters of animals are watched with wondering sympathy by the savage, as by the child. The beast is the very incarnation of familiar qualities of man; and such names as lion, bear, fox, owl, parrot, viper, worm, when we apply them as epithets to men, condense into a word some leading feature of a human life. Consistently with this, we see in looking over details of savage transmigration that the creatures often have an evident fitness to the character of the human beings whose souls are to pass into them, so that the savage philosopher's fancy of transferred souls offered something like an explanation of the likeness between beast and man. This comes more clearly into view among the more civilized races who have worked out the idea of transmigration into ethical schemes of retribution, where the appropriateness of the creatures chosen is almost as manifest to the modern critic as it could have been to the ancient believer. Perhaps the most graphic restoration of the state of mind in which the theological doctrine of metempsychosis was worked out in long-past ages, may be found in the writings of a modern theologian whose spiritualism often follows to the extreme the intellectual tracks of the lower races. In the spiritual world, says Emanuel Swedenborg, such persons as have opened themselves for the admission of the devil and acquired the nature of beasts, becoming foxes in cunning, etc., appear also at a distance in the proper shape of such beasts as they represent in disposition.[1] Lastly, one of the most notable points about the theory of transmigration is its close bearing upon a thought which lies very deep in the history of philosophy, the development-theory of organic life in successive stages. An elevation from the vegetable to the lower animal life, and thence onward through the higher animals to man, to say nothing of superhuman beings, does not here require even a succession of distinct individuals, but is brought by the theory of metempsychosis within the compass of the successive vegetable and animal lives of a single being.

Here a few words may be said on a subject which cannot be left out of sight, connecting as it does the two great branches of the doctrine of future existence, but which it is difficult to handle in definite terms, and much more to trace historically by comparing the views of lower and

---

[1] Swedenborg, 'The True Christian Religion,' 13. Compare the notion attributed to the followers of Basilides the Gnostic, of men whose souls are affected by spirits or dispositions as of wolf, ape, lion, or bear, wherefore their souls bear the properties of these, and imitate their deeds (Clem. Alex. Stromat. ii. c. 20).

higher races. This is the doctrine of a bodily renewal or resurrection. To the philosophy of the lower races it is by no means necessary that the surviving soul should be provided with a new body, for it seems itself to be of a filmy or vaporous corporeal nature, capable of carrying on an independent existence like other corporeal creatures. Savage descriptions of the next world are often such absolute copies of this, that it is scarcely possible to say whether the dead are or are not thought of as having bodies like the living; and a few pieces of evidence of this class are hardly enough to prove the lower races to hold original and distinct doctrines of corporeal resurrection.[1] Again, attention must be given to the practice, so common among low and high races, of preserving relics of the dead, from mere morsels of bone up to whole mummified bodies. It is well known that the departed soul is often thought apt to revisit the remains of the body. But how far the preservation of these remains may be connected with an idea of bodily resurrection, whether among the native races of America, or in ancient Egypt, or elsewhere, is a problem for which also the evidence available does not seem sufficient.[2] In discussing the closely allied doctrine of metempsychosis, I have described the theory of the soul's transmigration into a new human body as asserting in fact an earthly resurrection. From the same point of view, a bodily resurrection in Heaven or Hades is technically a transmigration of the soul. This is plain among the higher races, in whose religion these doctrines take at once clearer definition and more practical import. There are some distinct mentions of bodily resurrection in the Rig Veda: the dead is spoken of as glorified, putting on his body (tanu); and it is even promised that the pious man shall be born in

[1] See J. G. Müller, 'Amer. Urrel.' p. 208 (Caribs); but compare Rochefort, p. 429. Steller, 'Kamtschatka,' p. 269; Castrén, 'Finnische Mythologie,' p. 119.
[2] See for American evidence Brinton, 'Myths of New World,' p. 254, etc. For Egyptian evidence Birch's tr. of 'Book of Dead' in Bunsen's 'Egypt.' vol. vi.; Wilkinson, etc.

the next world with his entire body (sarvatanû). In Brahminism and Buddhism, the re-births of souls in bodies to inhabit heavens and hells are simply included as particular cases of transmigration. The question of an old Persian doctrine of resurrection, thought by some to be related to the late Jewish doctrine, is obscure.[1] In early Christianity, the conception of bodily resurrection is developed with especial strength and fulness in the Pauline doctrine. For an explicit interpretation of this doctrine, such as commended itself to the minds of later theologians, it is instructive to cite the remarkable passage of Origen, where he speaks of "corporeal matter, of which matter, in whatever quality placed, the soul always has use, now indeed carnal, but afterwards indeed subtler and purer, which is called spiritual."[2]

Passing from these metaphysical doctrines of civilized theology, we now take up a series of beliefs higher in practical moment, and more clearly conceived in savage thought. There may well have been, and there may still be, low races destitute of any belief in a Future State. Nevertheless, prudent ethnographers must often doubt accounts of such, for this reason, that the savage who declares that the dead live no more, may merely mean to say that they are dead. When the East African is asked what becomes of his buried ancestors, the "old people," he can reply that "they are ended," yet at the same time he fully admits that their ghosts survive.[3] In an account of the religious ideas of the Zulus, taken down from a native, it is explicitly stated that Unkulunkulu the Old-Old-One said that people "were to die and never rise again," and that he allowed them "to die and rise no more."[4] Knowing so thoroughly as we now do the

[1] Aryan evidence in 'Rig-Veda,' x. 14, 8; xi. 1, 8; Manu, xii. 16—22; Max Müller, 'Todtenbestattung,' pp. xii. xiv.; 'Chips,' vol. i. p. 47; Muir in 'Journ. As. Soc. Bengal,' vol. i. 1865, p. 306; Ward, 'Hindoos,' vol. ii. pp. 332, 347, 357; Haug, 'Parsees,' p. 266; see Alger, 'Future Life.'

[2] Origen, De. Princip. ii. 3, 2: "materiæ corporalis, cujus materiæ anima usum semper habet, in qualibet qualitate positæ, nunc quidem carnali, postmodum vero subtiliori et puriori, quæ spiritalis appellatur."

[3] Burton, 'Central Africa,' vol. ii. p. 345.

[4] Callaway, 'Rel. of Amazulu,' p. 84.

theology of the Zulus, whose ghosts not only survive in the under-world, but are the very deities of the living, we can put the proper sense to these expressions. But without such information, we might have mistaken them for denials of the soul's existence after death. This objection may even apply to one of the most formal denials of a future life ever placed on record among an uncultured race, a poem of the Dinka tribe of the White Nile, concerning Dendid the Creator:

" On the day when Dendid made all things,
   He made the sun;
And the sun comes forth, goes down, and comes again.
   He made the moon;
And the moon comes forth, goes down, and comes again:
   He made the stars;
And the stars come forth, go down, and come again:
   He made man;
And man comes forth, goes down into the ground, and comes no more."

It is to be remarked, however, that the close neighbours of these Dinka, the Bari, believe that the dead do return to live again on earth, and the question arises whether it is the doctrine of bodily resurrection, or the doctrine of the surviving ghost-soul, that the Dinka poem denies. The missionary Kaufmann says that the Dinka do not believe the immortality of the soul, that they think it but a breath, and with death all is over; Brun-Rollet's contrary authority goes to prove that they do believe in another life; both leave it an open question whether they recognize the existence of surviving ghosts.[1] The case is, like various others of the same kind, incomplete.

Looking at the religion of the lower races as a whole, we shall at least not be ill-advised in taking as one of its general and principal elements the doctrine of the soul's Future Life. But here it is needful to explain, to limit, and to reserve, lest modern theological ideas should lead us to misconstrue more primitive beliefs. In such enquiries the

[1] Kaufmann, 'Schilderungen aus Centralafrika,' p. 124; G. Lejean in 'Rev. des Deux Mondes,' Apr. 1, 1860, p. 760; see Brun-Rollet, 'Nil Blanc,' pp. 100, 234.

phrase "immortality of the soul" is to be avoided as misleading. It is doubtful how far the lower psychology entertains at all an absolute conception of immortality, for past and future fade soon into utter vagueness as the savage mind quits the present to explore them, the measure of months and years breaks down even within the narrow span of human life, and the survivor's thought of the soul of the departed dwindles and disappears with the personal memory that kept it alive. Even among races who distinctly accept the doctrine of the surviving soul, this acceptance is not unanimous. In savage as in civilized life, dull and careless natures ignore a world to come as too far off, while sceptical intellects are apt to reject its belief as wanting proof, or perhaps at most without closer scrutiny to prize its hope as a good influence in human life. Far from a life after death being held by all men as the destiny of all men, whole classes are formally excluded from it. In the Tonga islands, the future life was a privilege of caste, for while the chiefs and higher orders were to pass in divine ethereality to the happy land of Bolotu, the lower ranks were believed to be endowed only with souls that died with their bodies; and although some of these had the vanity to claim a place in paradise among their betters, the populace in general acquiesced in the extinction of their own plebeian spirits.[1] The Nicaraguans believed that if a man lived well, his soul would ascend to dwell among the gods, but if ill, it would perish with the body, and there would be an end of it.[2] Granted that the soul survives the death of the body, instance after instance from the records of the lower culture shows this soul to be regarded as a mortal being, liable like the body itself to accident and death. The Greenlanders pitied the poor souls who must pass in winter or in storm the dreadful mountain where the dead descend to reach the other world, for then a

[1] Mariner, 'Tonga Is.' vol. ii. p. 136.
[2] Oviedo, 'Nicaragua,' p. 50. For similar statements, see Martins, 'Ethnog. Amer.' vol. i. p. 247; Smith's 'Virginia' in Pinkerton, vol. xiii. p. 41; Meiners, vol. ii. p. 760.

soul is like to come to harm, and die the other death where there is nothing left, and this is to them the dolefullest thing of all.[1] Thus the Fijians tell of the fight which the ghost of a departed warrior must wage with the soul-killing Samu and his brethren; this is the contest for which the dead man is armed by burying the war-club with his corpse, and if he conquers, the way is open for him to the judgment-seat of Ndengei, but if he is wounded, his doom is to wander among the mountains, and if killed in the encounter he is cooked and eaten by Samu and his brethren. But the souls of unmarried Fijians will not even survive to stand this wager of battle; such try in vain to steal at low water round to the edge of the reef past the rocks where Nangananga, destroyer of wifeless souls, sits laughing at their hopeless efforts, and asking them if they think the tide will never flow again, till at last the rising flood drives the shivering ghosts to the beach, and Nangananga dashes them in pieces on the great black stone, as one shatters rotten firewood.[2] Such, again, were the tales told by the Guinea negroes of the life or death of departed souls. Either the great priest before whom they must appear after death would judge them, sending the good in peace to a happy place, but killing the wicked a second time with the club that stands ready before his dwelling; or else the departed shall be judged by their god at the river of death, to be gently wafted by him to a pleasant land if they have kept feasts and oaths and abstained from forbidden meats, but if not, to be plunged into the river by the god, and thus drowned and buried in eternal oblivion.[3] Even common water can drown a negro ghost, if we may believe the story of the Matamba widows having themselves ducked in the river or pond to drown off the souls of their

---

[1] Cranz, 'Grönland,' p. 259.

[2] Williams, 'Fiji,' vol. i. p. 244. See 'Journ. Ind. Archip.' vol. iii. p. 113 (Dayaks). Compare wasting and death of souls in depths of Hades, Taylor, 'New Zealand,' p. 232.

[3] Bosman, 'Guinea' in Pinkerton, vol. xvi. p. 401. See also Waitz, 'Anthropologie,' vol. ii. p. 191 (W. Afr.); Callaway, 'Rel. of Amazulu,' p. 355.

departed husbands, who might still be hanging about them, clinging closest to the best loved wives. After this ceremony, they went and married again.[1] From such details it appears that the conception of some souls suffering extinction at death or dying a second death, a thought still as heretofore familiar to speculative theology, is not unknown in the lower culture.

The soul, as recognized in the philosophy of the lower races, may be defined as an ethereal surviving being, conceptions of which preceded and led up to the more transcendental theory of the immaterial and immortal soul, which forms part of the theology of higher nations. It is principally the ethereal surviving soul of early culture that has now to be studied in the religions of savages and barbarians and the folklore of the civilized world. That this soul should be looked on as surviving beyond death is a matter scarcely needing elaborate argument. Plain experience is there to teach it to every savage; his friend or his enemy is dead, yet still in dream or open vision he sees the spectral form which is to his philosophy a real objective being, carrying personality as it carries likeness. This thought of the soul's continued existence is, however, but the gateway into a complex region of belief. The doctrines which, separate or compounded, make up the scheme of future existence among particular tribes, are principally these: the theories of lingering, wandering, and returning ghosts, and of souls dwelling on or below or above the earth in a spirit-world, where existence is modelled upon the earthly life, or raised to higher glory, or placed under reversed conditions, and lastly, the belief in a division between happiness and misery of departed souls, by a retribution for deeds done in life, determined in a judgment after death.

"All argument is against it; but all belief is for it," said Dr. Johnson of the apparition of departed spirits. The doctrine that ghost-souls of the dead hover among the

---

[1] Cavazzi, 'Congo, Matamba, et Angola,' lib. i. 270. See also Liebrecht in 'Zeitschr. für Ethnologie,' vol. v., p. 96, (Tartary, Scandinavia, Greece).

living is indeed rooted in the lowest levels of savage culture, extends through barbaric life almost without a break, and survives largely and deeply in the midst of civilization. From the myriad details of travellers, missionaries, historians, theologians, spiritualists, it may be laid down as an admitted opinion, as wide in distribution as it is natural in thought, that the two chief haunting-grounds of the departed soul are the scenes of its fleshly life and the burial place of its body. As in North America the Chickasaws believed that the spirits of the dead in their bodily shape moved about among the living in great joy; as the Aleutian islanders fancied the souls of the departed walking unseen among their kindred, and accompanying them in their journeys by sea and land; as Africans think that souls of the dead dwell in their midst, and eat with them at meal times; as Chinese pay their respects to kindred spirits present in the hall of ancestors;[1] so multitudes in Europe and America live in an atmosphere that swarms with ghostly shapes—spirits of the dead, who sit over against the mystic by his midnight fire, rap and write in spirit-circles, and peep over girls' shoulders as they scare themselves into hysterics with ghost-stories. Almost throughout the vast range of animistic religion, we shall find the souls of the departed hospitably entertained by the survivors on set occasions, and manes-worship, so deep and strong among the faiths of the world, recognises with a reverence not without fear and trembling those ancestral spirits which, powerful for good or ill, manifest their presence among mankind. Nevertheless death and life dwell but ill together, and from savagery onward there is recorded many a device by which the survivors have sought to rid themselves of household ghosts. Though the unhappy savage custom of deserting houses after a decease may mostly be connected with other causes, such as horror or abnegation of all things belonging to the dead, there are cases where it

[1] Schoolcraft, 'Indian Tribes,' part i. p. 310; Bastian, 'Psychologie,' pp. 111, 193; Doolittle, 'Chinese,' vol. i. p. 235.

appears that the place is simply abandoned to the ghost. In Old Calabar it was customary for the son to leave his father's house to decay, but after two years he might rebuild it, the ghost being thought by that time to have departed;[1] the Hottentots abandoned the dead man's house, and were said to avoid entering it lest the ghost should be within;[2] the Yakuts let the hut fall in ruins where any one had expired, thinking it the habitation of demons;[3] the Karens were said to destroy their villages to escape the dangerous neighbourhood of departed souls.[4] Such proceedings, however, scarcely extend beyond the limits of savagery, and only a feeble survival of the old thought lingers on into civilization, where from time to time a haunted house is left to fall in ruins, abandoned to a ghostly tenant who cannot keep it in repair. But even in the lowest culture we find flesh holding its own against spirit, and at higher stages the householder rids himself with little scruple of an unwelcome inmate. The Greenlanders would carry the dead out by the window, not by the door, while an old woman, waving a firebrand behind, cried "piklerrukpok!" *i. e.*, "there is nothing more to be had here!"[5]; the Hottentots removed the dead from the hut by an opening broken out on purpose, to prevent him from finding the way back;[6] the Siamese, with the same intention, break an opening through the house wall to carry the coffin through, and then hurry it at full speed thrice round the house;[7] the Siberian Chuwashes fling a red-hot stone

[1] Bastian, 'Mensch,' vol. ii. p. 323.
[2] Kolben, p. 579.
[3] Billings, p. 125.
[4] Bastian, 'Oestl. Asien.' vol. i. p. 145; Cross, l. c., p. 311. For other cases of desertion of dwellings after a death, possibly for the same motive, see Bourien, 'Tribes of Malay Pen.' in 'Tr. Eth. Soc.' vol. iii. p. 82; Polack, 'M. of New Zealanders,' vol. i. pp. 204, 216; Steller, 'Kamtschatka,' p. 271. But the Todas say that the buffaloes slaughtered and the hut burnt at the funeral are transferred to the spirit of the deceased in the next world; Shortt in 'Tr. Eth. Soc.' vol. vii. p. 247. See Waitz, vol. iii. p. 199.
[5] Egede, 'Greenland,' p. 152; Cranz, p. 300.
[6] Bastian, 'Mensch,' vol. ii. p. 323; see pp. 329, 363.
[7] Bowring, 'Siam,' vol. i. p. 122; Bastian, 'Oestl. Asien.' vol. iii. p. 258.

after the corpse is carried out, for an obstacle to bar the soul from coming back;[1] so Brandenburg peasants pour out a pail of water at the door after the coffin, to prevent the ghost from walking; and Pomeranian mourners returning from the churchyard leave behind the straw from the hearse, that the wandering soul may rest there, and not come back so far as home.[2] In the ancient and mediæval world, men habitually invoked supernatural aid beyond such material shifts as these, calling in the priest to lay or banish intruding ghosts, nor is this branch of the exorcist's art even yet forgotten. There is, and always has been, a prevalent feeling that disembodied souls, especially such as have suffered a violent or untimely death, are baneful and malicious beings. As Meiners suggests in his 'History of Religions,' they were driven unwillingly from their bodies, and have carried into their new existence an angry longing for revenge. No wonder that mankind should so generally agree that if the souls of the dead must linger in the world at all, their fitting abode should be not the haunts of the living but the resting-places of the dead.

After all, it scarcely seems to the lower animistic philosophy that the connexion between body and soul is utterly broken by death. Various wants may keep the soul from its desired rest, and among the chief of these is when its mortal remains have not had the funeral rites. Hence the deep-lying belief that the ghosts of such will walk. Among some Australian tribes the "ingna," or evil spirits, human in shape, but with long tails and long upright ears, are mostly souls of departed natives, whose bodies were left to lie unburied or whose death the avenger of blood did not expiate, and thus they have to prowl on the face of the earth, and about the place of death, with no gratification

---

[1] Castrén, 'Finn. Myth.' p. 120.
[2] Wuttke, 'Volksaberglaube,' pp. 213—17. Other cases of taking out the dead by a gap made on purpose: Arbousset and Daumas, p. 502 (Bushmen); Magyar, p. 351 (Kimbunda); Moffat, p. 307 (Bechuanas); Waitz, vol. iii. p. 199 (Ojibwas);—their motive is not clear.

but to harm the living.[1] In New Zealand, the ideas were to be found that the souls of the dead were apt to linger near their bodies, and that the spirits of men left unburied, or killed in battle and eaten, would wander; and the bringing such malignant souls to dwell within the sacred burial-enclosure was a task for the priest to accomplish with his charms.[2] Among the Iroquois of North America the spirit also stays near the body for a time, and "unless the rites of burial were performed, it was believed that the spirits of the dead hovered for a time upon the earth, in a state of great unhappiness. Hence their extreme solicitude to procure the bodies of the slain in battle."[3] Among Brazilian tribes, the wandering shadows of the dead are said to be considered unresting till burial.[4] In Turanian regions of North Asia, the spirits of the dead who have no resting-place in earth are thought of as lingering above ground, especially where their dust remains.[5] South Asia has such beliefs: the Karens say that the ghosts who wander on earth are not the spirits of those who go to Plu, the land of the dead, but of infants, of such as died by violence, of the wicked, and of those who by accident have not been buried or burned;[6] the Siamese fear as unkindly spirits the souls of such as died a violent death or were not buried with the proper rites, and who, desiring expiation, invisibly terrify their descendants.[7] Nowhere in the world had such thoughts a stronger hold than in classic antiquity, where it was the most sacred of duties to give the body its funeral rites, that the shade should not flit moaning near the gates of Hades, nor wander in the dismal crowd along the banks

[1] Oldfield in 'Tr. Eth. Soc.' vol. iii. pp. 228, 236, 245.
[2] Taylor, 'New Zealand,' p. 221; Schirren, p. 91; see Turner, 'Polynesia,' p. 233.
[3] Morgan, 'League of Iroquois,' p. 174.
[4] J. G. Müller, p. 286.
[5] Castrén, 'Finn. Myth.' p. 126.
[6] Cross in 'Journ. Amer. Or. Soc.' vol. iv. p. 309; Mason in 'Journ. As. Soc. Bengal,' 1865, part ii. p. 203. See also J. Anderson, 'Exp. to W. Yunnan,' pp. 126, 131 (Shans).
[7] Bastian, 'Psychologie,' pp. 51, 99—101.

of Acheron.[1] An Australian or a Karen would have taken in the full significance of the fatal accusation against the Athenian commanders, that they abandoned the bodies of their dead in the sea-fight of Arginousai. The thought is not unknown to Slavonic folklore: "Ha! with the shriek the spirit flutters from the mouth, flies up to the tree, from tree to tree, hither and thither till the dead is burned."[2] In mediæval Europe the classic stories of ghosts that haunt the living till laid by rites of burial pass here and there into new legends, where, under a changed dispensation, the doleful wanderer now asks Christian burial in consecrated earth.[3] It is needless to give here elaborate details of the world-wide thought that when the corpse is buried, exposed, burned, or otherwise disposed of after the accepted custom of the land, the ghost accompanies its relics. The soul stays near the Polynesian or the American Indian burial-place; it dwells among the twigs and listens joyfully to the singing birds in the trees where Siberian tribes suspend their dead; it lingers by the Samoyed's scaffolded coffin; it haunts the Dayak's place of burial or burning; it inhabits the little soul-hut above the Malagasy grave, or the Peruvian house of sun-dried bricks; it is deposited in the Roman tomb (animamque sepulchro condimus); it comes back for judgment into the body of the later Israelite and the Moslem; it inhabits, as a divine ancestral spirit, the palace-tombs of the old classic and new Asiatic world; it is kept down by the huge cairn raised over Antar's body lest his mighty spirit should burst forth, by the iron nails with which the Cheremiss secures the corpse in its coffin, by the stake that pins down the suicide's body at the four-cross way. And through all the changes of religious thought from first to last in the course of human history, the hovering ghosts of the dead make the midnight burial-ground a

---

[1] Lucian. De Luctu. See Pauly, 'Real. Encyclop.' and Smith, 'Dic. of Gr. and Rom. Ant.' s.v. 'inferi.'
[2] Hanusch, 'Slaw. Myth.' p. 277.
[3] Calmet, vol. ii. ch. xxxvi.; Brand, vol. iii. p. 67.

place where men's flesh creeps with terror. Not to discuss here the general subject of the funeral rites of mankind, of which only part of the multifarious details are directly relevant to the present purpose, a custom may be selected which is admirably adapted for the study of animistic religion, at once from the clear conception it gives of the belief in disembodied souls present among the living, and from the distinct line of ethnographic continuity in which it may be traced onward from the lower to the higher culture. This is the custom of Feasts of the Dead.

Among the funeral offerings described in the last chapter, of which the purpose more or less distinctly appears to be that the departed soul shall take them away in some ghostly or ideal manner, or that they shall by some means be conveyed to him in his distant spirit-home, there are given supplies of food and drink. But the feasts of the dead with which we are now concerned are given on a different principle; they are, so to speak, to be consumed on the premises. They are set out in some proper place, especially near the tombs or in the dwelling-houses, and there the souls of the dead come and satisfy themselves. In North America, among Algonquins who held that one of a man's two souls abides with the body after death, the provisions brought to the grave were intended for the nourishment of this soul; tribes would make offerings to ancestors of part of any dainty food, and an Indian who fell by accident into the fire would believe that the spirits of his ancestors pushed him in for neglecting to make due offerings.[1] The minds of the Hurons were filled with fancies not less lifelike than this. It seemed to them that the dead man's soul, in his proper human figure, walked in front of the corpse as they carried it to the burial-ground, there to dwell till the great feast of the dead; but meanwhile it would come and walk by night in the village, and eat the remnants in the kettles, wherefore some would not eat of these, nor touch the food

[1] Charlevoix, 'Nouvelle France,' vol. vi. p. 75; Schoolcraft, 'Indian Tribes,' part i. pp. 39, 83; part iv. p. 65; Tanner's 'Narr.' p. 293.

at funeral feasts—though some indeed would eat all.[1] In Madagascar, the elegant little upper chamber in King Radama's mausoleum was furnished with a table and two chairs, and a bottle of wine, a bottle of water, and two tumblers were placed there conformably with the ideas entertained by most of the natives, that the ghost of the departed monarch might occasionally visit the resting-place of his body, meet with the spirit of his father, and partake of what he was known to be fond of in his lifetime.[2] The Wanika of East Africa set a cocoa-nut shell full of rice and tembo near the grave for the "koma" or shade, which cannot exist without food and drink.[3] In West Africa the Efik cook food and leave it on the table in the little shed or "devil-house" near the grave, and thither not only the spirit of the deceased, but the spirits of the slaves sacrificed at his funeral, come to partake of it.[4] Farther south, in the Congo district, the custom has been described of making a channel into the tomb to the head or mouth of the corpse, whereby to send down month by month the offerings of food and drink.[5]

Among rude Asiatic tribes, the Bodo of North-East India thus celebrate the last funeral rites. The friends repair to the grave, and the nearest of kin to the deceased, taking an individual's usual portion of food and drink, solemnly presents it to the dead with these words, "Take and eat, heretofore you have eaten and drunk with us, you can do so no more; you were one of us, you can be so no longer; we come no more to you, come you not to us." Thereupon each of the party breaks off a bracelet of thread put on his wrist for this purpose, and casts it on the grave, a speaking symbol of breaking the bond of fellowship, and "next the party

[1] Brebeuf in 'Rel. des Jes.' 1636, p. 104.
[2] Ellis, 'Madagascar,' vol. i. pp. 253, 364. See Taylor, 'New Zealand,' p. 220.
[3] Krapf, 'E. Afr.' p. 150.
[4] T. J. Hutchinson, p. 206.
[5] Cavazzi, 'Congo, etc.' book i. p. 264. So in ancient Greece, Lucian. Charon, 22.

proceed to the river and bathe, and having thus lustrated themselves, they repair to the banquet and eat, drink, and make merry as though they never were to die."[1] With more continuance of affection, Naga tribes of Assam celebrate their funeral feasts month by month, laying food and drink on the graves of the departed.[2] In the same region of the world, the Kol tribes of Chota Nagpur are remarkable for their pathetic reverence for their dead. When a Ho or Munda has been burned on the funeral pile, collected morsels of his bones are carried in procession with a solemn, ghostly, sliding step, keeping time to the deep-sounding drum, and when the old woman who carries the bones on her bamboo tray lowers it from time to time, then girls who carry pitchers and brass vessels mournfully reverse them to show that they are empty; thus the remains are taken to visit every house in the village, and every dwelling of a friend or relative for miles, and the inmates come out to mourn and praise the goodness of the departed; the bones are carried to all the dead man's favourite haunts, to the fields he cultivated, to the grove he planted, to the threshing-floor where he worked, to the village dance-room where he made merry. At last they are taken to the grave, and buried in an earthen vase upon a store of food, covered with one of those huge stone slabs which European visitors wonder at in the districts of the aborigines in India. Beside these, monumental stones are set up outside the village to the memory of men of note; they are fixed on an earthen plinth where the ghost, resting in its walks among the living, is supposed to sit shaded by the pillar. The Kheriahs have collections of these monuments in the little enclosures round their houses, and offerings and libations are constantly made at them. With what feelings such rites are celebrated may be judged from this Ho dirge:—

"We never scolded you; never wronged you;
Come to us back!

---

[1] Hodgson, 'Abor. of India,' p. 180.   [2] 'Journ. Ind. Archip.' vol. ii. p. 235.

We ever loved and cherished you; and have lived long together
   Under the same roof;
   Desert it not now!
The rainy nights, and the cold blowing days, are coming on;
   Do not wander here!
Do not stand by the burnt ashes; come to us again!
You cannot find shelter under the peepul, when the rain comes down.
The saul will not shield you from the cold bitter wind.
   Come to your home!
It is swept for you, and clean; and we are there who loved you ever;
And there is rice put for you; and water;
   Come home, come home, come to us again!"

Among the Kol tribes this kindly hospitality to ancestral souls passes on into the belief and ceremony of full manes-worship: votive offerings are made to the "old folks" when their descendants go on a journey, and when there is sickness in the family it is generally they who are first propitiated.[1] Among Turanian races of North Asia, the Chuwash put food and napkins on the grave, saying, "Rise at night and eat your fill, and there ye have napkins to wipe your mouths!" while the Cheremiss simply said, "That is for you, ye dead, there ye have food and drink!" In this region we hear of offerings continued year after year, and even of messengers sent back by a horde to carry offerings to the tombs of their forefathers in the old land whence they had emigrated.[2]

Details of this ancient rite are to be traced from the level of these rude races far upward in civilization. South-East Asia is full of it, and the Chinese may stand as its representative. He keeps his coffined parent for years, serving him with meals as if alive. He summons ancestral souls with prayer and beat of drum to feed on the meat and drink set out on special days when they are thought to return home. He even gives entertainments for the benefit of

---

[1] Tickell in 'Journ. As. Soc. Bengal,' vol. ix. p. 795; Dalton, *ibid.* 1866, part ii. p. 153, etc.; and in 'Tr. Eth. Soc.' vol. vi. p. 1, etc.; Latham, 'Descr. Eth.' vol. ii. p. 415, etc.

[2] Bastian, 'Psychologie,' p. 62; Castrén, 'Finn. Myth.' p. 121.

destitute and unfortunate souls in the lower regions, such as those of lepers and beggars. Lanterns are lighted to show them the way, a feast is spread for them, and with characteristic fancy, some victuals are left over for any blind or feeble spirits who may be late, and a pail of gruel is provided for headless souls, with spoons for them to put it down their throats with. Such proceedings culminate in the so-called Universal Rescue, now and then celebrated, when a little house is built for the expected visitors, with separate accommodation and bath-rooms for male and female ghosts.[1] The ancient Egyptian would set out his provision of cakes and trussed ducks on reed scaffolds in the tomb, or would even keep the mummy in the house to be present as a guest at the feast, σύνδειπνον καὶ συμπότην ἐποιήσατο, as Lucian says.[2] The Hindu, as of old, offers to the dead the funeral cakes, places before the door the earthen vessels of water for him to bathe in, of milk for him to drink, and celebrates at new and full moon the solemn presentation of rice-cakes made with ghee, with its attendant ceremonies so important for the soul's release from its twelvemonth's sojourn with Yama in Hades, and its transition to the Heaven of the Pitaras, the Fathers.[3] In the classic world such rites were represented by funeral feasts and oblations of food.[4]

In Christian times there manifests itself that interesting kind of survival which, keeping up the old ceremony in form, has adapted its motive to new thoughts and feelings. The classic funeral oblations became Christian, the silicernium was succeeded by the feast held at the martyr's tomb. Faustus inveighs against the Christians for carrying on the ancient rites: "Their sacrifices indeed ye have turned into love-feasts, their idols into martyrs whom with like vows ye

---

[1] Doolittle, 'Chinese,' vol. i. p. 173, etc.; vol. ii. p. 91, etc.; Meiners, vol. i. p. 306.
[2] Wilkinson, 'Ancient Eg.' vol. ii. p. 362; Lucian. De Luctu, 21.
[3] Manu, iii.; Colebrooke, 'Essays,' vol. i. p. 161, etc.; Pictet, 'Origines Indo-Europ.' part ii. p. 600; Ward, 'Hindoos,' vol. ii. p. 332.
[4] Pauly, 'Real-Encyclop.' s. v. "funus;" Smith's 'Dic.' s. v. "funus." See Meiners, vol. i. pp. 305—19.

worship; ye appease the shades of the dead with wine and meals, ye celebrate the Gentiles' solemn days with them, such as calends and solstices,—of their life certainly ye have changed nought,"[1] and so forth. The story of Monica shows how the custom of laying food on the tomb for the manes passed into the ceremony, like to it in form, of setting food and drink to be sanctified by the sepulchre of a Christian saint. Saint-Foix, who wrote in the time of Louis XIV., has left us an account of the ceremonial after the death of a King of France, during the forty days before the funeral when his wax effigy lay in state. They continued to serve him at meal-times as though still alive, the officers laid the table, and brought the dishes, the maître d'hôtel handed the napkin to the highest lord present to be presented to the king, a prelate blessed the table, the basins of water were handed to the royal arm-chair, the cup was served in its due course, and grace was said in the accustomed manner, save that there was added to it the De Profundis.[2] Spaniards still offer bread and wine on the tombs of those they love, on the anniversary of their decease.[3] The conservative Eastern Church still holds to ancient rite. The funeral feast is served in Russia, with its table for the beggars, laden with fish-pasties and bowls of shchi and jugs of kvas, its more delicate dinner for friends and priests, its incense and chants of "everlasting remembrance"; and even the repetition of the festival on the ninth, and twentieth, and fortieth day are not forgotten. The offerings of saucers of kutiya or kolyvo are still made in the church; this used to be of parboiled wheat and was deposited over the body, it is now made of boiled rice and raisins, sweetened with honey. In their usual mystic fashion, the Greek Christians now explain away into symbolism this remnant of primitive offering to the dead: the honey is heavenly sweetness, the

---

[1] Augustin. contra Faustum, xx. 4; De Civ. Dei, viii. 27. See Beausobre, vol. ii. pp. 633, 685.

[2] Saint-Foix, 'Essais Historiques sur Paris,' in 'Œuvres,' vol. iv. p. 147, etc.

[3] Lady Herbert, 'Impressions of Spain,' p. 8.

shrivelled raisins will be full beauteous grapes, the grain typifies the resurrection, "that which thou sowest is not quickened except it die."[1]

In the calendar of many a people, differing widely as they may in race and civilization, there are to be found special yearly festivals of the dead. Their rites are much the same as those performed on other days for individuals; their season differs in different districts, but seems to have particular associations with harvest-time and the fall of the year, and with the year's end as reckoned at midwinter or in early spring.[2] The Karens make their annual offerings to the dead in the "month of shades," that is, December;[3] the Kocch of North Bengal every year at harvest-home offer fruits and a fowl to deceased parents;[4] the Barea of East Africa celebrate in November the feast of Thiyot, at once a feast of general peace and merry-making, of thanksgiving for the harvest, and of memorial for the deceased, for each of whom a little pot-full of beer is set out two days, to be drunk at last by the survivors;[5] in West Africa we hear of the feast of the dead at the time of yam-harvest;[6] at the end of the year the Haitian negroes take food to the graves for the shades to eat, "manger zombi," as they say.[7] The Roman Feralia and Lemuralia were held in February

---

[1] H. C. Romanoff, 'Rites and Customs of Greco-Russian Church,' p. 249; Ralston, 'Songs of the Russian People,' p. 135, 320; St. Clair and Brophy, 'Bulgaria,' p. 77; Brand, 'Pop. Ant.' vol. i. p. 115.

[2] Beside the accounts of annual festivals of the dead cited here, see the following:—Santos, Ethiopia, in Pinkerton, vol. xvi. p. 685 (Sept.); Brasseur, 'Mexique,' vol. iii. pp. 23, 522, 528 (Aug., Oct., Nov.); Rivero and Tschudi, 'Peru,' p. 134 (Peruvian feast dated as Nov. 2 in coincidence with All Souls', but this reckoning is vitiated by confusion of seasons of N. and S. hemisphere, see J. G. Müller, p. 389; moreover, the Peruvian feast may have been originally held at a different date, and transferred, as happened elsewhere, to the Spanish All Souls'); Doolittle, 'Chinese,' vol. ii. pp. 44, 62 (esp. Apr.); Caron, 'Japan' in Pinkerton, vol. vii. p. 629 (Aug.).

[3] Mason, 'Karens,' l. c. p. 238.

[4] Hodgson, 'Abor. of India,' p. 147.

[5] Munzinger, 'Ost. Afrika,' p. 473.

[6] Waitz, vol. ii. p. 194.

[7] G. D'Alaux in 'Rev. des Deux Mondes,' May 15, 1852, p. 768.

and May.[1] In the last five or ten days of their year the Zoroastrians hold their feasts for departed relatives, when souls come back to the world to visit the living, and receive from them offerings of food and clothing.[2] The custom of setting empty seats at the St. John's Eve feast, for the departed souls of kinsfolk, is said to have lasted on in Europe to the seventeenth century. Spring is the season of the time-honoured Slavonic rite of laying food on the graves of the dead. The Bulgarians hold a feast in the cemetery on Palm Sunday, and, after much eating and drinking, leave the remains upon the graves of their friends, who, they are persuaded, will eat them during the night. In Russia such scenes may still be watched on the two appointed days called Parents' Days. The higher classes have let the rite sink to prayer at the graves of lost relatives, and giving alms to the beggars who flock to the cemeteries. But the people still "howl" for the dead, and set out on their graves a handkerchief for a tablecloth, with gingerbread, eggs, curd-tarts, and even vodka, on it; when the weeping is over, they eat up the food, especially commemorating the dead in Russian manner by partaking of his favourite dainty, and if he were fond of a glass, the vodka is sipped with the ejaculation, "The Kingdom of Heaven be his! He loved a drink, the deceased!"[3] When Odilo, Abbot of Cluny, at the end of the tenth century, instituted the celebration of All Souls',[4] he set on foot one of

[1] Ovid. Fast. ii. 533; v. 420.
[2] Bleek, 'Avesta,' vol. ii. p. 31; vol. iii. p. 86; Alger, p. 137.
[3] Hanusch, 'Slaw. Myth.' pp. 374, 408; St. Clair and Brophy, 'Bulgaria,' p. 77; Romanoff, 'Greco-Roman Church,' p. 255.
[4] Petrus Damianus, 'Vita S. Odilonis,' in the Bollandist 'Acta Sanctorum, Jan. 1, has the quaint legend attached to the new ordinance. An island hermit dwelt near a volcano, where souls of the wicked were tormented in the flames. The holy man heard the officiating demons lament that their daily task of new torture was interfered with by the prayers and alms of devout persons leagued against them to save souls, and especially they complained of the monks of Cluny. Thereupon the hermit sent a message to Abbot Odilo, who carried out the work to the efficacy of which he had received such perfect spiritual testimony, by decreeing that Nov. 2, the day after All Saints', should be set apart for services for the departed.

those revivals which have so often given the past a new lease of life. The Western Church at large took up the practice, and round it, on the second of November, there naturally gathered surviving remnants of the primitive rite of banquets to the dead. The accusation against the early Christians, that they appeased the shades of the dead with feasts like the Gentiles, would not be beside the mark now, fifteen hundred years later. All Souls' Day keeps up, within the limits of Christendom, a commemoration of the dead which combines some touches of pathetic imagination with relics of savage animism scarcely to be surpassed in Africa or the South Sea Islands. In Italy the day is given to feasting and drinking in honour of the dead, while skulls and skeletons in sugar and paste form appropriate children's toys. In Tyrol, the poor souls released from purgatory fire for the night may come and smear their burns with the melted fat of the "soul-light" on the hearth, or cakes are left for them on the table, and the room is kept warm for their comfort. Even in Paris the souls of the departed come to partake of the food of the living. In Brittany the crowd pours into the churchyard at evening, to kneel bareheaded at the graves of dead kinsfolk, to fill the hollow of the tombstone with holy water, or to pour libations of milk upon it. All night the church bells clang, and sometimes a solemn procession of the clergy goes round to bless the graves. In no household that night is the cloth removed, for the supper must be left for the souls to come and take their part, nor must the fire be put out, where they will come to warm themselves. And at last, as the inmates retire to rest, there is heard at the door a doleful chant—it is the souls, who, borrowing the voices of the parish poor, have come to ask the prayers of the living.[1]

If we ask how the spirits of the dead are in general sup-

---

[1] Bastian, 'Mensch,' vol. ii. p. 336. Meiners, vol. i. p. 316; vol. ii. p. 290. Wuttke, 'Deutsche Volksaberglaube,' p. 216. Cortet, 'Fêtes Religieuses,' p. 233; 'Westminster Rev.' Jan. 1860; Hersart de la Villemarqué, 'Chants de la Bretagne,' vol. ii. p. 307.

posed to feed on the viands set before them, we come upon difficult questions, which will be met with again in discussing the theory of sacrifice. Even where the thought is certainly that the departed soul eats, this thought may be very indefinite, with far less of practical intention in it than of childish make-believe. Now and then, however, the sacrificers themselves offer closer definitions of their meaning. The idea of the ghost actually devouring the material food is not unexampled. Thus, in North America, Algonquin Indians considered that the shadow-like souls of the dead can still eat and drink, often even telling Father Le Jeune that they had found in the morning meat gnawed in the night by the souls. More recently, we read that some Potawatomis will leave off providing the supply of food at the grave if it lies long untouched, it being concluded that the dead no longer wants it, but has found a rich hunting-ground in the other world.[1] In Africa, again, Father Cavazzi records of the Congo people furnishing their dead with supplies of provisions, that they could not be persuaded that souls did not consume material food.[2] In Europe the Esths, offering food for the dead on All Souls', are said to have rejoiced if they found in the morning that any of it was gone.[3] A less gross conception is that the soul con-

[1] Le Jeune in 'Rel. des Jés.' 1634, p. 16 ; Waitz, vol. iii. p. 195.
[2] Cavazzi, 'Congo,' etc. book i. 265.
[3] Grimm, 'D. M.' p. 865, but not so in the account of the Feast of the Dead in Boecler, 'Ehsten Abergl. Gebr.' (ed. Kreutzwald), p. 89. Compare Martius, 'Ethnog. Amer.' vol. i. p. 345 (Gês). The following passage from a spiritualist journal, "The Medium," Feb. 9, 1872, shows this primitive notion curiously surviving in modern England. "Every time we sat at dinner, we had not only spirit-voices calling to us, but spirit-hands touching us ; and last evening, as it was his farewell, they gave us a special manifestation, unasked for and unlooked for. He sitting at the right hand of me, a vacant chair opposite to him began moving, and, in answer to whether it would have some dinner, said "Yes." I then asked it to select what it would take, when it chose *croquets des pommes de terre* (a French way of dressing potatoes, about three inches long and two wide. I will send you one that you may see it.) I was desired to put this on the chair, either in a tablespoon or on a plate. I placed it in a tablespoon, thinking that probably the plate might be broken. In a few seconds I was told that it was eaten, and looking, found the half of it gone, with the marks showing the teeth." (Note to 2nd ed.)

sumes the steam or savour of the food, or its essence or spirit. It is said to have been with such purpose that the Maoris placed food by the dead man's side, and some also with him in the grave.[1] The idea is well displayed among the natives in Mexican districts, where the souls who come to the annual feast are described as hovering over and smelling the food set out for them, or sucking out its nutritive quality.[2] The Hindu entreats the manes to quaff the sweet essence of the offered food; thinking on them, he slowly sets the dish of rice before the Brahmans, and while they silently eat the hot food, the ancestral spirits take their part of the feast.[3] At the old Slavonic meals for the dead, we read of the survivors sitting in silence and throwing morsels under the table, fancying that they could hear the spirits rustle, and see them feed on the smell and steam of the viands. One account describes the mourners at the funeral banquet inviting in the departed soul, thought to be standing outside the door, and every guest throwing morsels and pouring drink under the table, for him to refresh himself. What lay on the ground was not picked up, but was left for friendless and kinless souls. When the meal was over, the priest rose from table, swept out the house, and hunted out the souls of the dead "like fleas," with these words, "Ye have eaten and drunken, souls, now go, now go!"[4] Many travellers have described the imagination with which the Chinese make such offerings. It is that the spirits of the dead consume the impalpable essence of the food, leaving behind its coarse material substance, where fore the dutiful sacrificers, having set out sumptuous feasts for ancestral souls, allow them a proper time to satisfy their appetite, and then fall to themselves.[5] The Jesuit Father Christoforo Borri suggestively translates the native idea into his own scholastic phraseology. In Cochin China,

[1] Taylor, 'New Zealand,' p. 220, see 104.
[2] Brasseur, 'Mexique,' vol. iii. p. 24.
[3] Colebrooke, 'Essays,' vol. i. p. 163, etc.; Manu. iii.
[4] Hanusch, 'Slaw. Myth,' p. 408; Hartknoch, 'Preussen,' part i. p. 187.
[5] Doolittle, 'Chinese,' vol. ii. pp. 33, 48; Meiners, vol. i. p. 318.

according to him, people believed " that the souls of the dead have need of corporeal sustenance and maintenance, wherefore several times a year, according to their custom, they make splendid and sumptuous banquets, children to their deceased parents, husbands to their wives, friends to their friends, waiting a long while for the dead guest to come and sit down at table to eat." The missionaries argued against this proceeding, but were met by ridicule of their ignorance, and the reply "that there were two things in the food, one the substance, and the other the accidents of quantity, quality, smell, taste, and the like. The immaterial souls of the dead, taking for themselves the substance of the food, which being immaterial is food suited to the incorporeal soul, left only in the dishes the accidents which corporeal senses perceive; for this the dead had no need of corporeal instruments, as we have said." Thereupon the Jesuit proceeds to remark, as to the prospect of conversion of these people, "it may be judged from the distinction they make between the accidents and the substance of the food which they prepare for the dead," that it will not be very difficult to prove to them the mystery of the Eucharist.[1] Now to peoples among whom prevails the rite of feasts of the dead, whether they offer the food in mere symbolic pretence, or whether they consider the souls really to feed on it in this spiritual way (as well as in the cases inextricably mixed up with these, where the offering is spiritually conveyed away to the world of spirits), it can be of little consequence what becomes of the gross material food. When the Kafir sorcerer, in cases of sickness, declares that the shades of ancestors demand a particular cow, the beast is slaughtered and left shut up for a time for the shades to eat, or for its spirit to go to the land of shades, and then is taken out to be eaten by the sacrificers.[2] So, in more civilized Japan, when the survivors have placed

---

[1] Borri, 'Relatione della Nuova Missione della Comp. di Giesu,' Rome, 1631, p. 208; and in Pinkerton, vol. ix. p. 822, etc.
[2] Grout, 'Zulu-Land,' p. 140; see Callaway, 'Rel. of Amazulu,' p. 11.

their offering of unboiled rice and water in a hollow made for the purpose in a stone of the tomb, it seems to them no matter that the poor or the birds really carry off the grain.[1]

Such rites as these are especially exposed to dwindle in survival. The offerings of meals and feasts to the dead may be traced at their last stage into mere traditional ceremonies, at most tokens of affectionate remembrance of the dead, or works of charity to the living. The Roman Feralia in Ovid's time were a striking example of such transition, for while the idea was recognized that the ghosts fed upon the offerings, "nunc posito pascitur umbra cibo," yet there were but "parva munera," fruits and grains of salt, and corn soaked in wine, set out for their meal in the middle of the road. "Little the manes ask, the pious thought stands instead of the rich gift, for Styx holds no greedy gods : "—

> " Parva petunt manes. Pietas pro divite grata est
>   Munere. Non avidos Styx habet ima deos.
> Tegula porrectis satis est velata coronis,
>   Et sparsæ fruges, parcaque mica salis,
> Inque mero mollita ceres, violæque solutæ :
>   Hæc habeat media testa relicta via.
> Nec majora veto. Sed et his placabilis umbra est."[2]

Still farther back, in old Chinese history, Confucius had been called on to give an opinion as to the sacrifices to the dead. Maintainer of all ancient rites as he was, he stringently kept up this, "he sacrificed to the dead as if they were present," but when he was asked if the dead had knowledge of what was done or no, he declined to answer the question; for if he replied yes, then dutiful descendants would injure their substance by sacrifices, and if no, then undutiful children would leave their parents unburied. The evasion was characteristic of the teacher who expressed his theory

[1] Caron, 'Japan,' vol. vii. p. 629; see Turpin, 'Siam,' ibid. vol. ix. p. 590.
[2] Ovid. Fast. ii. 533.

of worship in this maxim, "to give oneself earnestly to the duties due to men, and, while respecting spiritual beings, to keep aloof from them, may be called wisdom." It is said that in our own time the Taepings have made a step beyond Confucius; they have forbidden the sacrifices to the spirits of the dead, yet keep up the rite of visiting their tombs on the customary day, for prayer and the renewal of vows.[1] How funeral offerings may pass into commemorative banquets and feasts to the poor, has been shown already. If we seek in England for vestiges of the old rite of funeral sacrifice, we may find a lingering survival into modern centuries, doles of bread and drink given to the poor at funerals, and "soul-mass cakes" which peasant girls perhaps to this day beg for at farmhouses with the traditional formula,

"Soul, soul, for a soul cake,
Pray you, mistress, a soul cake."[2]

Were it not for our knowledge of the intermediate stages through which these fragments of old custom have come down, it would seem far-fetched indeed to trace their origin back to the savage and barbaric times of the institution of feasts of departed souls.

[1] Legge, 'Confucius" pp. 101–2, 130; Bunsen, 'God in History,' p. 271.
[2] Brand, 'Pop. Ant.' vol. i. p. 392, vol. ii. p. 289.

CHAPTER XIII.

ANIMISM—(*continued*).

Journey of the Soul to the Land of the Dead—Visits by the Living to the Regions of Departed Souls—Connexion of such legends with myths of Sunset: the Land of the Dead thus imagined as in the West—Realization of current religious ideas, whether of savage or civilized theology, in narratives of visits to the Regions of Souls—Localization of the Future Life—Distant earthly region: Earthly Paradise, Isles of the Blest—Subterranean Hades or Sheol—Sun, Moon, Stars—Heaven—Historical course of belief as to such localization—Nature of Future Life—Continuance-theory, apparently original, belongs especially to the lower races—Transitional theories—Retribution-theory, apparently derived, belongs especially to the higher races—Doctrine of Moral Retribution as developed in the higher culture—Survey of Doctrine of Future State, from savage to civilized stages—Its practical effect on the sentiment and conduct of Mankind.

THE departure of the dead man's soul from the world of living men, its journey to the distant land of spirits, the life it will lead in its new home, are topics on which the lower races for the most part hold explicit doctrines. When these fall under the inspection of a modern ethnographer, he treats them as myths; often to a high degree intelligible and rational in their origin, consistent and regular in their structure, but not the less myths. Few subjects have aroused the savage poet's mind to such bold and vivid imagery as the thought of the hereafter. Yet also a survey of its details among mankind displays in the midst of variety a regular recurrence of episode that brings the ever-recurring question, how far is this correspondence due to transmission of the same thought from tribe to tribe, and how far to similar but independent development in distant lands?

From the savage state up into the midst of civilization, the comparison may be carried through. Low races and high, in region after region, can point out the very spot whence the flitting souls start to travel toward their new home. At the extreme western cape of Vanua Levu, a calm and solemn place of cliff and forest, the souls of the Fijian dead embark for the judgment-seat of Ndengei, and thither the living come in pilgrimage, thinking to see there ghosts and gods.[1] The Baperi of South Africa will venture to creep a little way into their cavern of Marimatlé, whence men and animals came forth into the world, and whither souls return at death.[2] In Mexico the cavern of Chalchatongo led to the plains of paradise, and the Aztec name of Mictlan, "Land of the Dead," now Mitla, keeps up the remembrance of another subterranean temple which opened the way to the sojourn of the blessed.[3] In the kingdom of Prester John, Maundevile tells of an entrance to the infernal regions: "Sum men clepen it the vale enchanted, some clepen it the vale of develes, and some clepen it the vale perilous. In that vale heren men often tyme grete tempestes and thonders, and grete murmures and noyses, alle dayes and nyghtes, and gret noyse, as it were soun of taboures and of nakeres and trompes, as though it were of a gret feste. This valle is alle fulle of develes, and hathe ben alleways; and men seyn there that it is on of the entrees of helle."[4] North German peasants still remember on the banks of the swampy Drömling the place of access to the land of departed souls.[5] To us Englishmen the shores of lake Avernus, trodden daily by our tourists, are more familiar than the Irish analogue of the place, Lough Derg, with its cavern entrance of St. Patrick's Purgatory leading down to the awful world below. The mass of mystic details

[1] Williams, 'Fiji,' vol. i. p. 239; Seemann, 'Viti,' p. 398.
[2] Arbousset and Daumas, p. 347; Casali's, p. 247.
[3] Brasseur, 'Mexique,' vol. iii. p. 20, etc.
[4] Sir John Maundevile, 'Voiage.'
[5] Wuttke, 'Volksaberglaube,' p. 215. Other cases in Bastian, 'Mensch,' vol. ii. pp. 58, 369, etc.

need not be repeated here of the soul's dread journey by caverns and rocky paths and weary plains, over steep and slippery mountains, by frail bark or giddy bridge across gulfs or rushing rivers, abiding the fierce onset of the soul-destroyer or the doom of the stern guardian of the other world. But before describing the spirit-world which is the end of the soul's journey, let us see what the proof is which sustains the belief in both. The lower races claim to hold their doctrines of the future life on strong tradition, direct revelation, and even personal experience. To them the land of souls is a discovered country, from whose bourne many a traveller returns.

Among the legendary visits to the world beyond the grave, there are some that seem pure myth, without a touch of real personal history. Ojibwa, the eponymic hero of his North American tribe, as one of his many exploits descended to the subterranean world of departed spirits, and came up again to earth.[1] When the Kamchadals were asked how they knew so well what happens to men after death, they could answer with their legend of Haetsh the first man. He died and went down into the world below, and a long while after came up again to his former dwelling, and there, standing above by the smoke-hole, he talked down to his kindred in the house and told them about the life to come; it was then that his two daughters whom he had left below followed him in anger and smote him so that he died a second time, and now he is chief in the lower world, and receives the Italmen when they die and rise anew.[2] Thus, again, in the great Finnish epic, the Kalewala, one great episode is Wainamoinen's visit to the land of the dead. Seeking the last charm-words to build his boat, the hero travelled with quick steps week after week through bush and wood till he came to the Tuonela river, and saw before him the island of Tuoni the god of death. Loudly he called to Tuoni's daughter to bring the ferry-boat across:—

[1] Schoolcraft, 'Algic. Res.' vol. ii. pp. 32, 64, and see ante, vol. i. p. 312.
[2] Steller, 'Kamtschatka,' p. 271; Klemm, 'C. G.' vol. ii. p. 312.

> " She, the virgin of Manala,
> She, the washer of the clothing,
> She, the wringer of the linen,
> By the river of Tuonela,
> In the under-world Manala,
> Spake in words, and this their meaning.
> This their answer to the hearer :—
> ' Forth the boat shall come from hither,
> When the reason thou hast given
> That hath brought thee to Manala,
> Neither slain by any sickness,
> Nor by Death dragged from the living,
> Nor destroyed by other ending.' "

Wainamoinen replies with lying reasons. Iron brought him, he says, but Tuoni's daughter answers that no blood drips from his garment; Fire brought him, he says, but she answers that his locks are unsinged, and at last he tells his real mission. Then she ferries him over, and Tuonetar the hostess brings him beer in the two-eared jug, but Wainamoinen can see the frogs and worms within and will not drink, for it was not to drain Manala's beer-jug he had come. He lay in the bed of Tuoni, and meanwhile they spread the hundred nets of iron and copper across the river that he might not escape; but he turned into a reed in the swamp, and as a snake crept through the meshes :—

> " Tuoni's son with hooked fingers
> Iron-pointed hooked fingers
> Went to draw his nets at morning—
> Salmon-trout he found a hundred,
> Thousands of the little fishes,
> But he found no Wainamoinen,
> Not the old friend of the billows.
> Then the ancient Wainamoinen,
> Come from out of Tuoni's kingdom,
> Spake in words, and this their meaning,
> This their answer to the hearer :—
> ' Never mayst thou, God of goodness,
> Never suffer such another
> Who of self-will goes to Mana,
> Thrusts his way to Tuoni's kingdom.

Many they who travel thither,
Few who thence have found the home-way,
From the houses of Tuoni
From the dwellings of Manala.' "[1]

It is enough to name the familiar classic analogues of these mythic visits to Hades,—the descent of Dionysos to bring back Semele, of Orpheus to bring back his beloved Eurydike, of Herakles to fetch up the three-headed Kerberos at the command of his master Eurystheus; above all, the voyage of Odysseus to the ends of the deep-flowing Ocean, to the clouded city of Kimmerian men, where shining Helios looks not down with his rays, and deadly night stretches always over wretched mortals,—thence they passed along the banks to the entrance of the land where the shades of the departed, quickened for a while by the taste of sacrificial blood, talked with the hero and showed him the regions of their dismal home.[2]

The scene of the descent into Hades is in very deed enacted day by day before our eyes, as it was before the eyes of the ancient myth-maker, who watched the sun descend to the dark under-world, and return at dawn to the land of living men. These heroic legends lie in close-knit connexion with episodes of solar myth. It is by the simplest poetic adaptation of the Sun's daily life, typifying Man's life in dawning beauty, in mid-day glory, in evening death, that mythic fancy even fixed the belief in the religions of the world, that the Land of Departed Souls lies in the Far West or the World Below. How deeply the myth of the Sunset has entered into the doctrine of men concerning a Future State, how the West and the Under-World have become by mere imaginative analogy Regions of the Dead, how the quaint day-dreams of savage poets may pass into

[1] Kalewala, Rune xvi.; see Schiefner's German Translation, and Castrén, 'Finn. Myth,' pp. 128, 134. A Slavonic myth in Hanusch, p 412.
[2] Homer. Odyss. xi. On the vivification of ghosts by sacrifice of blood, and on libations of milk and blood, see Meiners, vol. i. p. 315, vol. ii. p. 89; J. G. Müller, p. 85; Rochholz, 'Deutscher Glaube und Brauch,' vol. i. p. 1, etc.

honoured dogmas of classic sages and modern divines,—all this the crowd of details here cited from the wide range of culture stand to prove.

Moreover, visits from or to the dead are matters of personal experience and personal testimony. When in dream or vision the seer beholds the spirits of the departed, they give him tidings from the other world, or he may even rise and travel thither himself, and return to tell the living what he has seen among the dead. It is sometimes as if the traveller's material body went to visit a distant land, and sometimes all we are told is that the man's self went, but whether in body or in spirit is a mere detail of which the story keeps no record. Mostly, however, it is the seer's soul which goes forth, leaving his body behind in ecstasy, sleep, coma, or death. Some of these stories, as we trace them on from savage into civilized times, are no doubt given in good faith by the visionary himself, while others are imitations of these genuine accounts.[1] Now such visions are naturally apt to reproduce the thoughts with which the seer's mind was already furnished. Every idea once lodged in the mind of a savage, a barbarian, or an enthusiast, is ready thus to be brought back to him from without. It is a vicious circle; what he believes he therefore sees, and what he sees he therefore believes. Beholding the reflexion of his own mind like a child looking at itself in a glass, he humbly receives the teaching of his second self. The Red Indian visits his happy hunting-grounds, the Tongan his shadowy island of Bolotu, the Greek enters Hades and looks on the Elysian Fields, the Christian beholds the heights of Heaven and the depths of Hell.

Among the North American Indians, and especially the Algonquin tribes, accounts are not unusual of men whose spirits, travelling in dreams or in the hallucinations of extreme illness to the land of the dead, have returned to reanimate their bodies, and tell what they have seen.

[1] See for example, various details in Bastian, 'Mensch,' vol. ii. pp. 369-75, etc.

Their experiences have been in great measure what they were taught in early childhood to expect, the journey along the path of the dead, the monstrous strawberry at which the jebi-ug or ghosts refresh themselves, but which turned to red rock at the touch of their spoons, the bark offered them for dried meat and great puff-balls for squashes, the river of the dead with its snake-bridge or swinging log, the great dog standing on the other side, the villages of the dead beyond.[1] The Zulus of our own day tell of men who have gone down by holes in the ground into the underworld, where mountains and rivers and all things are as here above, and where a man may find his kindred, for the dead live in their villages, and may be seen milking their cattle, which are the cattle killed on earth and come to life anew. The Zulu Umpengula, who told one of these stories to Dr. Callaway, remembered when he was a boy seeing an ugly little hairy man called Uncama, who once, chasing a porcupine that ate his mealies, followed it down a hole in the ground into the land of the dead. When he came back to his home on earth he found that he had been given up for dead himself, his wife had duly burnt and buried his mats and blankets and vessels, and the wondering people at sight of him again shouted the funeral dirge. Of this Zulu Dante it used to be continually said, "There is the man who went to the underground people."[2] One of the most characteristic of these savage narratives is from New Zealand. This story, which has an especial interest from the reminiscence it contains of the gigantic extinct Moa, and which may be repeated at some length as an illustration of the minute detail and life-like reality which such visionary legends assume in barbaric life, was told to Mr. Shortland by a servant of his named Te Wharewera. An aunt of this

---

[1] See vol. i. p. 481; also below, p. 52, note. Tanner's 'Narr.' p. 290; Schoolcraft, 'Indian Tribes,' part iii. p. 233; Keating, vol. ii. p. 154; Loskiel, part i. p. 35; Smith, 'Virginia,' in Pinkerton, vol. xiii. p. 14. See Cranz, 'Grönland,' p. 269.

[2] Callaway, 'Zulu Tales,' vol. i. pp. 316-20.

man died in a solitary hut near the banks of Lake Rotorua. Being a lady of rank she was left in her hut, the door and windows were made fast, and the dwelling was abandoned, as her death had made it tapu. But a day or two after, Te Wharewera with some others paddling in a canoe near the place at early morning saw a figure on the shore beckoning to them. It was the aunt come to life again, but weak and cold and famished. When sufficiently restored by their timely help, she told her story. Leaving her body, her spirit had taken flight toward the North Cape, and arrived at the entrance of Reigna. There, holding on by the stem of the creeping akeake-plant, she descended the precipice, and found herself on the sandy beach of a river. Looking round, she espied in the distance an enormous bird, taller than a man, coming towards her with rapid strides. This terrible object so frightened her, that her first thought was to try to return up the steep cliff; but seeing an old man paddling a small canoe towards her she ran to meet him, and so escaped the bird. When she had been safely ferried across, she asked the old Charon, mentioning the name of her family, where the spirits of her kindred dwelt. Following the path the old man pointed out, she was surprised to find it just such a path as she had been used to on earth; the aspect of the country, the trees, shrubs, and plants were all familiar to her. She reached the village, and among the crowd assembled there she found her father and many near relations; they saluted her, and welcomed her with the wailing chant which Maoris always address to people met after long absence. But when her father had asked about his living relatives, and especially about her own child, he told her she must go back to earth, for no one was left to take care of his grandchild. By his orders she refused to touch the food that the dead people offered her, and in spite of their efforts to detain her, her father got her safely into the canoe, crossed with her, and parting gave her from under his cloak two enormous sweet potatoes to plant at home for his grandchild's especial eating. But as she began to climb the precipice again, two pursuing infant spirits pulled her back, and she only escaped by flinging the roots at them, which they stopped to eat, while she scaled the rock by help of the akeake-stem, till she reached the earth and flew back to where she had left her body. On returning to life she found herself in darkness, and what had passed seemed as a dream, till she perceived that she was deserted and the door fast, and concluded that she had really died and come to life again. When morning dawned, a faint light entered by the crevices of the shut-up house, and she saw on the floor near her a calabash partly full of red ochre mixed with water; this she eagerly drained to the dregs, and then feeling a little stronger, succeeded in opening the door and crawling down to the beach, where her friends soon after found her. Those who listened to her tale firmly believed the reality of her adventures, but it was much regretted that she had not brought back at least one of the huge sweet-potatoes, as evidence of her visit to the land of spirits.[1] Races of North Asia[2] and West Africa[3] have in like manner their explorers of the world beyond the grave.

Classic literature continues the series. Lucian's graphic

---

[1] Shortland, 'Traditions of New Zealand,' p. 150. The idea in this Maori story, that the living who tastes the food of the dead may not return, appears again among the Sioux of North America. Ahak-tah ('Male Elk') seems to die, but after two days comes down from the funeral-scaffold where his body had been laid, and tells his tale. His soul had travelled by the path of braves through the beautiful land of great trees and gay loud-singing birds, till he reached the river, and saw the homes of the spirits of his forefathers on the shore beyond. Swimming across, he entered the nearest house, where he found his uncle sitting in a corner. Very hungry, he noticed some wild rice n a bark dish. "I asked my uncle for some rice to eat, but he did not give it to me. Had I eaten of the food for spirits, I never should have returned to earth." Eastman, 'Dacotah,' p. 177. The analogy of this with the Homeric episode of the lotus-eaters may be deep.

[2] Castrén, 'Finn. Myth,' p. 139, etc.

[3] Bosman, 'Guinea,' Letter 19, in Pinkerton, vol. xvi. p. 501; Burton, 'Dahome,' vol. ii. p. 158. For modern visits to hell and heaven by Christianized negro visionaries in America, see Macrae, 'Americans at Home,' vol. ii. p. 91.

tales represent the belief of their age, if not of their author. His Eukrates looks down the chasm into Hades, and sees the dead reclining on the asphodel in companies of kinsfolk and friends; among them he recognizes Sokrates with his bald head and pot-belly, and also his own father, dressed in the clothes he was buried in. Then Kleodemos caps this story with his own, how when he was sick, on the seventh day when his fever was burning like a furnace, everyone left him, and the doors were shut. Then there stood before him an all-beauteous youth in a white garment, who led him through a chasm into Hades, as he knew by seeing Tantalos and Tityos and Sisyphos; and bringing him to the court of judgment, where were Aiakos and the Fates and the Erinyes, the youth set him before Pluto the King, who sat reading the names of those whose day of life was over. But Pluto was angry, and said to the guide, "This one's thread is not run out, that he should depart, but bring me Demylos the coppersmith, for he is living beyond the spindle." So Kleodemos came back to himself free from his fever, and announced that Demylos, who was a sick neighbour, would die; and accordingly a little while after there was heard the cry of the mourners wailing for him.[1] Plutarch's stories, told more seriously, are yet one in type with the mocking Lucian's. The wicked, pleasure-seeking Thespesios lies three days as dead, and revives to tell his vision of the world below. One Antyllos was sick, and seemed to the doctors to retain no trace of life; till, waking without sign of insanity, he declared that he had been indeed dead, but was ordered back to life, those who brought him being severely chidden by their lord, and sent to fetch Nikander instead, a well-known currier, who was accordingly taken with a fever, and died on the third day.[2] Such stories, old and new, are current among the Hindus at this day. A certain man's soul, for instance, is carried to the

[1] Lucian. Philopseudes, pp. 21-5.
[2] Plutarch. De Sera Numinis Vindicta, xxii.; and in Euseb. Præp. Evang. xi. 36.

realm of Yama by mistake for a namesake, and is sent back in haste to regain his body before it is burnt; but in the meanwhile he has a glimpse of the hideous punishments of the wicked, and of the glorious life of those who had mortified the flesh on earth, and of suttee-widows now sitting in happiness by their husbands.[1] Mutatis mutandis these tales reappear in Christian mythology, as when Gregory the Great records that a certain nobleman named Stephen died, who was taken to the region of Hades, and saw many things he had heard before but not believed; but when he was set before the ruler there presiding, he sent him back, saying that it was this Stephen's neighbour— Stephen the smith—whom he had commanded to be brought; and accordingly the one returned to life, and the other died.[2]

The thought of human visitors revealing the mysteries of the world beyond the grave, which indeed took no slight hold on Christian belief, attached itself in a remarkable way to the doctrine of Christ's descent into Hades. This dogma had so strongly established itself by the end of the 4th century, that Augustine could ask, " Quis nisi infidelis negaverit fuisse apud inferos Christum ? "[3] A distinct statement of the dogma was afterwards introduced into the symbol commonly called the "Apostles' Creed:" " Descendit ad inferos," " Descendit ad inferna," " He descended into hell."[4] The Descent into Hades, which had the theological use of providing a theory of salvation applicable to the saints of the old covenant, imprisoned in the limbo of the fathers, is narrated in full in the apocryphal Gospel of Nicodemus, and is made there to rest upon a legend which belongs to the present group of human visits to the other world. It is related that two sons of Simeon,

[1] Ward, 'Hindoos,' vol. ii. p. 63.
[2] Gregor. Dial. iv. 36. See Calmet, vol. ii. ch. 49.
[3] Augustin. Epist. clxiv. 2.
[4] See Pearson, 'Exposition of the Creed;' Bingham, 'Ant. Chr. Ch.' book x. ch. iii. Art. iii. of the Church of England was reduced to its present state by Archbp. Parker's revision.

named Charinus and Leucius, rose from their tombs at the Resurrection, and went about silently and prayerfully among men, till Annas and Caiaphas brought them into the synagogue, and charged them to tell of their raising from the dead. Then, making the sign of the cross upon their tongues, the two asked for parchment and wrote their record. They had been set with all their fathers in the depths of Hades, when on a sudden there appeared the colour of the sun like gold, and a purple royal light shining on them; then the patriarchs and prophets, from Adam to Simeon and John the Baptist, rejoicing proclaimed the coming of the light and the fulfilment of the prophecies; Satan and Hades wrangled in strife together; in vain the brazen gates were shut with their iron bars, for the summons came to open the gates that the king of glory may come in, who hath broken the gates of brass and cut the bars of iron asunder; then the mighty Lord broke the fetters and visited them who sat in darkness and the shadow of death; Adam and his righteous children were delivered from Hades, and led into the glorious grace of Paradise.[1]

Dante, elaborating in the 'Divina Commedia' the conceptions of paradise, purgatory, and hell familiar to the actual belief of his age, describes them once more in the guise of a living visitor to the land of the dead. Echoes in mediæval legend of such exploring expeditions to the world below still linger faintly in the popular belief of Europe. It has been thus with St. Patrick's Purgatory,[2] the cavern in the island of Lough Derg, in the county Donegal, which even in the seventeenth century O'Sullevan could describe first and foremost in his 'Catholic History' as "the greatest of all memorable things of Ireland." Mediæval visits to the other world were often made in the spirit. But

[1] Codex Apocr. N. T. Evang. Nicod. ed. Giles. 'Apocryphal Gospels,' etc. tr. by A. Walker; 'Gospel of Nicodemus.' The Greek and Latin texts differ much.

[2] The following details mostly from T. Wright, 'St. Patrick's Purgatory' (an elaborate critical dissertation on the mediæval legends of visits to the other world).

like Ulysses, Wainamoinen, and Dante, men could here make the journey in body, as did Sir Owain and the monk Gilbert. When the pilgrim had spent fifteen days in prayer and fasting in the church, and had been led with litanies and sprinkling of holy water to the entrance of the purgatory, and the last warnings of the monks had failed to turn him from the venture, the door was closed upon him, and if found next morning, he could tell the events of his awful journey—how he crossed the narrow bridge that spans the river of death, how he saw the hideous torments of hell, and approached the joys of paradise. Sir Owain, one of King Stephen's knights, went thither in penance for his life of violence and rapine, and this was one of the scenes he beheld in purgatory:—

> "There come develes other mony mo,
> And badde the knygth with hem to go,
> And ladde him into a fowle contreye,
> Where ever was nygth and never day,
> For hit was derke and wonther colde:
> Yette was there never man so bolde,
> Hadde he never so mony clothes on,
> But he wolde be colde as ony stone.
> Wynde herde he none blowe,
> But faste hit frese bothe hye and lowe.
> They browgte him to a felde full brode,
> Overe suche another never he yode,
> For of the lengthe none ende he knewe;
> Thereover algate he moste nowe.
> As he wente he herde a crye,
> He wondered what hit was, and why,
> He syg ther men and wymmen also
> That lowde cryed, for hem was woo.
> They leyen thykke on every londe,
> Faste nayled bothe fote and honde
> With nayles glowyng alle of brasse:
> They ete the erthe so wo hem was;
> Here face was nayled to the grownde.
> 'Spare,' they cryde, 'a lytylle stounde.'
> The develes wolde hem not spare:
> To hem peyne they thowgte yare."

When Owain had seen the other fields of punishment, with their fiery serpents and toads, and the fires where sinners were hung up by their offending members, and roasted on spits, and basted with molten metal, and turned about on a great wheel of fire, and when he had passed the Devil's Mouth over the awful bridge, he reached the fair white glassy wall of the Earthly Paradise, reaching upward and upward, and saw before him the beautiful gate, whence issued a ravishing perfume. Then he soon forgot his pains and sorrows.

> " As he stode, and was so fayne,
> Hym thowgth ther come hym agayne
> A swyde fayr processyour
> Of alle manere menne of relygyoun,
> Fayre vestementes they hadde on,
> So ryche syg he never none.
> Myche joye hym thowgte to se
> Bysshopes yn here dygnité;
> Ilkone wente other be and be,
> Every man yn his degré.
> He syg ther monkes and chanones,
> And freres with newe shavene crownes;
> Ermytes he saw there amonge,
> And nonnes with fulle mery songe;
> Persones, prestes, and vycaryes;
> They made fulle mery melodyes.
> He syg ther kynges and emperoures,
> And dukes that had casteles and toures,
> Erles and barones fele,
> That some tyme hadde the worldes wele.
> Other folke he syg also,
> Never so mony as he dede thoo.
> Wymmen he syg ther that tyde:
> Myche was the joye ther on every syde:
> For alle was joye that with hem ferde,
> And myche solempnyté he herde."

The procession welcomed Owain, and led him about, showing him the beauties of that country:—

> " Hyt was grene, and fulle of flowres
> Of mony dyvers colowres;

> Hyt was grene on every syde,
> As medewus are yn someres tyde,
> Ther were trees growyng fulle grene
> Fulle of fruyte ever more, y wene;
> For ther was frwyte of mony a kynde,
> Such yn the londe may no mon fynde.
> Ther they have the tree of lyfe,
> Theryn ys myrthe, and never stryfe;
> Frwyte of wysdom also ther ys,
> Of the whyche Adam and Eve dede amysse:
> Other manere frwytes ther were fele,
> And alle manere joye and wele.
> Moche folke he syg ther dwelle,
> There was no tongue that mygth hem telle;
> Alle were they cloded yn ryche wede,
> What cloth hit was he kowthe not rede.
>
> . . . .
>
> There was no wronge, but ever rygth,
> Ever day and nevere nygth.
> They shone as brygth and more clere
> Than ony sonne yn the day doth here."

The poem, in fifteenth-century English, from which these passages are taken, is a version of the original legend of earlier date, and as such contrasts with a story really dating from early in the fifteenth century—William Staunton's descent into Purgatory, where the themes of the old sincerely-believed visionary lore are fading into moral allegory, and the traveller sees the gay gold and silver collars and girdles burning into the wearer's flesh, and the jags that men were clothed in now become adders and dragons, sucking and stinging them, and the fiends drawing down the skin of women's shoulders into pokes, and smiting into their heads with burning hammers their gay chaplets of gold and jewels turned to burning nails, and so forth. Late in this fifteenth century, St. Patrick's Purgatory fell into discredit, but even the destruction of the entrance-building, in 1497, by Papal order, did not destroy the ideal road. About 1693, an excavation on the spot brought to light a window with iron stanchions; there was a cry for holy water to keep the spirits from breaking out from prison, and the priest smelt brimstone from the dark cavity below, which, however, unfortunately turned out to be a cellar. In still later times, the yearly pilgrimage of tens of thousands of votaries to the holy place has kept up this interesting survival from the lower culture, whereby a communication may still be traced, if not from Earth to Hades, at least from the belief of the New Zealander to that of the Irish peasant.

To study and compare the ideal regions where man has placed the abodes of departed souls is not an unprofitable task. True, geography has now mapped out into mere earth and water the space that lay beyond the narrower sea and land known to the older nations, and astronomy no longer recognizes the flat earth trodden by men as being the roof of subterranean halls, nor the sky as being a solid firmament, shutting out men's gaze from strata or spheres of empyræan regions beyond. Yet if we carry our minds back to the state of knowledge among the lower races, we shall not find it hard to understand the early conceptions as to the locality of the regions beyond the grave. They are no secrets of high knowledge made known to sages of old; they are the natural fancies which childlike ignorance would frame in any age. The regularity with which such conceptions repeat themselves over the world bears testimony to the regularity of the processes by which opinion is formed among mankind. At the same time, the student who carefully compares them will find in them a perfect illustration of an important principle, widely applicable to the general theory of the formation of human opinion. When a problem has presented itself to mankind at large, susceptible of a number of solutions about equally plausible, the result is that the several opinions thus produced will be found lying scattered in country after country. The problem here is, given the existence of souls of the dead who from time to time visit the living, where is the home of these ghosts? Why men in one district should have preferred the earth, in another the under-world, in another the sky, as the abode of departed souls, is a question often difficult to answer. But we may at least see how again and again the question was taken in hand, and how out of the three or four available answers some people adopted one, some another, some several at once. Primitive theologians had all the world before them where to choose their place of rest for the departed, and they used to the full their speculative liberty.

Firstly, when the land of souls is located on the surface of the earth, there is choice of fit places among wild and cloudy precipices, in secluded valleys, in far-off plains and islands. In Borneo, Mr. St. John visited the heaven of the Idaan race, on the summit of Kina Balu, and the native guides, who feared to pass the night in this abode of spirits, showed the traveller the moss on which the souls of their ancestors fed, and the footprints of the ghostly buffaloes that followed them. On Gunung Danka, a mountain in West Java, there is such another 'Earthly Paradise.' The Sajira who dwell in the district indeed profess themselves Mohammedans, but they secretly maintain their old belief, and at death or funeral they enjoin the soul in solemn form to set aside the Moslem Allah, and to take the way to the dwelling-place of his own forefathers' souls:—

'Step up the bed of the river, and cross the neck of land,
Where the aren trees stand in a clump, and the pinangs in a row,
Thither direct thy steps, Laillah being set aside."

Mr. Jonathan Rigg had lived ten years among these people, and knew them well, yet had never found out that their paradise was on this mountain. When at last he heard of it, he made the ascent, finding on the top only a few river-stones, forming one of the balai, or sacred cairns, common in the district. But the popular belief, that a tiger would devour the chiefs who permitted a violation of the sacred place, soon received the sort of confirmation which such beliefs receive everywhere, for a tiger killed two children a

few days later, and the disaster was of course ascribed to
Mr. Rigg's profanation.[1] The Chilians said that the soul
goes westward over the sea to Gulcheman, the dwelling-
place of the dead beyond the mountains; life, some said,
was all pleasure there, but others thought that part would
be happy and part miserable.[2] Hidden among the mountains of Mexico lay the joyous garden-land of Tlalocan,
where maize, and pumpkins, and chilis, and tomatos never
failed, and where abode the souls of children sacrificed to
Tlaloc, its god, and the souls of such as died by drowning
or thunderstroke, or by leprosy or dropsy, or other acute
disease.[3] A survival of such thought may be traced into
mediæval civilization, in the legends of the Earthly Paradise, the fire-girt abode of saints not yet raised to highest
bliss, localized in the utmost East of Asia, where earth
stretches up towards heaven.[4] When Columbus sailed westward across the Atlantic to seek the "new heaven and a
new earth" he had read of in Isaiah, he found them, though
not as he sought. It is a quaint coincidence that he found
there also, though not as he sought it, the Earthly Paradise
which was another main object of his venturous quest. The
Haitians described to the white men their Coaibai, the
paradise of the dead, in the lovely Western valleys of their
island, where the souls hidden by day among the cliffs came
down at night to feed on the delicious fruit of the mamey-
trees, of which the living ate but sparingly, lest the souls of
their friends should want.[5]

Secondly, there are Australians who think that the spirit
of the dead hovers a while on earth, and goes at last toward

[1] St. John, 'Far East,' vol. i. p. 278. Rigg in 'Journ. Ind. Archip.' vol.
iv. p. 119. See also Ellis, 'Polyn. Res.' vol. i. p. 397; Bastian, 'Oestl.
Asien,' vol. i. p. 83; Irving, 'Astoria,' p. 142.

[2] Molina, 'Chili,' vol. ii. p. 89.

[3] Brasseur, 'Mexique,' vol. iii. p. 496; Sahagun, iii. App. c. 2, x. c. 29;
Clavigero, vol. ii. p. 5.

[4] See Wright, l. c. etc.; Alger, p. 391; Maundevile, etc.

[5] 'History of Colon,' ch. 61; Pet. Martyr. Dec. i. lib. ix.; Irving, 'Life
of Columbus,' vol. ii. p. 121.

the setting sun, or westward over the sea to the island of souls, the home of his fathers. Thus these rudest savages have developed two thoughts which we meet with again and again far onward in the course of culture—the thought of an island of the dead, and the thought that the world of departed souls is in the West, whither the Sun descends at evening to his daily death.[1] Among the North American Indians, when once upon a time an Algonquin hunter left his body behind and visited the land of souls in the sunny south, he saw before him beautiful trees and plants, but found he could walk right through them. Then he paddled in the canoe of white shining stone across the lake where wicked souls perish in the storm, till he reached the beautiful and happy island where there is no cold, no war, no bloodshed, but the creatures run happily about, nourished by the air they breathe.[2] Tongan legend says that, long ago, a canoe returning from Fiji was driven by stress of weather to Bolotu, the island of gods and souls lying in the ocean north-west of Tonga. That island is larger than all theirs together, full of all finest fruits and loveliest flowers, that fill the air with fragrance, and come anew the moment they are plucked; birds of beauteous plumage are there, and hogs in plenty, all immortal save when killed for the gods to eat, and then new living ones appear immediately to fill their places. But when the hungry crew of the canoe landed, they tried in vain to pluck the shadowy bread-fruit, they walked through unresisting trees and houses, even as the souls of chiefs who met them walked unchecked through their solid bodies. Counselled to hasten home from this land of no earthly food, the men sailed to Tonga, but the deadly air of Bolotu had infected them, and they soon all died.[3]

[1] Stanbridge in 'Tr. Eth. Soc.' vol. i. p. 299; G. F. Moore, 'Vocab. W. Austr.' p. 83; Bonwick, 'Tasmanians,' p. 181.

[2] Schoolcraft, 'Indian Tribes,' part i. p. 321; see part iii. p. 229.

[3] Mariner, 'Tonga Is.' vol. ii. p. 107. See also Burton, 'W. and W. fr. W. Africa,' p. 154 (Gold Coast).

ANIMISM. 63

Such ideas took strong hold on classic thought, in the belief in a paradise in the Fortunate Islands of the far Western Ocean. Hesiod in the 'Works and Days' tells of the half-gods of the Fourth Age, between the Age of Bronze and the Age of Iron. When death closed on this heroic race, Zeus granted them at the ends of Earth a life and home, apart from man and far from the immortals. There Kronos reigns over them, and they dwell careless in the Islands of the Happy, beside deep-eddying Ocean—blest heroes, for whom the grain-giving field bears, thrice blooming yearly, the honey-sweet fruit:—

"'Ενθ' ήτοι τοὺς μὲν θανάτου τέλος ἀμφεκάλυψε·
Τοῖς δὲ δίχ' ἀνθρώπων βίοτον καὶ ἤθε' ὀπάσσας
Ζεὺς Κρονίδης κατένασσε πατὴρ ἐς πείρατα γαίης,
Τηλοῦ ἀπ' ἀθανάτων· τοῖσιν Κρόνος ἐμβασιλεύει·
Καὶ τοὶ μὲν ναίουσιν ἀκηδέα θυμὸν ἔχοντες
Ἐν μακάρων νήσοισι παρ' Ὠκεανὸν βαθυδίνην,
Ὄλβιοι ἥρωες, τοῖσιν μελιηδέα καρπὸν
Τρὶς ἔτεος θάλλοντα φέρει ζείδωρος ἄρουρα."

These Islands of the Blest, assigned as the abode of blessed spirits of the dead, came indeed to be identified with the Elysian Fields. Thus Pindar sings of steadfast souls, who through three lives on either side have endured free from injustice; then they pass by the road of Zeus to the tower of Kronos, where the ocean breezes blow round the islands of the happy, blazing with golden flowers of land and water. Thus, also, in the famous hymn of Kallistratos in honour of Harmodios and Aristogeiton, who slew the tyrant Hipparchos:—

"Φίλταθ' Ἁρμόδι', οὔ τί πω τέθνηκας
Νήσοις δ' ἐν μακάρων σε φασὶν εἶναι,
Ἵνα περ ποδώκης Ἀχιλλεὺς,
Τυδείδην τε φασὶ τὸν ἐσθλὸν Διομήδεα." [1]

This group of legends has especial interest to us Englishmen, who ourselves dwell, it seems, on such an island of the

[1] Hesiod. Opera et Dies, 165. Pindar, Olymp. ii. antistr. 4. Callistrat. Hymn. in Ilgen, Scolia Græca, 10. Strabo, iii. 2, 13; Plin. iv. 36.

dead. It is not that we or our country are of a more ghostly nature than others, but the idea is geographical, we are dwellers in the region of the setting sun, the land of death. The elaborate account by Procopius, the historian of the Gothic War, dates from the 6th century. The island of Brittia, according to him, lies opposite the mouths of the Rhine, some 200 stadia off, between Britannia and Thule, and on it dwell three populous nations, the Angles, Frisians, and Britons. (By Brittia, it appears, he means our Great Britain, his Britannia being the coast-land from modern Brittany to Holland, and his Thule being Scandinavia.) In the course of his history it seems to him needful to record a story, mythic and dreamlike as he thinks, yet which numberless men vouch for as having been themselves witnesses by eye and ear to its facts. This story is that the souls of the departed are conveyed across the sea to the island of Brittia. Along the mainland coast are many villages, inhabited by fishermen and tillers of the soil and traders to this island in their vessels. They are subject to the Franks, but pay no tribute, having from of old had to do by turns the burdensome service of transporting the souls. Those on duty for each night stay at home till they hear a knocking at the doors, and a voice of one unseen calling them to their work. Then without delay rising from their beds, compelled by some unknown power they go down to the beach, and there they see boats, not their own but others, lying ready but empty of men. Going on board and taking the oars, they find that by the burden of the multitude of souls embarked, the vessel lies low in the water, gunwale under within a finger's breadth. In an hour they are at the opposite shore, though in their own boats they would hardly make the voyage in a night and day. When they reach the island, the vessel becomes empty, till it is so light that only the keel touches the waves. They see no man on the voyage, no man at the landing, but a voice is heard that proclaims the name and rank and parentage of each newly arrived passenger, or if women, those of their husbands. Traces of this remarkable legend seem to have survived, thirteen centuries later, in that endmost district of the Britannia of Procopius which still keeps the name of Bretagne. Near Raz, where the narrow promontory stretches westward into the ocean, is the 'Bay of Souls' (boé ann anavo); in the commune of Plouguel the corpse is taken to the churchyard, not by the shorter road by land, but in a boat by the "Passage de l'Enfer," across a little arm of the sea; and Breton folklore holds fast to the legend of the Curé de Braspar, whose dog leads over to Great Britain the souls of the departed, when the wheels of the soul-car are heard creaking in the air. These are but mutilated fragments, but they seem to piece together with another Keltic myth, told by Macpherson in the last century, the voyage of the boat of heroes to Flath-Innis, Noble Island, the green island home of the departed, which lies calm amid the storms far in the Western Ocean. With full reason, also, Mr. Wright traces to the situation of Ireland in the extreme West its especial association with legends of descents to the land of shades. Claudian placed at the extremity of Gaul the entrance where Ulysses found a way to Hades—

"Est locus extremum qua pandit Gallia litus,
Oceani praetentus aquis, ubi fertur Ulysses," etc.

No wonder that this spot should have been since identified with St. Patrick's Purgatory, and that some ingenious etymologist should have found in the name of "Ulster" a corruption of "Ulyssisterra," and a commemoration of the hero's visit.[1]

Thirdly, the belief in a subterranean Hades peopled by the ghosts of the dead is quite common among the lower races. The earth is flat, say the Italmen of Kamchatka,

[1] Procop. De Bello Goth. iv. 20; Plut. Fragm. Comm. in Hesiod. 2; Grimm, 'D. M.' p. 793; Hersart de Villemarqué, vol. i. p. 136; Souvestre, 'Derniers Bretons,' p. 37; Jas. Macpherson, 'Introd. to Hist. of Great Britain and Ireland,' 2d. ed. London, 1772, p. 180; Wright, 'St. Patrick's Purgatory,' pp. 64, 129.

for if it were round, people would fall off; it is the wrong side of another heaven, which covers another earth below, whither the dead will go down to their new life, and so, as Steller says, their mundane system is like a tub with three bottoms.[1] In North America, the Tacullis held that the soul goes after death into the bowels of the earth, whence it can come back in human shape to visit friends.[2] In South America, Brazilian souls travel down to the world below in the West, and Patagonian souls will depart to enjoy eternal drunkenness in the caves of their ancestral deities.[3] The New Zealander who says "The sun has returned to Hades" (kua hoki mai te Ra ki te Rua), simply means that it has set. When a Samoan islander dies, the host of spirits that surround the house, waiting to convey his soul away, set out with him crossing the land and swimming the sea, to the entrance of the spirit-world. This is at the westernmost point of the westernmost island, Savaii, and there one may see the two circular holes or basins where souls descend, chiefs by the bigger and plebeians by the smaller, into the regions of the underworld. There below is a heaven, earth, and sea, and people with real bodies, planting, fishing, cooking, as in the present life; but at night their bodies become like a confused collection of fiery sparks, and in this state during the hours of darkness they come up to revisit their former abodes, retiring at dawn to the bush or to the lower regions.[4] For the state of thought on this subject among rude African tribes, it is enough to cite the Zulus, who at death will descend to live in Hades among their ancestors, the "Abapansi," the "people underground."[5] From among rude Asiatic tribes, let us take example from the Karens.

[1] Steller, 'Kamtschatka,' p. 269.
[2] Harmon, 'Journal,' p. 299 ; see Lewis and Clarke, p. 139 (Mandans).
[3] J. G. Müller, 'Amer. Urrelig.' pp. 140, 287 ; see Humboldt and Bonpland, 'Voy.' vol. iii. p. 132; Falkner, 'Patagonia,' p. 114.
[4] Taylor, 'New Zealand,' p. 232 ; Turner, 'Polynesia,' p. 235.
[5] Callaway, 'Zulu Tales,' vol. i. p. 317, etc.; Arbousset and Daumas, p. 474. See also Burton, 'Dahome,' vol. ii. p. 157.

They are not quite agreed where Plu, the land of the dead, is situate; it may be above the earth or beyond the horizon. But the dominant and seemingly indigenous opinion is that it is below the earth. When the sun sets on earth, it rises in the Karen Hades, and when it sets in Hades it rises in this world. Here, again, the familiar belief of the European peasant is found; the spirits of the dead may come up from the land of shades by night, but at daybreak must return.[1]

Such ideas, developed by uncultured races, may be followed up in various detail, through the stage of religion represented by the Mexican and Peruvian nations, into higher ranges of culture. The Roman Orcus was in the bowels of the earth, and when the 'lapis manalis,' the stone that closed the mouth of the world below, was moved away on certain solemn days, the ghosts of the dead came up to the world above, and partook of the offerings of their friends.[3] Among the Greeks, the land of Hades was in the world below, nor was the thought unknown that it was the sunset-realm of the Western god ($\pi\rho\grave{o}s\ \dot{\epsilon}\sigma\pi\acute{\epsilon}\rho o\upsilon\ \theta\epsilon o\hat{\upsilon}$). What Hades seemed like to the popular mind, Lucian thus describes:—
"The great crowd, indeed, whom the wise call 'idiots,' believing Homer and Hesiod, and the other myth-makers about these things, and setting up their poetry as a law, have supposed a certain deep place under the earth, Hades, and that it is vast, and roomy, and gloomy, and sunless, and how thought to be lighted up so as to behold every one within, I know not."[4] In the ancient Egyptian doctrine of the future life, modelled as it was on solar myth, Amenti, the western region of the departed, is an under-world or Hades; the dead passes the gate of the setting sun to traverse the roads of darkness, and behold his father Osiris; and with a

[1] Mason, 'Karens,' l. c. p. 195; Cross, l. c. p. 313. Turanian examples in Castrén, 'Finn. Myth.' p. 119.

[2] See below, pp. 79, 85.

[3] Festus, s. v. "manalis," etc.

[4] Sophocl. Œdip. Tyrann. 178; Lucian. De Luctu, 2. See classic details in Pauly, 'Real-Encyclop.' art. 'inferi.'

like solar thought the Egyptian priests, representing in symbolic ceremony the scenes of the other world, carried the corpse in the sacred boat across to the burial-place, on the western side of the sacred lake.[1] So, too, the cavernous Sheol of the Israelites, the shadowy region of departed souls, lay deep below the earth. Through the great Aryan religious systems, Brahmanism, Zarathustrism, Buddhism, and onward into the range of Islam and of Christianity, subterranean hells of purgatory or punishment make the doleful contrast to heavens of light and glory.

It is, however, a point worthy of special notice that the conception of hell as a fiery abyss, so familiar to the religions of the higher civilization, is all but unknown to savage thought, so much so that if met with, its genuineness is doubtful. Captain John Smith's 'History of Virginia,' published in 1624, contains two different accounts of the Indians' doctrine of a future life. Smith's own description is of a land beyond the mountains, toward sunset, where chiefs and medicine-men in paint and feathers shall smoke, and sing, and dance with their forefathers, while the common people have no life after death, but rot in their graves. Heriot's description is of tabernacles of the gods to which the good are taken up to perpetual happiness, while the wicked are carried to 'Popogusso,' a great pit which they think to be at the furthest parts of the world where the sun sets, and there burn continually.[2] Now knowing so much as we do of the religion of the Algonquins, to whom these Virginians belonged, we may judge that while the first account is genuinely native, though perhaps not quite correctly understood, the second was borrowed by the Indians from the white men themselves. Yet even here the touch of solar myth is manifest, and the description of the fiery abyss in the region of sunset may be compared with one

---

[1] Birch in Bunsen's 'Egypt,' vol. v.; Wilkinson, 'Ancient Eg.' vol. ii. p. 368; Alger, p. 101.

[2] Smith, Virginia, in Pinkerton, vol. xiii. pp. 14, 41; vol. xii. p. 604; see below, p. 95.

from our own country, in the Anglo-Saxon dialogue of Saturn and Solomon. "Saga me forhwan byth seo sunne read on æfen? Ic the secge, forthon heo locath on helle. —Tell me, why is the sun red at even? I tell thee, because she looketh on hell."[1] To the same belief belongs another striking mythic feature. The idea of volcanos being mouths of the under-world seems not unexampled among the lower races, for we hear of certain New Zealanders casting their dead down into a crater.[2] But in connexion with the thought of a gehenna of fire and brimstone, Vesuvius, Etna, and Hecla had spiritual as well as material terrors to the mind of Christendom, for they were believed to be places of purgatory or the very mouths of the pit where the souls of the damned were cast down.[3] The Indians of Nicaragua used in old times to offer human sacrifices to their volcano Masaya, flinging the corpses into the crater, and in later years, after the conversion of the country, we hear of Christian confessors sending their penitents to climb the mountain, and (as a glimpse of hell) to look down upon the molten lava.[4]

Fourthly, in old times and new, it has come into men's minds to fix upon the sun and moon as abodes of departed souls. When we have learnt from the rude Natchez of the Mississippi and the Apalaches of Florida that the sun is the bright dwelling of departed chiefs and braves, and have traced like thoughts on into the theologies of Mexico and Peru, then we may compare these savage doctrines with Isaac Taylor's ingenious supposition in his 'Physical Theory of Another Life,'—the sun of each planetary system is the house of the higher and ultimate spiritual corporeity, and the centre of assembly to those who have passed on the planets their preliminary era of corruptible organization. Or perhaps some may prefer the Rev. Tobias Swinden's

[1] Thorpe, 'Analecta Anglo-Saxonica,' p. 115.
[2] Schirren, p. 151. See Taylor, 'N. Z.' p. 525.
[3] Meiners, vol. ii. p. 781; Maury, 'Magie,' etc. p. 170.
[4] Oviedo, 'Nicaragua,' p. 160; Brinton, p. 288.

book, published in the last century, and translated into French and German, which proved the sun to be hell, and its dark spots gatherings of damned souls.[1] And when in South America the Saliva Indians have pointed out the moon, their paradise where no mosquitos are, and the Guaycurus have shown it as the home of chiefs and medicine-men deceased, and the Polynesians of Tokelau in like manner have claimed it as the abode of departed kings and chiefs, then these pleasant fancies may be compared with that ancient theory mentioned by Plutarch, that hell is in the air and elysium in the moon,[2] and again with the mediæval conception of the moon as the seat of hell, a thought elaborated in profoundest bathos by Mr. M. F. Tupper:

"I know thee well, O Moon, thou cavern'd realm,
Sad Satellite, thou giant ash of death,
Blot on God's firmament, pale home of crime,
Scarr'd prison-house of sin, where damned souls
Feed upon punishment. Oh, thought sublime,
That amid night's black deeds, when evil prowls
Through the broad world, thou, watching sinners well,
Glarest o'er all, the wakeful eye of—Hell!"

Skin for skin, the brown savage is not ill matched in such speculative lore with the white philosopher.

Fifthly, as Paradise on the face of the earth, and Hades beneath it where the sun goes down, are regions whose existence is asserted or not denied by savage and barbaric science, so it is with Heaven. Among the examples which display for us the real course of knowledge among mankind, and the real relation which primitive bears to later culture, the belief in the existence of a firmament is one of the most

[1] J. G. Müller, 'Amer. Urrel.' p. 138, see also 220 (Caribs), 402 (Peru), 505, 660 (Mexico); Brinton, 'Myths of New World,' p. 233; Taylor, 'Physical Theory,' ch. xvi.; Alger, 'Future Life,' p. 590; see also above, p. 16, note.

[2] Humboldt and Bonpland, 'Voy.' vol. v. p. 90; Martius, 'Ethnog. Amer.' vol. i. p. 233; Turner, 'Polynesia,' p. 531; Plutarch. De Facie in Orbe Lunæ; Alger, l. c.; Bastian, 'Psychologie,' pp. 80, 89 (souls in stars.)

instructive. It arises naturally in the minds of children still, and in accordance with the simplest childlike thought, the cosmologies of the North American Indians [1] and the South Sea Islanders [2] describe their flat earth arched over by the solid vault of heaven. Like thoughts are to be traced on through such details as the Zulu idea that the blue heaven is a rock encircling the earth, inside which are the sun, moon, and stars, and outside which dwell the people of heaven; the modern negro's belief that there is a firmament stretched above like a cloth or web; the Finnish poem which tells how Ilmarinen forged the firmament of finest steel, and set in it the moon and stars.[3] The New Zealander, with his notion of a solid firmament, through which the waters can be let down on earth through a crack or hole from the reservoir of rain above, could well explain the passage in Herodotus concerning that place in North Africa where, as the Libyans said, the sky is pierced, as well as the ancient Jewish conception of a firmament of heaven, "strong as a molten mirror," with its windows through which the rain pours down in deluge from the reservoirs above, windows which late Rabbinical literature tells us were made by taking out two stars.[4] In nations where the theory of the firmament prevails, accounts of bodily journeys or spiritual ascents to heaven are in general meant not as figure, but as fact. Among the lower races, the tendency to localize the region of departed souls above the sky seems less strong than that which leads them to place their world of the dead on or below the earth's surface. Yet some well-marked descriptions of a savage

[1] See Schoolcraft, 'Indian Tribes,' part i. pp. 269, 311; Smith, 'Virginia, in Pinkerton, vol. xiii. p. 54; Waitz, vol. iii. p. 223; Squier, 'Abor. Mon. of N. Y.' p. 156; Catlin, 'N. A. Ind.' vol. i. p. 180.

[2] Mariner, 'Tonga Is.' vol. ii. p. 134; Turner, 'Polynesia,' p. 103; Taylor, 'New Zealand,' pp. 101, 114, 256.

[3] Callaway, 'Rel. of Amazulu,' p. 393; Burton, 'W. and W. fr. W. Afr.' p. 454; Castrén, 'Finn. Myth.' p. 295.

[4] Herodot. iv. 158, see 185, and Rawlinson's note. See Smith's 'Dic. of the Bible,' s. v. "firmament." Eisenmenger, part i. p. 408.

Heaven are on record, the following, and others to be cited presently. Even some Australians seem to think of going up to the clouds at death, to eat and drink, and hunt and fish as here.[1] In North America, the Winnebagos placed their paradise in the sky, where souls travel along that "Path of the Dead" which we call the Milky Way; and, working out the ever-recurring solar idea, the modern Iroquois speak of the soul going upward and westward, till it comes out on the beauteous plains of heaven, with people and trees and things as on earth.[2] In South America the Guarayos, representatives in some sort of the past condition of the Guarani race, worship Tamoi the Grandfather, the Ancient of Heaven; he was their first ancestor, who lived among them in old days and taught them to till the ground; then rising to heaven in the East he disappeared, having promised to be the helper of his people on earth, and to transport them, when they died, from the top of a sacred tree into another life, where they shall find their kindred and have hunting in plenty, and possess all that they possessed on earth; therefore it is that the Guarayos adorn their dead, and burn their weapons for them, and bury them with their faces to the East, whither they are to go.[3] Among American peoples whose culture rose to a higher level than that of these savage tribes, we hear of the Peruvian Heaven, the glorious "Upper World," and of the temporary abode of Aztec warriors on heavenly wooded plains, where the sun shines when it is night on earth, wherefore it was a Mexican saying that the sun goes at evening to lighten the dead.[4] What thoughts of heaven were in the minds of the old Aryan poets, this hymn from the Rig-Veda may show:—

[1] Eyre, 'Australia,' vol. ii. p. 367.
[2] Schoolcraft, 'Indian Tribes,' part iv. p. 240 (but compare part v. p. 403); Morgan, 'Iroquois,' p. 176; Sproat, 'Savage Life,' p. 209.
[3] D'Orbigny, 'L'Homme Américain,' vol. ii. pp. 319, 328; see Martius, vol. i. p. 485 (Jumanas).
[4] J. G. Müller, p. 403; Brasseur, 'Mexique,' vol. iii. p. 496; Kingsborough, 'Mexico,' Cod. Letellier, fol. 20.

"Where there is eternal light, in the world where the sun is placed, in that immortal imperishable world place me, O Soma!
Where king Vaivasvata reigns, where the secret place of heaven is, where these mighty waters are, there make me immortal!
Where life is free, in the third heaven of heavens, where the worlds are radiant, there make me immortal!
Where wishes and desires are, where the place of the bright sun is, where there is freedom and delight, there make me immortal!
Where there is happiness and delight, where joy and pleasure reside, where the desires of our desire are attained, there make me immortal!"[1]

In such bright vague thoughts from the poet's religion of nature, or in cosmic schemes of ancient astronomy, with their artificial glories of barbaric architecture exaggerated in the skies, or in the raptures of mystic vision, or in the calmer teaching of the theologic doctrine of a future life, descriptions of realms of blessed souls in heaven are to be followed through the religions of the Brahman, the Buddhist, the Parsi, the later Jew, the Moslem, and the Christian.

For the object, not of writing a handbook of religions, but of tracing the relation which the religion of savages bears to the religion of cultured nations, these details are enough to show the general line of human thought regarding the local habitations of departed souls. It seems plain from the most cursory inspection of these various localizations, however much we may consider them as inherited or transmitted from people to people in the complex movements of theological history, that they are at any rate not derived from any single religion accepted among ancient or primæval men. They bear evident traces of independent working out in the varied definition of the region of souls, as on earth among men, on earth in some distant country, below the earth, above or beyond the sky. Similar ideas of this kind are found in different lands, but this simi-

[1] Max Müller, 'Chips,' vol. i. p. 46; Roth in 'Zeitschr. d. Deutsch. Morgenl. Ges.' vol. iv. p. 427.

larity seems in large measure due to independent recurrence of thoughts so obvious. Not less is independent fancy compatible with the ever-recurring solar myth in such ideas, placing the land of Death in the land of Evening or of Night, and its entrance in the gates of Sunset. Barbaric poets of many a distant land must have gazed into the West to read the tale of Life and Death, and tell it of Man. If, however, we look more closely into the stages of intellectual history to which these theories of the Future World belong, it will appear that the assignment of the realm of departed souls to the three great regions, Earth, Hades, Heaven, has not been uniform. Firstly, the doctrine of a land of souls on Earth belongs widely and deeply to savage culture, but dwindles in the barbaric stage, and survives but feebly into the mediæval. Secondly, the doctrine of a subterranean Hades holds as large a place as this in savage belief, and has held it firmly along the course of higher religions, where, however, this under-world is looked on less and less as the proper abode of the dead, but rather as the dismal place of purgatory and hell. Lastly, the doctrine of a Heaven, floored upon a firmament, or placed in the upper air, seems in early savage belief less common than the other two, but yields to neither of them in its vigorous retention by the thought of modern nations. These local theories appear to be taken, firstly and mostly, in the most absolute literal sense, and although, under the influence of physical science, much that was once distinctly-meant philosophy has now passed among theologians into imagery and metaphor, yet at low levels of knowledge the new canons of interpretation find little acceptance, and even in modern Europe the rude cosmology of the lower races in no small measure retains its place.

Turning now to consider the state of the departed in these their new homes, we have to examine the definitions of the Future Life which prevail through the religions of mankind. In these doctrines there is much similarity caused by the spreading of established beliefs into new countries, and also much similarity that is beyond what such transmission can account for. So there is much variety due to local colour and circumstance, and also much variety beyond the reach of such explanation. The main causes of both similarity and variety seem to lie far deeper, in the very origin and inmost meaning of the doctrines. The details of the future life, among the lower races and upwards, are no heterogeneous mass of arbitrary fancies. Classified, they range themselves naturally round central ideas, in groups whose correspondence seems to indicate the special course of their development. Amongst the pictures into which this world has shaped its expectations of the next, two great conceptions are especially to be discerned. The one is that the future life is, as it were, a reflexion of this; in a new world, perhaps of dreamy beauty, perhaps of ghostly gloom, men are to retain their earthly forms and their earthly conditions, to have around them their earthly friends, to possess their earthly property, to carry on their earthly occupations. The other is that the future life is a compensation for this, where men's conditions are re-allotted as the consequence, and especially as the reward or punishment, of their earthly life. The first of these two ideas we may call (with Captain Burton) the "continuance-theory," contrasting with it the second as the "retribution-theory." Separately or combined, these two doctrines are the keys of the subject, and by grouping typical examples under their two headings, it will be possible to survey systematically man's most characteristic schemes of his life beyond the grave.

To the doctrine of Continuance belongs especially the savage view of the spirit-land, that it is as the dream-land where the souls of the living so often go to visit the souls of the dead. There the soul of the dead Karen, with the souls of his axe and cleaver, builds his house and cuts his rice; the shade of the Algonquin hunter hunts souls of beaver and elk, walking on the souls of his snow-shoes over the soul of the snow; the fur-wrapped Kamchadal drives his dog-sledge; the Zulu milks his cows and drives his cattle to kraal; South American tribes live on, whole or mutilated, healthy or sick, as they left this world, leading their old lives, and having their wives with them again, though indeed, as the Araucanians said, they have no more children, for they are but souls.[1] Soul-land is dream-land in its shadowy unreal pictures, for which, nevertheless, material reality so plainly furnished the models, and it is dream-land also in its vivid idealization of the soberer thoughts and feelings of waking life,

> " There was a time when meadow, grove, and stream,
> The earth, and every common sight,
> To me did seem
> Apparell'd in celestial light,
> The glory and the freshness of a dream."

Well might the Mohawk Indian describe the good land of paradise, as he had seen it in a dream. The shade of the Ojibwa follows a wide and beaten path that leads toward the West, he crosses a deep and rapid water, and reaching a country full of game and all things the Indian covets, he joins his kindred in their long lodge.[2] So, on the southern continent, the Bolivian Yuracarés will go, all of them, to a future life where there will be plenty of hunting, and Brazilian forest-tribes will find a pleasant forest full of calabash-trees and game, where the souls of the dead will live happily in company.[3] The Greenlanders hoped that their souls—pale, soft, disembodied forms which the living could not grasp—would lead a life better than that of earth, and never ceasing. It might be in heaven, reached by the rainbow, where the souls pitch their tents round the great

[1] Cross, 'Karens,' l. c. pp. 309, 313; Le Jeune in 'Rel. des Jes.' 1634, p. 16; Steller, 'Kamtschatka,' p. 272; Callaway, 'Zulu Tales,' vol. i. p. 316; Klemm, 'Cultur-Gesch.' vol. ii. pp. 310, 315; J. G. Müller, 'Amer. Urrel.' pp. 139, 286.
[2] Bastian, 'Psychologie,' p. 224; Schoolcraft, 'Indian Tribes,' part ii. p. 135.
[3] D'Orbigny, 'L'Homme Américain,' vol. i. p. 364; Spix and Martius, 'Brasilien,' vol. i. p. 383; De Laet, Novus Orbis, xv. 2.

lake rich in fish and fowl, the lake whose waters above the firmament overflowing make rain on earth, and if its banks broke, there would be another deluge. But gaining the most and best of their living from the depths of the sea, they were also apt to think the land of Torngarsuk to be below the sea or earth, and to be entered by the deep holes in the rocks. Perpetual summer is there, ever beauteous sunshine, and no night, good water and superfluity of birds and fish, seals and reindeer to be caught without difficulty, or found alive seething in a great kettle.[1] In the Kimbunda country of South-West Africa, souls live on in "Kalunga" the world where it is day when it is night here; and with plenty of food and drink, and women to serve them, and hunting and dancing for pastime, they lead a life which seems a corrected edition of this.[2] When we compare these pictures of the future life with such as have expressed the longings of more cultured nations, there appear indeed different details, but the principle is ever the same—the idealization of earthly good. The Norseman's ideal is sketched in the few broad touches which show him in Walhalla, where he and the other warriors without number ride forth arrayed each morning and hew each other on Odin's plain, till the slain have been "chosen" as in earthly battle, and meal-tide comes, and slayers and slain mount and ride home to feast on the everlasting boar, and drink mead and ale with the Æsir.[3] To understand the Moslem's mind, we must read the two chapters of the Koran where the Prophet describes the faithful in the garden of delights, reclining on their couches of gold and gems, served by children ever young, with bowls of liquor whose fumes will not rise into the drinkers' heads, living among the thornless lotus-trees and bananas loaded to the ground, feasting on the fruits they love and the meat of the rarest birds, with the houris near them with beautiful black eyes, like

[1] Cranz, 'Grönland,' p. 258.
[2] Magyar, 'Süd-Afrika,' p. 336.
[3] Edda: 'Gylfaginning.'

pearls in the shell, where no idle or wicked speech is heard, but only the words "Peace, Peace."

"They who fear the judgment of God shall have two gardens.
Which of the benefits of God will ye deny?
Adorned with groves.
Which of the benefits of God will ye deny?
In each of them shall spring two fountains.
Which of the benefits of God will ye deny?
In each of them shall grow two kinds of fruits.
Which of the benefits of God will ye deny?
They shall lie on carpets brocaded with silk and embroidered with gold; the fruits of the two gardens shall be near, easy to pluck.
Which of the benefits of God will ye deny?
There shall be young virgins with modest looks, unprofaned by man or jinn.
Which of the benefits of God will ye deny?
They are like jacinth and coral.
Which of the benefits of God will ye deny?
What is the recompence of good, if not good?
Which of the benefits of God will ye deny?" etc.[1]

With these descriptions of Paradise idealized on secular life, it is interesting to compare others which bear the impress of a priestly caste, devising a heaven after their manner. We can almost see the faces of the Jewish rabbis settling their opinions about the high schools in the firmament of heaven, where Rabbi Simeon ben Yochai and the great Rabbi Eliezer teach Law and Talmud as they taught when they were here below, and masters and learners go prosing on with the weary old disputations of cross question and crooked answer that pleased their souls on earth.[2] Nor less suggestively do the Buddhist heavens reflect the minds of the ascetics who devised them. As in their thoughts sensual pleasure seemed poor and despicable in comparison with mystic inward joy, rising and rising till consciousness fades in trance, so, above their heavens of millions of years of mere divine happiness, they raised other ranges of heavens where sensual pain and pleasure cease, and enjoy-

[1] 'Koran,' ch. lv. lvi.
[2] Eisenmenger, 'Entdecktes Judenthum,' part i. p. 7.

ment becomes intellectual, till at a higher grade even bodily form is gone, and after the last heaven of "Neither-consciousness-nor-unconsciousness" there follows Nirwâna, as ecstasy passes into swoon.[1]

But the doctrine of the continuance of the soul's life has another and a gloomier side. There are conceptions of an abode of the dead characterized not so much by dreaminess as by ghostliness. The realm of shades, especially if it be a cavern underground, has seemed a dim and melancholy place to the dwellers in this "white world," as the Russian calls the land of the living. One description of the Hurons tells how the other world, with its hunting and fishing, its much-prized hatchets and robes and necklaces, is like this world, yet day and night the souls groan and lament.[2] Thus the region of Mictlan, the subterranean land of Hades whither the general mass of the Mexican nation, high and low, expected to descend from the natural death-bed, was an abode looked forward to with resignation, but scarcely with cheerfulness. At the funeral the survivors were bidden not to mourn too much, the dead was reminded that he had passed and suffered the labours of this life, transitory as when one warms himself in the sun, and he was bidden to have no care or anxiety to return to his kinsfolk now that he has departed for ever and aye, for his consolation must be that they too will end their labours, and go whither he has gone before.[3] Among the Basutos, where the belief in a future life in Hades is general, some imagine in this underworld valleys ever green, and herds of hornless speckled cattle owned by the dead; but it seems more generally thought that the shades wander about in silent calm, experiencing neither joy nor sorrow. Moral retribution there is none.[4] The Hades of the West African seems no

[1] Hardy, 'Manual of Buddhism,' pp. 5, 24; Köppen, 'Rel. des Buddha,' vol. i. p. 235, etc.
[2] Brebeuf in 'Rel. des Jes.' 1636, p. 105.
[3] Sahagun, 'Hist. de Nueva España,' book iii. appendix ch. i., in Kingsborough, vol. vii.; Brasseur, vol. iii. p. 571.
[4] Casalis, 'Basutos,' pp. 247, 254.

ecstatic paradise, to judge by Captain Burton's description:
"It was said of the old Egyptians that they lived rather in
Hades than upon the banks of the Nile. The Dahomans
declare that this world is man's plantation, the next is his
home,—a home which, however, no one visits of his own
accord. They of course own no future state of rewards
and punishments: there the King will be a King, and the
slave a slave for ever. Ku-to-men, or Deadman's land, the
Dahoman's other but not better world, is a country of
ghosts, of umbræ, who, like the spirits of the nineteenth
century in Europe, lead a quiet life, except when by means
of mediums they are drawn into the drawing-rooms of the
living." With some such hopeless expectation the neighbours of the Dahomans, the Yorubas, judge the life to come
in their simple proverb that "A corner in this world is
better than a corner in the world of spirits."[1] The Finns,
who feared the ghosts of the departed as unkind, harmful
beings, fancied them dwelling with their bodies in the grave,
or else, with what Castrén thinks a later philosophy, assigned
them their dwelling in the subterranean Tuonela. Tuonela
was like this upper earth, the sun shone there, there was no
lack of land and water, wood and field, tilth and meadow,
there were bears and wolves, snakes and pike, but all things
were of a hurtful, dismal kind, the woods dark and swarming with wild beasts, the water black, the cornfields bearing
seed of snakes' teeth, and there stern pitiless old Tuoni,
and his grim wife and son with the hooked fingers with iron
points, kept watch and ward over the dead lest they should
escape.[2] Scarce less dismal was the classic ideal of the
dark realm below, whither the shades of the dead must go
to join the many gone before ($\dot{\epsilon}$ς πλεόνων ἱκέσθαι; penetrare
ad plures; andare tra i più). The Roman Orcus holds the
pallid souls, rapacious Orcus, sparing neither good nor bad.

[1] Burton, 'Dahome,' vol. ii. p. 156; 'Tr. Eth. Soc.' vol. iii. p. 403; 'Wit and Wisdom from W. Afr.' pp. 280, 449; see J. G. Müller, p. 140.
[2] Castrén, 'Finn. Myth.' p. 126, etc.; Kalewala, Rune xv. xvi. xlv. etc.; Meiners, vol. ii. p. 780.

Gloomy is the Greek land of Hades, dark dwelling of the images of departed mortals, where the shades carry at once their living features and their dying wounds, and glide and cluster and whisper, and lead the shadow of a life. Like the savage hunter on his ghostly prairie, the great Orion still bears his brazen mace, still chases over the meadows of asphodel the flying beasts he slew of yore in the lonely mountains. Like the rude African of to-day, the swift-footed Achilles scorns such poor, thin, shadowy life; rather would he serve a mean man upon earth than be lord of all the dead.

> "Truly, oxen and goodly sheep may be taken for booty,
> Tripods, too, may be bought, and the yellow beauty of horses;
> But from the fence of the teeth when once the soul is departed,
> Never cometh it back, regained by plunder or purchase." [1]

Where and what was Sheol, the dwelling of the ancient Jewish dead? Though its description is so suggested by the dark, quiet, inevitable cavern-tomb, that the two conceptions melt together in Hebrew poetic phrase, nevertheless Sheol is not a mere general term for burial-places. Nations to whom the idea of a subterranean region of departed spirits was a familiar thought, with familiar words to express it, quite naturally use these words in Biblical translation as the equivalents of Sheol. To the Greek Septuagint, *Sheol* was *Hades*, and for this the Coptic translators had their long-inherited Egyptian name of *Amenti*, while the Vulgate renders it as *Infernus*, the lower regions. The Gothic Ulfilas, translating the Hades of the New Testament, could use *Halja* in its old German sense of the dim shadowy home of the dead below the earth; and the corresponding word *Hell*, if this its earlier sense be borne in mind, fairly translates Sheol and Hades in the English version of the Old and New Testament, though the word has become misleading to uneducated ears by being used also in the sense of Gehenna, the place of torment. The

---

[1] Homer. Il. ix. 405; Odyss. xi. 218, 475; Virg. Æn. vi. 243, etc. etc.

early Hebrew historians and prophets, holding out neither the hope of everlasting glory nor the fear of everlasting agony as guiding motives for man's present life, lay down little direct doctrine of a future state, yet their incidental mentions justify the translators who regard Sheol as Hades. Sheol is a special locality where dead men go to their dead ancestors: "And Isaac gave up the ghost, and died, and was gathered unto his people . . . . and his sons Esau and Jacob buried him." Abraham, though not even buried in the land of his forefathers, is thus "gathered unto his people;" and Jacob has no thought of his body being laid with Joseph's body, torn by wild beasts in the wilderness, when he says, "I shall go down to my son mourning to Sheol" ("εἰς ᾅδου" in the LXX., "ἐπεσēt ἐàmenti" in the Coptic, "in infernum" in the Vulgate). Sheol (שאול from שאל) is, as its name implies, a cavernous recess, yet it is no mere surface-grave or tomb, but an under-world of awful depth: "High as Heaven, what doest thou? deeper than Sheol, what knowest thou?" "Though they dig into Sheol, thence shall mine hand take them; though they climb up to Heaven, thence will I bring them down." Thither Jew and Gentile go down: "What man liveth, and shall not see death? shall he deliver his soul from the hand of Sheol?" Asshur and all her company, Elam and all her multitude, the mighty fallen of the uncircumcised, lie there. The great king of Babylon must go down:—

"Sheol from beneath is moved because of thee, to meet thee at thy coming:
He rouseth for thee the mighty dead, all the great chiefs of the earth;
He maketh to rise up from their thrones, all the kings of the nations.
All of them shall accost thee, and shall say unto thee:
Art thou, even thou too, become weak as we? Art thou made like unto us?"

The rephaim, the "shades" of the dead, who dwell in Sheol, love not to be disturbed from their rest by the necromancer; "And Samuel said to Saul, why hast thou disquieted me to bring me up?" Yet their quiet is contrasted in a tone of sadness with the life on earth; "Whatsoever thy hand findeth to do, do it with thy might; for there is no work, nor device, nor knowledge, nor wisdom, in Sheol, whither thou goest."[1] Such thoughts of the life of the shades below did not disappear when, in the later years of the Jewish nation, the great change in the doctrine of the future life passed in so large a measure over the Hebrew mind, their earlier thoughts of ghostly continuance giving place to the doctrines of resurrection and retribution. The ancient ideas have even held their place on into Christian thought, in pictures like that of the Limbus Patrum, the Hades where Christ descended to set free the patriarchs.

The Retribution-theory of the future life comprises in a general way the belief in different grades of future happiness, especially in different regions of the other world allotted to men according to their lives in this. This doctrine of retribution is, as we have already seen, far from universal among mankind, many races recognizing the idea of a spirit outliving the body, without considering the fate of this spirit to depend at all upon the conduct of the living man. The doctrine of retribution indeed hardly seems an original part of the doctrine of the future life. On the contrary, if we judge that men in a primitive state of culture arrived at the notion of a surviving spirit, and that some races, but by no means all, afterwards reached the further stage of recognizing a retribution for deeds done in the body, this theory will not, so far as I know, be discountenanced by facts.[2] Even among the higher savages, however, a con-

[1] Gen. xxxv. 29; xxv. 8; xxxvii. 35; Job xi. 8; Amos ix. 2; Psalm lxxxix. 48; Ezek. xxxi., xxxii.; Isaiah xiv. 9, xxxviii. 10-18; 1 Sam. xxviii. 15; Eccles. ix. 10. This argument is mainly from Alger, 'Critical History of the Doctrine of a Future Life.' ch. viii.; see F. W. Farrar in Smith's 'Dic. of the Bible,' art. "hell."

[2] The doctrine of reversal, as in Kamschatka, where rich and poor will change places in the other world (Steller, pp. 269-72), is too exceptional in the lower culture to be generalized. See Steinhauser, 'Rel. des Negers.' l. c. p. 135. A Wolof proverb is "The more powerful one is in this world,

nexion between man's life and his happiness or misery after death is often held as a definite article of theology, and thence it is to be traced onward through barbaric religions, and into the very heart of Christianity. Yet the grounds of future reward and punishment are so far from uniform among the religions of the world, that they may differ widely within what is considered one and the same creed. The result is more definite than the cause, the end than the means. Men who alike look forward to a region of unearthly happiness beyond the grave, hope to reach that happy land by roads so strangely different, that the path of life which leads one nation to eternal bliss may seem to the next the very descent into the pit. In noticing among savage and barbaric peoples the qualifications which determine future happiness, we may with some distinctness define these as being excellence, valour, social rank, religious ordinance. On the whole, however, in the religions of the lower range of culture, unless where they may have been affected by contact with higher religions, the destiny of the soul after death seems comparatively seldom to turn on a judicial system of reward and punishment. Such difference as they make between the future conditions of different classes of souls, seems more often to belong to a remarkable intermediate doctrine, standing between the earlier continuance-theory and the later retribution-theory. The idea of the next life being similar to this seems to have developed into the idea that what gives prosperity and renown here will give it also there, so that earthly conditions carry on their contrasts into the changed world after death. Thus a man's condition after death will be a result of, rather than a compensation or retribution for, his condition during life. A comparison of doctrines held at various stages of culture may justify a tentative speculation as to their actual sequence in history, favouring the opinion that through such an intermediate stage the doctrine of simple

the more servile one will be in the next." (Burton, 'Wit and Wisdom, p. 28.)

future existence was actually developed into the doctrine of future reward and punishment, a transition which for deep import to human life has scarcely its rival in the history of religion.

The effect of earthly rank on the future life, as looked at by the lower races, brings out this intermediate stage in bold relief. Mere transfer from one life to another makes chiefs and slaves here chiefs and slaves hereafter, and this natural doctrine is very usual. But there are cases in which earthly caste is exaggerated into utter difference in the life to come. The aerial paradise of Raiatea, with its fragrant ever-blooming flowers, its throngs of youths and girls all perfection, its luxurious feasts and merrymakings, were for the privileged orders of Areois and chiefs who could pay the priests their heavy charges, but hardly for the common populace. This idea reached its height in the Tonga islands, where aristocratic souls would pass to take their earthly rank and station in the island paradise of Bolotu, while plebeian souls, if indeed they existed, would die with the plebeian bodies they dwelt in.[1] In Vancouver's Island, the Ahts fancied Quawteaht's calm sunny plenteous land in the sky as the resting-place of high chiefs, who live in one great house as the Creator's guests, while the slain in battle have another to themselves. But otherwise all Indians of low degree go deep down under the earth to the land of Chay-her, with its poor houses and no salmon and small deer, and blankets so small and thin, that when the dead are buried the friends often bury blankets with them, to send them to the world below with the departed soul.[2] The expectation of royal dignity in the life after death, distinct from the fate of ordinary mortals, comes well into view among the Natchez of Louisiana, where the sun-descended royal family would in some way return to the Sun; thus

---

[1] Ellis, 'Polyn. Res.' vol. i. pp. 245, 397; see also Turner, 'Polynesia,' p. 237 (Samoans); Mariner, 'Tonga Is.' vol. ii. p. 105.
[2] Sproat, 'Savage Life,' p. 209.

also in the mightier empire of Peru, where each sun-descended Inca, feeling the approach of death, announced to his assembled vassals that he was called to heaven to rest with his father the Sun.[1] But in the higher religions, the change in this respect from the doctrine of continuance to the doctrine of retribution is wonderful in its completeness. The story of that great lady who strengthened her hopes of future happiness by the assurance, " They will think twice before they refuse a person of my condition," is a mere jest to modern ears. Yet, like many another modern jest, it is only an archaism which in an older stage of culture had in it nothing ridiculous.

To the happy land of Torngarsuk the Great Spirit, says Cranz, only such Greenlanders come as have been valiant workers, for other ideas of virtue they have none; such as have done great deeds, taken many whales and seals, borne much hardship, been drowned at sea, or died in childbirth.[2] Thus Charlevoix says of the Indians further south, that their claim to hunt after death on the prairies of eternal spring is to have been good hunters and warriors here. Lescarbot, speaking of the belief among the Indians of Virginia that after death the good will be at rest and the wicked in pain, remarks that their enemies are the wicked and themselves the good, so that in their opinion they are after death much at their ease, and principally when they have well defended their country and slain their enemies.[3] So Jean de Lery said of the rude Tupinambas of Brazil, that they think the souls of such as had lived virtuously, that is to say, who have well avenged themselves and eaten many of their enemies, will go behind the great mountains and dance in beautiful gardens with the souls of their fathers, but the souls of the effeminate and worthless, who

---

[1] 'Rec. des Voy. au Nord,' vol. v. p. 23 (Natchez); Garcilaso de la Vega, 'Commentarios Reales,' lib. i. c. 23, tr. by C. R. Markham; Prescott, 'Peru,' vol. i. pp. 29, 83; J. G. Müller, p. 402, etc.

[2] Cranz, 'Grönland,' p. 259.

[3] Charlevoix, 'Nouvelle France,' vol. vi. p. 77; Lescarbot, 'Hist. de la Nouvelle France,' Paris, 1619, p. 679.

have not striven to defend their country, will go to Aygnan, the Evil Spirit, to incessant torments.[1] More characteristic, and probably more genuinely native than this, is the fancy of the Caribs, that the braves of their nation should go after death to happy islands, where all good fruits grow wild, there to spend their time in dancing and feasting, and to have their enemies the Arawaks for slaves; but the cowards who feared to go to war should go to serve the Arawaks, dwelling in their waste and barren lands beyond the mountains.[2]

The fate of warriors slain in battle is the subject of two singularly contrasted theories. We have elsewhere examined the deep-lying belief that if a man's body be wounded or mutilated, his soul will arrive in the same state in the other world. Perhaps it is some such idea of the soul being spoilt with the body by a violent death, that leads the Mintira of the Malay Peninsula, though not believing in a future reward and punishment, to exclude from the happy paradise of "Fruit Island" (Pulo Bua) the souls of such as die a bloody death, condemning them to dwell on "Red Land" (Tana Mera), a desolate barren place, whence they must even go to the fortunate island to fetch their food.[3] In North America, the idea is mentioned among the Hurons that the souls of the slain in war live in a band apart, neither they nor suicides being admitted to the spirit-villages of their tribe. A belief ascribed to certain Indians of California may be cited here, though less as a sample of real native doctrine than to illustrate that borrowing of Christian ideas which so often spoils such evidence for ethnological purposes. They held, it is said, that Niparaya, the Great Spirit, hates war, and will have no warriors in his paradise, but that his adversary Wac, shut up for rebellion in a great cave, takes thither to himself the

[1] Lery, 'Hist. d'un Voy. en Brésil,' p. 234; Coreal, 'Voi. aux Indes Occ.' vol. i. p. 224.
[2] Rochefort, 'Iles Antilles,' p. 430.
[3] 'Journ. Ind. Archip.' vol. i. p. 325.

slain in battle.[1] On the other hand, the thought which shows out in such bold relief in the savage mind, that courage is virtue, and battle and bloodshed the hero's noblest pursuit, leads naturally to a hope of glory for his soul when his body has been slain in fight. Such expectation was not strange in North America, to that Indian tribe, for instance, who talked of the Great Spirit walking in the moonlight on his island in Lake Superior, whither slain warriors will go to him to take their pleasure in the chace.[2] The Nicaraguans declared that men who died in their houses went underground, but the slain in war went to serve the gods in the east, where the sun comes from. This corresponds in part with the remarkable threefold contrast of the future life among their Aztec kinsfolk. Mictlan, the Hades of the general dead, and Tlalocan, the Earthly Paradise, reached by certain special and acute ways of death, have been mentioned here already. But the souls of warriors slain in battle or sacrificed as captives, and of women who died in childbirth, were transported to the heavenly plains; there the heroes, peeping through the holes in their bucklers pierced by arrows in earthly fight, watched the Sun arise and saluted him with shout and clash of arms, and at noon the mothers received him with music and dance to escort him on his western way.[3] In such wise, to the old Norseman, to die the "straw-death" of sickness or old age was to go down into the dismal loathly house of Hela the Death-goddess; if the warrior's fate on the field of battle were denied him, and death came to fetch him from a peaceful couch, yet at least he could have the scratch of the spear, Odin's mark, and so contrive to go with a blood-stained soul to the glorious Walhalla. Surely then if ever, says a

[1] Brebeuf in 'Rel. des Jes.' 1636, p. 104; see also Meiners, vol. ii. p. 769. J. G. Müller, pp. 89, 139.

[2] Chateaubriand, 'Voy. en Amérique' (Religion).

[3] Oviedo, 'Nicaragua,' p. 22; Torquemada, 'Monarquia Indiana,' book xiii. c. 48; Sahagun, book iii. app. ch. i.-iii. in Kingsborough, vol. vii. Compare Anderson, 'Exp. to W. Yunnan,' p. 125. (Shans, good men and mothers dying in childbirth to heaven, bad men and those killed by the sword to hell.)

modern writer, the kingdom of heaven suffered violence, and the violent took it by force.' Thence we follow the idea onward to the battle-fields of holy war, where the soldier earned with his blood the unfading crown of martyrdom, and Christian and Moslem were urged in mutual onset and upheld in agony by the glimpse of paradise opening to receive the slayer of the infidel.

Such ideas, current among the lower races as to the soul's future happiness or misery, do not seem, setting aside some exceptional points, to be thoughts adopted or degraded from doctrines of cultured nations. They seem rather to belong to the intellectual stratum in which they are found. If so, we must neither ignore nor exaggerate their standing in the lower ethics. "The good are good warriors and hunters," said a Pawnee chief; whereupon the author who mentions the saying remarks that this would also be the opinion of a wolf, if he could express it.[2] Nevertheless, if experience has led societies of savage men to fix on certain qualities, such as courage, skill, and industry, as being virtues, then many moralists will say that such a theory is not only ethical, but lying at the very foundation of ethics. And if these savage societies further conclude that such virtues obtain their reward in another world as in this, then their theories of future happiness and misery, destined for what in this sense they call good and bad men, may be fairly looked on as belonging to morality, though at no high stage of development. But many or most writers, when they mention morality, assume a narrower definition of it. This must be borne in mind in appreciating what is meant by the statements of several well-qualified ethnologists, who have, in more or less degree, denied a moral character to the future retribution as conceived in savage religion. Mr. Ellis, describing the Society Islanders, at least gives an explicit definition. When he tried to ascertain whether they connected a person's con-

---

[1] Alger, 'Future Life,' p. 93.
[2] Brinton, 'Myths of New World,' p. 300.

dition in a future state with his disposition and conduct in this, he never could learn that they expected in the world of spirits any difference in the treatment of a kind, generous, peaceful man, and that of a cruel, parsimonious, quarrelsome one.[1] This remark, it seems to me, applies to savage religion far and wide. Dr. Brinton, commenting on the native religions of America, draws his line in a somewhat different place. Nowhere, he says, was any well-defined doctrine that moral turpitude was judged and punished in the next world. No contrast is discoverable between a place of torments and a realm of joy; at the worst but a negative castigation awaited the liar, the coward, or the niggard.[2] Professor J. G. Müller, in his 'American Religions,' yet more pointedly denies any "ethical meaning" in the contrasts of the savage future life, and looks upon what he well calls its "light-side" and "shadow-side" not as recompensing earthly virtue and vice, but rather as carrying on earthly conditions in a new existence.[3]

The idea that admission to the happier region depends on the performance of religious rites and the giving of offerings, seems scarcely known to the lowest savages. It is worth while, however, to notice some statements which seem to mark its appearance at the level of high savagery or low barbarism. Thus in the Society Islands, though the destiny of man's spirit to the region of night or to elysium was irrespective of moral character, we hear of neglect of rites and offerings as being visited by the displeasure of deities.[4] In Florida, the belief of the Sun-worshipping people of Achalaque was thus described: those who had lived well, and well served the Sun, and given many gifts to the poor in his honour, would be happy after

[1] Ellis, 'Polyn. Res.' vol. i. p. 397; see also Williams, 'Fiji,' vol. i. p. 243.
[2] Brinton, p. 242, etc.
[3] J. G. Müller, 'Amer. Urrel.' pp. 87, 224. See also the opinions of Meiners, 'Gesch. der Religion,' vol. ii. p. 768; Wuttke, 'Gesch. des Heidenthums,' vol. i. p. 115.
[4] Ellis, l. c.; Moerenhout, 'Voyage,' vol. i. p. 433.

death and be changed into stars, whereas the wicked would be carried to a destitute and wretched existence among mountain precipices, where fierce wild beasts have their dens.[1] According to Bosman, the souls of Guinea negroes reaching the river of death must answer to the divine judge how they have lived; have they religiously observed the holy days dedicated to their god, have they abstained from all forbidden meats and kept their vows inviolate, they are wafted across to paradise; but if they have sinned against these laws they are plunged in the river and there drowned for ever.[2] Such statements among peoples at these stages of culture are not frequent, and perhaps not very valid as accounts of original native doctrine. It is in the elaborate religious systems of more organized nations, in modern Brahmanism and Buddhism, and degraded forms of Christianity, that the special adaptation of the doctrine of retribution to the purposes of priestcraft and ceremonialism has become a commonplace of missionary reports.

It is well not to speak too positively on a subject so difficult and doubtful as this of the history of the belief in future retribution. Careful criticism of the evidence is above all necessary. For instance, we have to deal with several statements recorded among low races, explicitly assigning reward or punishment to men after death, according as they were good or bad in life. Here the first thing to be done is to clear up, if possible, the question whether the doctrine of retribution may have been borrowed from some more cultured neighbouring religion, as the very details often show to have been the case. Examples of direct adoption of foreign dogmas on this subject are not uncommon in the world. When among the Dayaks of Borneo it is said that a dead man becomes a spirit and lives in the jungle, or haunts the place of burial or burning, or when some distant mountain-top is pointed to as the abode of spirits of departed friends, it is hardly needful to question

[1] Rochefort, 'Iles Antilles,' p. 378.
[2] Bosman, 'Guinea,' letter x.; in Pinkerton, vol. xvi. p. 401.

the originality of ideas so characteristically savage. But one of these Dayak tribes, burning the dead, says that "as the smoke of the funeral pile of a good man rises, the soul ascends with it to the sky, and that the smoke from the pile of a wicked man descends, and his soul with it is borne down to the earth, and through it to the regions below."[1] Did not this exceptional idea come into the Dayak's mind by contact with Hinduism? In Orissa, again, Khond souls have to leap across the black unfathomable river to gain a footing on the slippery Leaping Rock, where Dinga Pennu, the judge of the dead, sits writing his register of all men's daily lives and actions, sending virtuous souls to become blessed spirits, keeping back wicked ones and sending them to suffer their penalties in new births on earth.[2] Here the striking myth of the leaping rock is perfectly savage, but the ideas of a judgment, moral retribution, and transmigration, may have come from the Hindus of the plains, as the accompanying notion of the written book unquestionably did. Dr. Mason is no doubt right in taking as the indigenous doctrine of the Karens their notion of an underworld where the ghosts of the dead live on as here, while he sets down to Hindu influence the idea of Tha-ma, the judge of the dead (the Hindu Yama), as allotting their fate according to their lives, sending those who have done deeds of merit to heaven, those who have done wickedness to hell, and keeping in hades the neither good nor bad.[3] How the theory of moral retribution may be superposed on more primitive doctrines of the future life, comes remarkably into view in Turanian religion. Among the Lapps, Jabme-Aimo, the subterranean "home of the dead" below the earth, where the departed have their cattle and follow their livelihood like Lapps above, though they are a richer, wiser,

[1] St. John, 'Far East,' vol. i. p. 181; see Mundy, 'Narrative,' vol. i. p. 332.
[2] Macpherson, p. 92. Compare Moerenhout, l. c. (Tahiti).
[3] Mason, l. c. p. 195. See also De Brosses, 'Nav. aux Terres Australes,' vol. ii. p. 482 (Caroline Is.).

stronger folk, and also Saivo-Aimo, a yet happier "home of the gods," are conceptions thoroughly in the spirit of the lower culture. But in one account the subterranean abode becomes a place of transition, where the dead stay awhile, and then with bodies renewed are taken up to the Heaven-god, or if misdoers, are flung into the abyss. Castrén is evidently right in rejecting this doctrine as not native, but due to Catholic influence. So, at the end of the 16th Rune of the Finnish Kalewala, which tells of Wainamoinen's visit to the dismal land of the dead, there is put into the hero's mouth a second speech, warning the children of men to harm not the innocent, for sad payment is in Tuoni's dwelling—the bed of evil-doers is there, with its glowing red-hot stones below and its canopy of snakes above. But the same critic condemns this moral "tag," as a later addition to the genuine heathen picture of Manala, the under-world of the dead.[1] Nor did Christianity scorn to borrow details from the religions it abolished. The narrative of a mediæval visit to the other world would be incomplete without its description of the awful Bridge of Death; Acheron and Charon's bark were restored to their places in Tartarus by the visionary and the poet; the wailing of sinful souls might be heard as they were hammered white-hot in Vulcan's smithies; and the weighing of good and wicked souls, as we may see it figured on every Egyptian mummy-case, now passed into the charge of St. Paul and the Devil.[2]

The foregoing considerations having been duly weighed, it remains to call attention to the final problem, at what stage of religious history the full theological doctrine of judicial retribution and moral compensation in a future life may have arisen. It is hard, however, to define where this development takes place even at a barbaric stage of culture. Thus among the barbaric nations of West Africa, there

[1] Castrén, 'Finn. Myth,' pp. 136, 144. See Georgi, 'Reise in Russ. Reich,' vol. i. p. 278. Compare accounts of Purgatory among the North American Indians, apparently derived from missionaries, in Morgan, 'Iroquois,' p. 169; Waitz, vol. iii. p. 345.

[2] See T. Wright, 'St. Patrick's Purgatory.'

appear such beliefs as that in Nuffi, that criminals who escape their punishment here will receive it in the other world; the division of the Yoruba under-world into an upper and a lower region for the righteous and wicked; the Kru doctrine that only the good will rejoin their ancestors in heaven; the Oji doctrine that only the good will dwell after death in the heavenly house or city of the Deity whom they call the "Highest."[1] How far is all this to be taken as native conception, and how far as due to ages of Christian and Moslem intercourse, to which at any rate few will scruple to refer the last case?

In the lower ranges of civilization, some of the most remarkable doctrines of this class are recorded in North America. Thus they appear in connexion with the fancy of a river or gulf to be passed by the departing soul on its way to the land of the dead, one of the most remarkable traits of the mythology of the world. This seems in its origin a nature-myth, connected probably with the Sun's passage across the sea into Hades, and in many of its versions it appears as a mere episode of the soul's journey, without any moral sense attached to it. Brebeuf, the same early Jesuit missionary who says explicitly of the Hurons that there is no difference in their future life between the fate of the virtuous and the vicious, mentions also among them the tree-trunk that bridges the river of death; here the dead must cross, the dog that guards it attacks some souls, and they fall. Yet in other versions this myth has a moral sense attached to it, and the passage of the heaven-gulf becomes an ordeal to separate good and wicked. To take but one instance, there is Catlin's account of the Choctaw souls journeying far westward, to where the long slippery barkless pine-log, stretching from hill to hill, bridges over the deep and dreadful river; the good pass safely to a beauteous Indian paradise, the wicked fall into the abyss of waters, and go to the dark hungry wretched

---

[1] Waitz, vol. ii. pp. 171, 191; Bowen, 'Yoruba Lang.' p. xvi. See J. L. Wilson, p. 210.

land where they are henceforth to dwell.[1] This and many similar beliefs current in the religions of the world, which need not be particularised here, seem best explained as originally nature-myths, afterwards adapted to a religious purpose. A different conception was recorded so early as 1622, by Captain John Smith among the Massachusetts, whose name is still borne by the New England district they once inhabited: They say, at first there was no king but Kiehtan, that dwelleth far westerly above the heavens, whither all good men go when they die, and have plenty of all things. The bad men go thither also and knock at the door, but he bids them go wander in endless want and misery, for they shall not stay there.[2] Lastly, the Salish Indians of Oregon say that the good go to a happy hunting-ground of endless game, while the bad go to a place where there is eternal snow, hunger, and thirst, and are tantalised by the sight of game they cannot kill, and water they cannot drink.[3] If, now, looking at these records, it be granted that the idea of moral retribution involved in them is of genuine native origin, and that the goodness and wickedness for which men are to be rewarded and punished are moral qualities, however undeveloped in definition, this will amount to an admission that the doctrine of moral retribution at any rate appears within the range of savage theology. This by no means invalidates the view here put forward as to the historical development of this doctrine, but only shows at how early a stage it may have begun to take place. The general mass of evidence remains to show the savage doctrine of the future state, as originally involving no moral retribution, or arriving at this through transitional and rudimentary stages.[4]

[1] Brebeuf in 'Rel des Jes.' 1635, p. 35; 1636, p. 105. Catlin, 'N. A. Ind.' vol. ii. p. 127; Long's 'Exp.' vol. i. p. 180. See Brinton, p. 247; Waitz, vol. ii. p. 191, vol. iii. p. 197; and the collection of myths of the Heaven-Bridge and Heaven-Gulf in 'Early History of Mankind,' chap. xii.

[2] Smith, 'New England,' in Pinkerton, vol. xiii. p. 244.

[3] Wilson in 'Tr. Eth. Soc.' vol. iii. p. 303.

[4] The remarks on this subject in the 1st edition have been in some respects

In strong contrast with the schemes of savage future existence, I need but set before the reader's mind a salient point here and there in the doctrine of distinct and unquestionable moral retribution, as held in religions of the higher culture. The inner mystic doctrines of ancient Egypt may perhaps never be extracted now from the pictures and hieroglyphic formulas of the 'Book of the Dead.' But the ethnographer may satisfy himself of two important points as to the place which the Egyptian view of the future life occupies in the history of religion. On the one hand, the transmigration into animals, the connexion kept up between corpse and soul, the good and evil life beyond the tomb, the soul's passage into the dark western Hades or to the bright sun in heaven—all these are conceptions which connect the Egyptian religion with the religions of the ruder races of mankind. But on the other hand, the mixed ethical and ceremonial standard by which the dead are to be judged adapts these primitive and even savage thoughts to a higher social development, such as may be shown by fragments from that remarkable "negative confession" which the dead must make before Osiris and the forty-two judges in Amenti. "O ye Lords of Truth! let me know you! . . . Rub ye away my faults. I have not privily done evil against mankind. . . . I have not told falsehoods in the tribunal of Truth. . . . I have not done any wicked thing. I have not made the labouring man do more than his task daily. . . . I have not calumniated the slave to his master. . . . I have not murdered. . . . I have not done fraud to men. I have not changed the measures of the country. I have not injured the images of the gods. I have not taken scraps of the bandages of the dead. I have not committed adultery. I have not withheld milk from the mouths of sucklings. I have not hunted wild animals in the pasturages. I have not netted

modified, with particular reference to the argument of Prof. Calderwood, on 'Moral Philosophy and Savage Life' in 'Contemporary Review,' Jan. 1872. (Note to 2nd Ed.).

sacred birds. . . . I am pure! I am pure! I am pure!"[1]

The Vedic hymns, again, tell of endless happiness for the good in heaven with the gods, and speak also of the deep pit where the liars, the lawless, they who give no sacrifice, will be cast. The rival theories of continuance and retribution are seen in instructive coexistence in classic Greece and Rome. What seems the older belief holds its ground in the realm of Hades; that dim region of bodiless, smoke-like ghosts remains the home of the undistinguished crowd in the $\mu\acute{\epsilon}\sigma os$ $\beta\acute{\iota}os$, the "middle life." Yet at the same time the judgment-seat of Minos and Rhadamanthos, the joys of Elysium for the just and good, fiery Tartarus echoing with the wail of the wicked, represent the newer doctrine of a moral retribution.[2] The idea of purgatorial suffering, which hardly seems to have entered the minds of the lower races, expands in immense vigour in the great Aryan religions of Asia. In Brahmanism and Buddhism, the working out of good and evil actions into their necessary consequence of happiness and misery is the very key to the philosophy of life, whether life's successive transmigrations be in animal, or human, or demon births on earth, or in luxurious heaven-palaces of gold and jewels, or in the agonizing hells where Oriental fancy riots in the hideous inventory of torture—caldrons of boiling oil and liquid fire; black dungeons and rivers of filth; vipers, and vultures, and cannibals; thorns, and spears, and red-hot pincers, and whips of flame. To the modern Hindu, it is true, ceremonial morality seems to take the upper hand, and the question of happiness or misery after death turns rather on ablutions and fasts, on sacrifices and gifts to brahmans, than on purity and beneficence of life. Buddhism in South-

[1] Bunsen, 'Egypt's Place in Univ. Hist.' vol. iv. p. 618, etc.; Birch's Introduction to and translation of the 'Book of the Dead,' ibid. vol. v.; Wilkinson, 'Ancient Eg.' vol. v.

[2] For details see Max Müller, 'Chips,' vol. i. p. 47; Pauly, "Real Encyclop.' and Smith's 'Dic. of Biog. and Myth.'

East Asia, sadly degenerate from its once high estate, is apt to work out the doctrine of merit and demerit into debtor and creditor accounts kept in good and bad marks from day to day; to serve out so much tea in hot weather counts 1 to the merit-side, and putting a stop to one's women scolding for a month counts 1 likewise, but this may be balanced by the offence of letting them keep the bowls and plates dirty for a day, which counts 1 the wrong way; and it appears that giving wood for two coffins, which count 30 marks each, and burying four bones, at 10 marks a-piece, would just be balanced by murdering a child, which counts 100 to the bad.[1] It need hardly be said here that these two great religions of Asia must be judged rather in their records of long past ages, than in the lingering degeneration of their modern reality.

In the Khordah-Avesta, a document of the old Persian religion, the fate of good and wicked souls at death is pictured in a dialogue between Zarathustra (Zoroaster), and Ahura-Mazda and Anra-Mainyu (Ormuzd and Ahriman). Zarathustra asks, "Ahura-Mazda, Heavenly, Holiest, Creator of the corporeal world, Pure! When a pure man dies, where does his soul dwell during this night?" Then answers Ahura-Mazda: "Near his head it sits down, reciting the Gâthâ Ustavaiti, praying happiness for itself; 'Happiness be to the man who conduces to the happiness of each. May Ahura-Mazda create, ruling after his wish.'" On this night the soul sees as much joyfulness as the whole living world possesses; and so the second and the third night. When the lapse of the third night turns itself to light, then the soul of the pure man goes forward, recollecting itself by the perfume of plants. A wind blows to meet it from the mid-day regions, a sweet-scented one, more sweet-scented than the other winds, and the soul of the pure man receives it—'Whence blows this wind, the sweetest-scented which I ever have smelt with the nose?' Then comes to meet him

[1] 'Journ. Ind. Archip.' new ser. vol. ii. p. 210. See Bastian, 'Oestl. Asien.'

his own law (his rule of life) in the figure of a maiden, beautiful, shining, with shining arms, powerful, well-grown, slender, large-bosomed, with praiseworthy body, noble, with brilliant face, one of fifteen years, as fair in her growth as the fairest creatures. Then to her speaks the soul of the pure man, asking, 'What maiden art thou whom I have seen here as the fairest of maidens in body?' She answers, 'I am, O youth, thy good thoughts, words, and works, thy good law, the own law of thine own body. Thou hast made the pleasant yet pleasanter to me, the fair yet fairer, the desirable yet more desirable, the sitting in a high place sitting in a yet higher place.' Then the soul of the pure man takes the first step and comes to the first paradise, the second and third step to the second and third paradise, the fourth step and arrives at the Eternal Lights. To the soul speaks a pure one deceased before, asking it, 'How art thou, O pure deceased, come away from the fleshly dwellings, from the corporeal world hither to the invisible, from the perishable world hither to the imperishable. Hail! has it happened to thee long?' "Then speaks Ahura-Mazda: 'Ask not him whom thou askest, for he is come on the fearful way of trembling, the separation of body and soul. Bring him hither of the food, of the full fatness, that is the food for a youth who thinks, speaks, and does good, who is devoted to the good law after death—that is the food for a woman who especially thinks good, speaks good, does good, the following, obedient, pure after death.'" And now Zarathustra asks, when a wicked one dies, where his soul dwells? He is told how, running about near the head, it utters the prayer, Ke maúm:—" Which land shall I praise, whither shall I go praying, O Ahura-Mazda?" In this night it sees as much unjoyfulness as the whole living world; and so the second and the third night, and it goes at dawn to the impure place, recollecting itself by the stench. An evil-smelling wind comes towards him from the north, and with it the ugly hateful maiden who is his own wicked deeds, and the soul takes the fourth step into the darkness without beginning, and a wicked soul asks how long—woe to thee!—art thou come? and the mocking Anra-Mainyu, answering in words like the words of Ahura-Mazda to the good, bids food to be brought—poison, and mixed with poison, for them who think and speak and do evil, and follow the wicked law. The Parsi of our own time, following in obscure tradition the ancient Zoroastrian faith, before he prays for forgiveness for all that he ought to have thought, and said, and done, and has not, for all that he ought not to have thought, and said, and done, and has, confesses thus his faith of the future life :—" I am wholly without doubt in the existence of the good Mazadayaçnian faith, in the coming of the resurrection and the later body, in the stepping over the bridge Chinvat, in an invariable recompense of good deeds and their reward, and of bad deeds and their punishment."[1]

In Jewish theology, the doctrine of future retribution appears after the Babylonish captivity, not in ambiguous terms, but as the strongly-expressed and intensely-felt religious conviction it has since remained among the children of Israel. Not long afterward, it received the sanction of Christianity.

A broad survey of the doctrine of the Future Life among the various nations of the world shows at once how difficult and how important is a systematic theory of its development. Looked at ethnographically, the general relations of the lower to the higher culture as to the belief in future existence may be defined somewhat as follows :—If we draw a line dividing civilization at the junction of savagery and barbarism—about where the Carib and New Zealander ends and the Aztec or Tatar begins, we may see clearly the difference of prevalent doctrine on either side. On the savage side, the theory of hovering ghosts is strong, rebirth in human or animal bodies is often thought of, but above all there prevails the expectation of a new life, most

[1] Spiegel, 'Avesta,' ed. Bleek, vol. iii. pp. 136, 163 ; see vol. i. pp. xviii. 90, 141; vol. ii. p. 68.

often located in some distant earthly region, or less commonly in the under-world or on the sky. On the cultured side, the theory of hovering ghosts continues, but tends to subside from philosophy into folklore, the theory of re-birth is elaborated into great philosophic systems, but eventually dies out under the opposition of scientific biology, while the doctrine of a new life after death maintains its place with immense power in the human mind, although the dead have been ousted by geography from any earthly district, and the regions of heaven and hell are more and more spiritualized out of definite locality into vague expressions of future happiness and misery. Again, on the savage side we find the dominant idea to be a continuance of the soul in a new existence, like the present life, or idealized and exaggerated on its model; while on the cultured side the doctrine of judgment and moral retribution prevails with paramount, though not indeed absolute sway. What, then, has been the historical course of theological opinion, to have produced in different stages of culture these contrasted phases of doctrine?

In some respects, theories deriving savage from more civilized ideas are tenable. In certain cases, to consider a particular savage doctrine of the future state as a fragmentary, or changed, or corrupted outcome of the religion of higher races, seems as easy as to reverse this view by taking savagery as representing the starting-point. It is open to anyone to suppose that the doctrine of transmigration among American savages and African barbarians may have been degraded from elaborate systems of metempsychosis established among philosophic nations like the Hindus; that the North American and South African doctrine of continued existence in a subterranean world may be derived from similar beliefs held by races at the level of the ancient Greeks; that when rude tribes in the Old or New World assign among the dead a life of happiness to some, and of misery to others, this idea may have been inherited or adopted from cultured nations holding more strongly and systematically the doctrine of retribution. In such cases the argument is to a great extent the same, whether the lower race be considered degenerate descendants of a higher nation, or whether the simpler supposition be put forward that they have adopted the ideas of some more cultured people. These views ought to have full attention, for degenerate and borrowed beliefs form no small item in the opinions of uncivilized races. Yet this kind of explanation is more adapted to meet special cases than general conditions; it is rather suited to piecemeal treatment, than to comprehensive study, of the religions of mankind. Worked out on the large scale, it would endeavour to account for the doctrines of the savage world, as being a patchwork of fragments from various religions of high nations, transported by not easily-conceived means from their distant homes and set down in remote regions of the earth. It may be safely said that no hypothesis can account for the varied doctrines current among the lower tribes, without the admission that religious ideas have been in no small measure developed and modified in the districts where they are current.

Now this theory of development, in its fullest scope, combined with an accessory theory of degeneration and adoption, seems best to meet the general facts of the case. A hypothesis which finds the origin of the doctrine of the future life in the primitive animism of the lower races, and thence traces it along the course of religious thought, in varied developments fitted to exacter knowledge and forming part of loftier creeds, may well be maintained as in reasonable accordance with the evidence. Such a theory, as has been sufficiently shown in the foregoing chapters, affords a satisfactory explanation of the occurrence, in the midst of cultured religions, of intellectually low superstitions, such as that of offerings to the dead, and various others. These, which the development theory treats naturally as survivals from a low stage of education lingering on in a higher, are by no means so readily accounted for by the degeneration theory. There are more special arguments which favour the priority of the savage to the civilized phases of the doctrine of a future life. If savages did in general receive their views of another existence from the religious systems of cultured nations, these systems can hardly have been such as recognize the dominant doctrines of heaven and hell. For, as to the locality of the future world, savage races especially favour a view little represented in civilized belief, namely, that the life to come is in some distant earthly country. Moreover, the belief in a fiery abyss or Gehenna, which excites so intensely and lays hold so firmly of the imagination of the most ignorant men, would have been especially adapted to the minds of savages, had it come down to them by tradition from an ancestral faith. Yet, in fact, the lower races so seldom recognize such an idea, that even the few cases in which it occurs lie open to suspicion of not being purely native. The proposition that the savage doctrines descend from the more civilized seems thus to involve the improbable supposition, that tribes capable of keeping up traditions of Paradise, Heaven, or Hades, should nevertheless have forgotten or discarded a tradition of Hell. Still more important is the contrast between the continuance-theory and the retribution-theory of the future existence, in the sections of culture where they respectively predominate. On the one hand, the continuance-theory, with its ideas of a ghostly life like this, is directly vouched for by the evidence of the senses in dreams and visions of the dead, and may be claimed as part of the "Natural Religion," properly so called, of the lower races. On the other hand, the retribution-theory is a dogma which this evidence of apparitions could hardly set on foot, though capable of afterwards supporting it. Throughout the present study of animistic religion, it constantly comes into view that doctrines which in the lower culture are philosophical, tend in the higher to become ethical; that what among savages is a science of nature, passes among civilized nations into a moral engine. Herein lies the distinction of deepest import between the two great theories of the soul's existence after bodily death. According to a development theory of culture, the savage, unethical doctrine of continuance would be taken as the more primitive, succeeded in higher civilization by the ethical doctrine of retribution. Now this theory of the course of religion in the distant and obscure past is conformable with experience of its actual history, so far as this lies within our knowledge. Whether we compare the early Greek with the later Greek, the early Jew with the later Jew, the ruder races of the world in their older condition with the same races as affected by the three missionary religions of Buddhism, Mohammedanism, Christianity, the testimony of history vouches for the like transition towards ethical dogma.

In conclusion, though theological argument on the actual validity of doctrines relating to the future life can have no place here, it will be well not to pass by without further remark one great practical question which lies fairly within the province of Ethnography. How, in the various stages of culture, has the character and conduct of the living been affected by the thought of a life to come? If we take the savage beliefs as a starting-point, it will appear that these belong rather to speculative philosophy than to practical rule of life. The lower races hold opinions as to a future state because they think them true, but it is not surprising that men who take so little thought of a contingency three days off, should receive little practical impulse from vague anticipations of a life beyond the grave. Setting aside the consideration of possible races devoid of all thought of a future existence, there unquestionably has been and is a great mass of mankind whose lives are scarcely affected by such expectations of another life as they do hold. The doctrine of continuance, making death as it were a mere journey into a new country, can have little direct action on men's conduct, though indirectly it has indeed an enormous and disastrous influence on society, leading as it does to the slaughter of wives and slaves, and the destruction of property, for the use of the dead in the next world. If this world to come be thought a happier region, the looking forward to it makes men more willing to risk their lives in battle, promotes the habit of despatching the sick and aged into a better life, and encourages suicide when life is very hateful here. When the half-way house between continuance and retribution is reached, and the idea prevails that the manly virtues which give rank and wealth and honour here will lead hereafter to yet brighter glory; then this belief must add new force to the earthly motives which make bold warriors and mighty chiefs. But among men who expect to become hovering ghosts at death, or to depart to some gloomy land of shades, such expectation strengthens the natural horror and hatred of dissolution. They tend toward the modern African's state of mind, whose thought of death is that he shall drink no more rum, wear no more fine clothes, have no more wives. The negro of our own day would feel to the utmost the sense of those lines in the beginning of the Iliad, which describe the heroes' "souls" being cast down to Hades, but "themselves" left a prey to dogs and carrion birds.

Rising to the level of the higher races, we mark the thought of future existence taking a larger and larger place in the convictions of religion, the expectation of a judgment after death gaining in intensity and becoming, what it scarcely seems to the savage, a real motive in life. Yet this change is not to be measured as proceeding throughout in any direct proportion with the development of culture. The doctrine of the future life has hardly taken deeper and stronger root in the higher than in the middle levels of civilization. The Osiris-mummy carried round at Egyptian feasts symbolized not dissolution, but entrance into glory, to the nation who looked for their real life not on the banks of the Nile, but in the sunset regions of the mystic Amenti. The Moslem says that men sleep in life and wake in death; the Hindu likens the body which a soul has quitted to the bed he rises from in the morning. The story of the ancient

Getæ, who wept at births and laughed at funerals, embodies an idea of the relation of this life to the next which comes to the surface again and again in the history of religion, nowhere perhaps touched in with a lighter hand than in the Arabian Nights' tale where Abdallah of the Sea indignantly breaks off his friendship with Abdallah of the Land, when he hears that the dwellers on the land do not feast and sing when one of them dies, like the dwellers in the sea, but mourn and weep and tear their garments. Such thoughts lead on into the morbid asceticism that culminates in the life of the Buddhist saint, eating his food with loathing from the alms-bowl that he carries as though it held medicine, wrapping himself in grave-clothes from the cemetery, or putting on his disfigured robe as though it were a bandage to cover a sore, whose looking forward is to death for deliverance from the misery of life, whose dreamiest hope is that after an inconceivable series of successive existences he may find in utter dissolution and not-being a refuge even from heaven.

The belief in future retribution has been indeed a powerful engine in shaping the life of nations. Powerful both for good and evil, it has been made the servant-of-all-work of many faiths. Priesthoods have used it unscrupulously for their professional ends, to gain wealth and power for their own caste, to stop intellectual and social progress beyond the barriers of their consecrated systems. On the banks of the river of death, a band of priests has stood for ages to bar the passage against all poor souls who cannot satisfy their demands for ceremonies, and formulas, and fees. This is the dark side of the picture. On the bright side, as we study the moral standards of the higher nations, and see how the hopes and fears of the life to come have been brought to enforce their teachings, it is plain that through most widely differing religions the doctrine of future judgment has been made to further goodness and to check wickedness, according to the shifting rules by which men have divided right from wrong. The philosophic schools which from classic times onward have rejected the belief in a future existence, appear to have come back by a new road to the very starting point which perhaps the rudest races of men never quitted. At least this seems true as regards the doctrine of future retribution, which is alike absent from the belief of classes of men at the two extremes of culture. How far the moral standard of life may have been adjusted throughout the higher races with reference to a life hereafter, is a problem difficult of solution, so largely do unbelievers in this second life share ethical principles which have been more or less shaped under its influence. Men who live for one world or for two, have high motives of virtue in common; the noble self-respect which impels them to the life they feel worthy of them; the love of goodness for its own sake and for its immediate results; and beyond this, the desire to do good that shall survive the doer, who will not indeed be in the land of the living to see his work, but who can yet discount his expectations into some measure of present satisfaction. Yet he who believes that his thread of life will be severed once and for ever by the fatal shears, well knows that he wants a purpose and a joy in life, which belong to him who looks for a life to come. Few men feel real contentment in the expectation of vanishing out of conscious existence, henceforth, like the great Buddha, to exist only in their works. To remain incarnate in the memory of friends is something. A few great spirits may enjoy in the reverence of future ages a thousand years or so of "subjective immortality;" though as for mankind at large, the individual's personal interest hardly extends beyond those who have lived in his time, while his own memory scarce outlives the third and fourth generation. But over and above these secular motives, the belief in immortality extends its powerful influence through life, and culminates at the last hour, when, setting aside the very evidence of their senses, the mourners smile through their tears, and say it is not death but life.

CHAPTER XIV.

ANIMISM (*continued*).

Animism, expanding from the Doctrine of Souls to the wider Doctrine of Spirits, becomes a complete Philosophy of Natural Religion—Definition of Spirits similar to and apparently modelled on that of Souls—Transition-stage : classes of Souls passing into good and evil Demons—Manes-Worship—Doctrine of Embodiment of Spirits in human, animal, vegetable, and inert bodies—Demoniacal Possession and Obsession as causes of Disease and Oracle-inspiration—Fetishism—Disease-spirits embodied— Ghost attached to remains of Corpse—Fetish produced by a Spirit embodied in, attached to, or operating through, an Object—Analogues of Fetish-doctrine in Modern Science—Stock-and-Stone Worship—Idolatry —Survival of Animistic Phraseology in modern Language—Decline of Animistic theory of Nature.

THE general scheme of Animism, of which the doctrine of souls hitherto discussed forms part, thence expands to complete the full general philosophy of Natural Religion among mankind. Conformably with that early childlike philosophy in which human life seems the direct key to the understanding of nature at large, the savage theory of the universe refers its phenomena in general to the wilful action of pervading personal spirits. It was no spontaneous fancy, but the reasonable inference that effects are due to causes, which led the rude men of old days to people with such ethereal phantoms their own homes and haunts, and the vast earth and sky beyond. Spirits are simply personified causes. As men's ordinary life and actions were held to be caused by souls, so the happy or disastrous events which affect mankind, as well as the manifold physical operations of the outer world, were accounted for as caused by soul-like beings, spirits whose essential similarity of origin is evident through all their wondrous variety of power and function. Much that the primitive animistic view thus explains, has been indeed given over by more advanced education to the "metaphysical" and "positive" stages of thought. Yet animism is still plainly to be traced onward from the intellectual state of the lower races, along the course of the higher culture, whether its doctrines have been continued and modified into the accepted philosophy of religion, or whether they have dwindled into mere survivals in popular superstition. Though all I here undertake is to sketch in outline such features of this spiritualistic philosophy as I can see plainly enough to draw at all, scarcely attempting to clear away the haze that covers great parts of the subject, yet even so much as I venture on is a hard task, made yet harder by the responsibility attaching to it. For it appears that to follow the course of animism on from its more primitive stages, is to account for much of mediæval and modern opinion whose meaning and reason could hardly be comprehended without the aid of a development-theory of culture, taking in the various processes of new formation, abolition, survival, and revival. Thus even the despised ideas of savage races become a practically important topic to the modern world, for here, as usual, whatever bears on the origin of philosophic opinion, bears also on its validity.

At this point of the investigation, we come fully into sight of the principle which has been all along implied in the use of the word Animism, in a sense beyond its narrower meaning of the doctrine of souls. By using it to express the doctrine of spirits generally, it is practically asserted that the ideas of souls, demons, deities, and any other classes of spiritual beings, are conceptions of similar nature throughout, the conceptions of souls being the original ones of the series. It was best, from this point of view, to begin with a careful study of souls, which are the spirits proper to men, animals, and things, before extending the survey of the spirit-world to its fullest range. If it be admitted that souls and other spiritual beings are conceived of as essentially similar in their nature, it may be reasonably argued that the class of conceptions based on evidence most direct and accessible to ancient men, is the earlier and fundamental class. To grant this, is in effect to agree that the doctrine of souls, founded on the natural perceptions of primitive man, gave rise to the doctrine of spirits, which extends and modifies its general theory for new purposes, but in developments less authenticated and consistent, more fanciful and far-fetched. It seems as though the conception of a human soul, when once attained to by man, served as a type or model on which he framed not only his ideas of other souls of lower grade, but also his ideas of spiritual beings in general, from the tiniest elf that sports in the long grass up to the heavenly Creator and Ruler of the world, the Great Spirit.

The doctrines of the lower races fully justify us in classing their spiritual beings in general as similar in nature to the souls of men. It will be incidentally shown here, again and again, that souls have the same qualities attributed to them as other spirits, are treated in like fashion, and pass without distinct breaks into every part of the general spiritual definition. The similar nature of soul and other spirit is, in fact, one of the commonplaces of animism, from its rudest to its most cultured stages. It ranges from the native New Zealanders' and West Indians' conceptions of the "atua" and the "cemi," beings which require special definition to show whether they are human souls or demons or deities of some other class,[1] and so onward to the declaration of Philo Judæus, that souls demons, and angels differ indeed in name, but are in reality one,[2] and to the state of mind of the modern Roman Catholic priest, who is

[1] See Taylor, 'New Zealand,' p. 134; J. G. Müller 'Amerikanische Urreligionen,' p. 171.
[2] Philo Jud. de Gigantibus, iv.

cautioned in the rubric concerning the examination of a possessed patient, not to believe the demon if he pretends to be the soul of some saint or deceased person, or a good angel (neque ei credatur, si dæmon simularet se esse animam alicujus Sancti, vel defuncti, vel Angelum bonum).[1] Nothing can bring more broadly into view the similar nature of souls and other spiritual beings than the existence of a full transitional series of ideas. Souls of dead men are in fact considered as actually forming one of the most important classes of demons and deities.

It is quite usual for savage tribes to live in terror of the souls of the dead as harmful spirits. Thus Australians have been known to consider the ghosts of the unburied dead as becoming malignant demons.[2] New Zealanders have supposed the souls of their dead to become so changed in nature as to be malignant to their nearest and dearest friends in life;[3] the Caribs said that, of man's various souls, some go to the seashore and capsize boats, others to the forests to be evil spirits:[4] among the Sioux Indians the fear of the ghost's vengeance has been found to act as a check on murder;[5] of some tribes in Central Africa it may be said that their main religious doctrine is the belief in ghosts, and that the main characteristic of these ghosts is to do harm to the living.[6] The Patagonians lived in terror of the souls of their wizards, which become evil demons after death;[7] Turanian tribes of North Asia fear their shamans even more when dead than when alive, for they become a special class of spirits who are the hurtfullest in all nature, and who among the Mongols plague the living on

[1] Rituale Romanum : De Exorcizandis Obsessis a Dæmonio.
[2] Oldfield, 'Abor. of Australia' in 'Tr. Eth. Soc.' vol. iii. p. 236. See Bonwick, 'Tasmanians,' p. 181.
[3] Taylor, 'New Zealand,' p. 104.
[4] Rochefort, 'Iles Antilles,' p. 429.
[5] Schoolcraft, 'Indian Tribes,' part ii. p. 195; M. Eastman, 'Dahcotah,' p. 72.
[6] Burton, 'Central Afr.' vol. ii. p. 344; Schlegel, 'Ewe-Sprache,' p. xxv.
[7] Falkner, 'Patagonia,' p. 116; but cf. Musters, p. 180.

purpose to make them bring offerings.[1] In China it is held that the multitudes of wretched destitute spirits in the world below, such as souls of lepers and beggars, can sorely annoy the living; therefore at certain times they are to be appeased with offerings of food, scant and beggarly; and a man who feels unwell, or fears a mishap in business, will prudently have some mock-clothing and mock-money burnt for these "gentlemen of the lower regions."[2] Notions of this sort are widely prevalent in Indo-China and India; whole orders of demons there were formerly human souls, especially of people left unburied or slain by plague or violence, of bachelors or of women who died in childbirth, and who henceforth wreak their vengeance on the living. They may, however, be propitiated by temples and offerings, and thus have become in fact a regular class of local deities.[3] Among them may be counted the diabolic soul of a certain wicked British officer, whom native worshippers in the Tinnevelly district still propitiate by offering at his grave the brandy and cheroots he loved in life.[4] India even carries theory into practice by an actual manufacture of demons, as witness the two following accounts. A certain brahman, on whose lands a kshatriya raja had built a house, ripped himself up in revenge, and became a demon of the kind called brahmadasyu, who has been ever since the terror of the whole country, and is the most common village deity in Kharakpur.[5] Toward the close of the last century there were two brahmans, out of whose house a man had wrongfully, as they thought, taken forty rupees; whereupon one of the brahmans proceeded to cut off his own mother's

---

[1] Castrén, 'Finn. Myth.' p. 122.
[2] Doolittle, 'Chinese,' vol. i. p. 206.
[3] Bastian, 'Oestl. Asien,' vol. ii. pp. 129, 416; vol. iii. pp. 29, 257, 278; 'Psychologie,' pp. 77, 99; Cross, 'Karens,' l. c. p. 316; Elliot in 'Journ. Eth. Soc.' vol. i. p. 115; Buchanan, 'Mysore, etc.' in Pinkerton, vol. viii. p. 677.
[4] Shortt, 'Tribes of India,' in 'Tr. Eth. Soc.' vol. vii. p. 192; Tinling, 'Tour round India,' p. 19.
[5] Bastian, 'Psychologie,' p. 101.

head, with the professed view, entertained by both mother and son, that her spirit, excited by the beating of a large drum during forty days, might haunt, torment, and pursue to death the taker of their money and those concerned with him. Declaring with her last words that she would blast the thief, the spiteful hag deliberately gave up her life to take ghostly vengeance for those forty rupees.[1] By instances like these it appears that we may trace up from the psychology of the lower races the familiar ancient and modern European tales of baleful ghost-demons. The old fear even now continues to vouch for the old belief.

Happily for man's anticipation of death, and for the treatment of the sick and aged, thoughts of horror and hatred do not preponderate in ideas of deified ancestors, who are regarded on the whole as kindly patron spirits, at least to their own kinsfolk and worshippers. Manes-worship is one of the great branches of the religion of mankind. Its principles are not difficult to understand, for they plainly keep up the social relations of the living world. The dead ancestor, now passed into a deity, simply goes on protecting his own family and receiving suit and service from them as of old; the dead chief still watches over his own tribe, still holds his authority by helping friends and harming enemies, still rewards the right and sharply punishes the wrong. It will be enough to show by a few characteristic examples the general position of manes-worship among mankind, from the lower culture upward.[2] In the two Americas it appears not unfrequently, from the low savage level of the Brazilian Camacans, to the somewhat higher stage of northern Indian tribes whom we hear of as praying to the spirits of their forefathers for good weather or luck in hunting, and fancying when an Indian falls into the fire that the ancestral spirits pushed him in to punish

---

[1] Sir J. Shore in 'Asiatic Res.' vol. iv. p. 331.
[2] For some collections of details of manes-worship, see Meiners, 'Geschichte der Religionen,' vol. i. book 3; Bastian, 'Mensch,' vol. ii. pp. 402-11; 'Psychologie,' pp. 72-114.

neglect of the customary gifts, while the Natchez of Louisiana are said to have even gone so far as to build temples for dead men.[1] Turning to the dark races of the Pacific, we find the Tasmanians laying their sick round a corpse on the funeral pile, that the dead might come in the night and take out the devils that caused the diseases; it is asserted in a general way of the natives, that they believed most implicitly in the return of the spirits of their departed friends or relations to bless or injure them as the case might be.[2] In Tanna, the gods are spirits of departed ancestors, aged chiefs becoming deities after death, presiding over the growth of yams and fruit trees, and receiving from the islanders prayer and offerings of first fruits.[3] Nor are the fairer Polynesians behind in this respect. Below the great mythological gods of Tonga and New Zealand, the souls of chiefs and warriors form a lower but active and powerful order of deities, who in the Tongan paradise intercede for man's benefit with the higher deities, who direct the Maori war parties on the march, hover over them and give them courage in the fight, and, watching jealously their own tribes and families, punish any violation of the sacred laws of tapu.[4] Thence we trace the doctrine into the Malay islands, where the souls of deceased ancestors are looked to for prosperity in life and help in distress.[5] In Madagascar, the worship of the spirits of the dead is remarkably associated with the Vazimbas, the aborigines of the island, who are said still to survive as a distinct race in the interior, and whose peculiar graves testify to their former occupancy of other districts. These graves, small in size, and distinguished by a cairn and an upright stone slab or altar,

[1] J. G. Müller, 'Amer. Urrel.' pp. 73, 173, 209, 261; Schoolcraft, 'Indian Tribes,' part i. p. 39, part iii. p. 237; Waitz, 'Anthropologie,' vol. iii. pp. 191, 204.
[2] Backhouse, 'Australia,' p. 105; Bonwick, 'Tasmanians,' p. 182.
[3] Turner, 'Polynesia,' p. 88.
[4] Mariner, 'Tonga Is.' vol. ii. p. 104; S. S. Farmer, p. 126; Shortland, 'Trads. of N. Z.' p. 81; Taylor, 'New Zealand,' p. 108.
[5] J. R. Forster, 'Observations,' p. 604; Marsden, 'Sumatra,' p. 258; 'Journ. Ind. Archip.' vol. ii. p. 234.

are places which the Malagasy regard with equal fear and veneration, and their faces become sad and serious when they even pass near. To take a stone or pluck a twig from one of these graves, to stumble against one in the dark, would be resented by the angry Vazimba inflicting disease, or coming in the night to carry off the offender to the region of ghosts. The Malagasy is thus enabled to account for every otherwise unaccountable ailment by his having knowingly or unknowingly given offence to some Vazimba. They are not indeed always malevolent, they may be placable or implacable, or partake of both characters. Thus it comes to pass, that at the altar-slab which long ago some rude native family set up for commemoration or dutiful offering of food to a dead kinsman, a barbaric supplanting race now comes to smear the burnt fat of sacrifice, and set up the heads of poultry and sheep and the horns of bullocks, that the mysterious tenant may be kind, not cruel, with his superhuman powers.[1]

On the continent of Africa, manes-worship appears with extremest definiteness and strength. Thus Zulu warriors, aided by the "amatongo," the spirits of their ancestors, conquer in the battle; but if the dead turn their backs on the living, the living fall in the fight, to become ancestral spirits in their turn. In anger the "itongo" seizes a living man's body and inflicts disease and death; in beneficence he gives health, and cattle, and corn, and all men wish. Even the little children and old women, of small account in life, become at death spirits having much power, the infants for kindness, the crones for malice. But it is especially the head of each family who receives the worship of his kin. Why it is naturally and reasonably so, a Zulu thus explains. "Although they worship the many Amatongo of their tribe, making a great fence around them for their protection; yet their father is far before all others when they worship the Amatongo. Their father is a great

[1] Ellis, 'Madagascar,' vol. i. pp. 123, 423. As to the connexion of the Vaximbas with the Mazimba of East Africa, see Waitz, vol. ii. p. 360, 426.

treasure to them even when he is dead. And those of his children who are already grown up know him thoroughly, his gentleness, and his bravery." "Black people do not worship all Amatongo indifferently, that is, all the dead of their tribe. Speaking generally, the head of each house is worshipped by the children of that house; for they do not know the ancients who are dead, nor their laud-giving names, nor their names. But their father whom they knew is the head by whom they begin and end in their prayer, for they know him best, and his love for his children; they remember his kindness to them whilst he was living; they compare his treatment of them whilst he was living, support themselves by it, and say, 'He will still treat us in the same way now he is dead. We do not know why he should regard others besides us; he will regard us only.'"[1] We shall see in another place how the Zulu follows up the doctrine of divine ancestors till he reaches a first ancestor of man and creator of the world, the primæval Unkulunkulu. In West Africa, manes-worship displays in contrast its two special types. On the one hand, we see the North Guinea negroes transferring the souls of the dead, according to their lives, to the rank of good and evil spirits, and if evil worshipping them the more zealously as fear is to their minds a stronger impulse than love. On the other hand, in Southern Guinea, we see the deep respect paid to the aged during life, passing into worship when death has raised them to yet higher influence. There the living bring to the images of the dead food and drink, and even a small portion of their profits gained in trade; they look especially to dead relatives for help in the trials of life, and "it is no uncommon thing to see large groups of men and women, in times of peril or distress, assembled along the brow of some commanding eminence, or along the skirts of some dense

---

[1] Callaway, 'Religious System of Amazulu,' part ii.; see also Arbousset and Daumas, p. 469; Casalis, 'Basutos,' pp. 248-54; Waitz, 'Anthropologie,' vol. ii. pp. 411, 419; Magyar, 'Reisen in Süd-Afrika,' pp. 21, 335 (Congo); Cavazzi, 'Congo,' lib. i.

forest, calling in the most piteous and touching tones upon the spirits of their ancestors."[1]

In Asia, manes-worship comes to the surface in all directions. The rude Veddas of Ceylon believe in the guardianship of the spirits of the dead; these, they say, are "ever watchful, coming to them in sickness, visiting them in dreams, giving them flesh when hunting;" and in every calamity and want they call for aid on the "kindred spirits," and especially the shades of departed children, the "infant spirits."[2] Among non-Hindu tribes of India, whose religions more or less represent præ-Brahmanic and præ-Buddhistic conditions, wide and deep traces appear of an ancient and surviving cultus of ancestors.[3] Among Turanian tribes spread over the northern regions of the Old World, a similar state of things may be instanced from the Mongols, worshipping as good deities the princely souls of Genghis Khan's family, at whose head stands the divine Genghis himself.[4] Nor have nations of the higher Asiatic culture generally rejected the time-honoured rite. In Japan the "Way of the Kami," better known to foreigners as the Sin-tu religion, is one of the officially recognized faiths, and in it there is still kept up in hut and palace the religion of the rude old mountain-tribes of the land, who worshipped their divine ancestors, the Kami, and prayed to them for help and blessing. To the time of these ancient Kami, say the modern Japanese, the rude stone implements belong which are found in the ground in Japan as elsewhere: to modern ethnologists, however, these bear witness not of divine but savage parentage.[5] In Siam the lower orders scruple to

[1] J. L. Wilson, 'W. Afr.' pp. 217, 388-93. See Waitz, vol. ii. pp. 181, 194.

[2] Bailey in 'Tr. Eth. Soc.' vol. ii. p. 301. Compare Taylor, 'New Zealand,' p. 153.

[3] Buchanan, 'Mysore,' in Pinkerton, vol. viii. pp. 674-7. See Macpherson, 'India,' p. 95 (Khonds); Hunter, 'Rural Bengal,' p. 183 (Santals).

[4] Castrén, 'Finn. Myth.' p. 122; Bastian, 'Psychologie,' p. 90. See Palgrave, 'Arabia,' vol. i. p. 373.

[5] Siebold, 'Nippon,' vol. i. p. 3, vol. ii. p. 51; Kempfer, 'Japan,' in Pinkerton, vol. vii. pp. 672, 680, 723, 755.

worship the great gods, lest through ignorance they should blunder in the complex ritual; they prefer to pray to the "theparak," a lower class of deities among whom the souls of great men take their places at death.[1] In China, as every one knows, ancestor-worship is the dominant religion of the land, and interesting problems are opened out to the Western mind by the spectacle of a great people who for thousands of years have been thus seeking the living among the dead. Nowhere is the connexion between parental authority and conservatism more graphically shown. The worship of ancestors, begun during their life, is not interrupted but intensified when death makes them deities. The Chinese, prostrate bodily and mentally before the memorial tablets that contain the souls of his ancestors, little thinks that he is all the while proving to mankind how vast a power unlimited filial obedience, prohibiting change from ancestral institutions, may exert in stopping the advance of civilization. The thought of the souls of the dead as sharing the happiness and glory of their descendants is one which widely pervades the world, but most such ideas would seem vague and weak to the Chinese, who will try hard for honours in his competitive examination with the special motive of glorifying his dead ancestors, and whose titles of rank will raise his deceased father and grandfather a grade above himself, as though, with us, Zachary Macaulay and Copley the painter should now have viscounts' coronets officially placed on their tombstones. As so often happens, what is jest to one people is sober sense to another. There are 300 millions of Chinese who would hardly see a joke in Charles Lamb reviling the stupid age that would not read him, and declaring that he would write for antiquity. Had he been a Chinese himself, he might have written his book in all seriousness for the benefit of his great-great-grandfather. Among the Chinese, manes-worship is no rite of mere affection. The living want the help of the ancestral spirits, who reward virtue and punish vice: "The exalted

---

[1] Bastian, 'Oestl. Asien.' vol. iii. p. 250.

ancestor will bring thee, O Prince, much good!"—"Ancestors and fathers will abandon you and give you up, and come not to help, and ye will die." If no help comes in time of need, the Chinese will reproach his ancestor, or even come to doubt his existence. Thus in a Chinese ode the sufferers in a dreadful drought cry, "Heu-tsi cannot or will not help. . . . Our ancestors have surely perished. . . . Father, mother, ancestors, how could you calmly bear this!" Nor does manes-worship stop short with direct family ties; it is naturally developed to produce, by deification of the heroic dead, a series of superior gods to whom worship is given by the public at large. Thus, according to legend, the War-god or Military Sage was once in human life a distinguished soldier, the Mechanics' god was a skilful workman and inventor of tools, the Swine-god was a hog-breeder who lost his pigs and died of sorrow, and the Gamblers' god, a desperate gamester who lost his all and died of want, is represented by a hideous image called a "devil gambling for cash," and in this shape receives the prayers and offerings of confirmed gamblers, his votaries. The spirits of San-kea Ta-te, and Chang-yuen-sze go to partake of the offerings set out in their temples, returning flushed and florid from their meal; and the spirit of Confucius is present in the temple, where twice a year the Emperor does sacrifice to him.[1]

The Hindu unites in some degree with the Chinese as to ancestor-worship, and especially as to the necessity of having a son by blood or adoption, who shall offer the proper sacrifices to him after death. "May there be born in our lineage," the manes are supposed to say, "a man to offer to us, on the thirteenth day of the moon, rice boiled in milk, honey and ghee." Offerings made to the divine manes, the " pitris " (patres, fathers) as they are called, preceded and followed by offerings to the greater deities, give to the worshipper merit

[1] Plath, 'Religion der alten Chinesen,' part i. p. 65, part ii. p. 89; Doolittle, 'Chinese,' vol. i. pp. vi. viii.; vol. ii. p. 373; 'Journ. Ind. Archip.' New Ser. vol. ii. p. 363; Legge, 'Confucius,' p. 92.

and happiness.[1] In classic Europe, apotheosis lies part within the limits of myth, where it was applied to fabled ancestors, and part within the limits of actual history, as where Julius and Augustus shared its honours with the vile Domitian and Commodus. The most special representatives of ancestor-worship in Europe were perhaps the ancient Romans, whose word "manes" has become the recognized name for ancestral deities in modern civilized language; they embodied them as images, set them up as household patrons, gratified them with offerings and solemn homage, and counting them as or among the infernal gods, inscribed on tombs D. M., "Diis Manibus."[2] The occurrence of this D. M. in Christian epitaphs is an often-noticed case of religious survival.

Although full ancestor-worship is not practised in modern Christendom, there remains even now within its limits a well-marked worship of the dead. A crowd of saints, who were once men and women, now form an order of inferior deities, active in the affairs of men and receiving from them reverence and prayer, thus coming strictly under the definition of manes. This Christian cultus of the dead, belonging in principle to the older manes-worship, was adapted to answer another purpose in the course of religious transition in Europe. The local gods, the patron gods of particular ranks and crafts, the gods from whom men sought special help in special needs, were too near and dear to the inmost heart of præ-Christian Europe to be done away with without substitutes. It proved easier to replace them by saints who could undertake their particular professions, and even succeed them in their sacred dwellings. The system of spiritual division of labour was in time worked out with wonderful minuteness in the vast array of professional saints, among whom the most familiar to modern English ears are St. Cecilia, patroness of musicians; St. Luke, patron

---

[1] Manu, book iii.
[2] Details in Pauly, 'Real-Encyclop.' s. v. 'inferi;' Smith's 'Dic. of Gr. and Rom. Biog. and Myth. ;' Meiners, Hartung, etc.

of painters; St. Peter, of fishmongers; St. Valentine, of lovers; St. Sebastian, of archers; St. Crispin, of cobblers; St. Hubert, who cures the bite of mad dogs; St. Vitus, who delivers madmen and sufferers from the disease which bears his name; St. Fiacre, whose name is now less known by his shrine than by the hackney-coaches called after him in the seventeenth century. Not to dwell here minutely on an often-treated topic, it will be enough to touch on two particular points. First, as to the direct historical succession of the Christian saint to the heathen deity, the following are two very perfect illustrations. It is well known that Romulus, mindful of his own adventurous infancy, became after death a Roman deity propitious to the health and safety of young children, so that nurses and mothers would carry sickly infants to present them in his little round temple at the foot of the Palatine. In after ages the temple was replaced by the church of St. Theodorus, and there Dr. Conyers Middleton, who drew public attention to its curious history, used to look in and see ten or a dozen women, each with a sick child in her lap, sitting in silent reverence before the altar of the saint. The ceremony of blessing children, especially after vaccination, may still be seen there on Thursday mornings.[1] Again, Sts. Cosmas and Damianus, according to Maury, owe their recognized office to a similar curious train of events. They were martyrs who suffered under Diocletian, at Ægææ in Cilicia. Now this place was celebrated for the worship of Æsculapius, in whose temple incubation, *i. e.*, sleeping for oracular dreams, was practised. It seems as though the idea was transferred on the spot to the two local saints, for we next hear of them as appearing in a dream to the Emperor Justinian, when he was ill at Byzantium. They cured him, he built them a temple, their cultus spread far and wide, and they frequently appeared to the sick to show them what they should do. Legend settled that Cosmas and Damianus were physicians while they lived on earth,

[1] Middleton, 'Letter from Rome;' Murray's 'Handbook of Rome.'

and at any rate they are patron-saints of the profession of medicine to this day.[1] Second, as to the actual state of hagiolatry in modern Europe, it is obvious on a broad view that it is declining among the educated classes. Yet modern examples may be brought forward to show ideas as extreme as those which prevailed more widely a thousand years ago. In the Church of the Jesuit College at Rome lies buried St. Aloysius Gonzaga, on whose festival it is customary especially for the college students to write letters to him, which are placed on his gaily decorated and illuminated altar, and afterwards burnt unopened. The miraculous answering of these letters is vouched for in an English book of 1870. To the same year belongs an English tract commemorating a late miraculous cure. An Italian lady afflicted with a tumour and incipient cancer of the breast was exhorted by a Jesuit priest to recommend herself to the Blessed John Berchmans, a pious Jesuit novice from Belgium, who died in 1621, and was beatified in 1865. Her adviser procured for her "three small packets of dust gathered from the coffin of this saintly innocent, a little cross made of the boards of the room the blessed youth occupied, as well as some portion of the wadding in which his venerable head was wrapped." During nine days' devotion the patient accordingly invoked the Blessed John, swallowed small portions of his dust in water, and at last pressed the cross to her breast so vehemently that she was seized with sickness, went to sleep, and awoke without a symptom of the complaint. And when Dr. Panegrossi the physician beheld the incredible cure, and heard that the patient had addressed herself to the Blessed Berchmans, he bowed his head, saying, "When such physicians interfere, *we* have nothing more to say!"[2] To sum up the whole

---

[1] L. F. Alfred Maury, 'Magie, etc.' p. 249; 'Acta Sanctorum,' 27 Sep.; Gregor. Turon. De Gloria Martyr. i. 98.

[2] J. R. Beste, 'Nowadays at Home and Abroad,' London, 1870, vol. ii. p. 44; 'A New Miracle at Rome; being an Account of a Miraculous Cure etc. etc.' London (Washbourne), 1870.

history of manes-worship, it is plain that in our time the dead still receive worship from far the larger half of mankind, and it may have been much the same ever since the remote periods of primitive culture in which the religion of the manes probably took its rise.

It has now been seen that the theory of souls recognizes them as capable either of independent existence, or of inhabiting human, animal, or other bodies. On the principle here maintained, that the general theory of spirits is modelled on the theory of souls, we shall be able to account for several important branches of the lower philosophy of religion, which without such explanation may appear in great measure obscure or absurd. Like souls, other spirits are supposed able either to exist and act flitting free about the world, or to become incorporate for more or less time in solid bodies. It will be well at once to get a secure grasp of this theory of Embodiment, for without it we shall be stopped every moment by a difficulty in understanding the nature of spirits, as defined in the lower animism. The theory of embodiment serves several highly important purposes in savage and barbarian philosophy. On the one hand it provides an explanation of the phenomena of morbid exaltation and derangement, especially as connected with abnormal utterance, and this view is so far extended as to produce an almost general doctrine of disease. On the other hand, it enables the savage either to "lay" a hurtful spirit in some foreign body, and so get rid of it, or to carry about a useful spirit for his service in a material object, to set it up as a deity for worship in the body of an animal, or in a block or stone or image or other thing, which contains the spirit as a vessel contains a fluid: this is the key to strict fetishism, and in no small measure to idolatry. In briefly considering these various branches of the Embodiment-theory, there may be conveniently included certain groups of cases often impossible to distinguish apart. These cases belong theoretically rather to obsession than possession, the spirits not actually inhabiting the bodies, but hanging or hovering about them and affecting them from the outside.

As in normal conditions the man's soul, inhabiting his body, is held to give it life, to think, speak, and act through it, so an adaptation of the self-same principle explains abnormal conditions of body or mind, by considering the new symptoms as due to the operation of a second soul-like being, a strange spirit. The possessed man, tossed and shaken in fever, pained and wrenched as though some live creature were tearing or twisting him within, pining as though it were devouring his vitals day by day, rationally finds a personal spiritual cause for his sufferings. In hideous dreams he may even sometimes see the very ghost or nightmare-fiend that plagues him. Especially when the mysterious unseen power throws him helpless on the ground, jerks and writhes him in convulsions, makes him leap upon the bystanders with a giant's strength and a wild beast's ferocity, impels him, with distorted face and frantic gesture, and voice not his own nor seemingly even human, to pour forth wild incoherent raving, or with thought and eloquence beyond his sober faculties to command, to counsel, to foretell—such a one seems to those who watch him, and even to himself, to have become the mere instrument of a spirit which has seized him or entered into him, a possessing demon in whose personality the patient believes so implicitly that he often imagines a personal name for it, which it can declare when it speaks in its own voice and character through his organs of speech; at last, quitting the medium's spent and jaded body, the intruding spirit departs as it came. This is the savage theory of dæmoniacal possession and obsession, which has been for ages, and still remains, the dominant theory of disease and inspiration among the lower races. It is obviously based on an animistic interpretation, most genuine and rational in its proper place in man's intellectual history, of the actual symptoms of the cases. The general doctrine of disease-spirits and oracle-spirits appears to have its earliest, broadest, and most consistent position within the limits of savagery. When we have gained a clear idea of it in this its original home, we shall be able to trace it along from grade to grade of civilization, breaking away piecemeal under the influence of new medical theories, yet sometimes expanding in revival, and at least in lingering survival holding its place into the midst of our modern life. The possession-theory is not merely known to us by the statements of those who describe diseases in accordance with it. Disease being accounted for by attack of spirits, it naturally follows that to get rid of these spirits is the proper means of cure. Thus the practices of the exorcist appear side by side with the doctrine of possession, from its first appearance in savagery to its survival in modern civilization, and nothing could display more vividly the conception of a disease or a mental affection as caused by a personal spiritual being than the proceedings of the exorcist who talks to it, coaxes or threatens it, makes offerings to it, entices or drives it out of the patient's body, and induces it to take up its abode in some other. That the two great effects ascribed to such spiritual influence in obsession and possession, namely, the infliction of ailments and the inspiration of oracles, are not only mixed up together but often run into absolute coincidence, accords with the view that both results are referred to one common cause. Also that the intruding or invading spirit may be either a human soul or may belong to some other class in the spiritual hierarchy, countenances the opinion that the possession-theory is derived from, and indeed modelled on, the ordinary theory of the soul acting on the body. In illustrating the doctrine by typical examples from the enormous mass of available details, it will hardly be possible to discriminate among the operating spirits, between those which are souls and those which are demons, nor to draw an exact line between obsession by a demon outside and possession by a demon inside, nor between the condition of the demon-tormented patient and the demon-actuated doctor, seer, or priest. In a word, the confusion of these conceptions in the savage mind only fairly represents their intimate connexion in the Possession-theory itself.

In the Australian-Tasmanian district, disease and death are ascribed to more or less defined spiritual influences; descriptions of a demon working a sorcerer's wicked will by coming slyly behind his victim and hitting him with his club on the back of his neck, and of a dead man's ghost angered by having his name uttered, and creeping up into the utterer's body to consume his liver, are indeed peculiarly graphic details of savage animism.[1] The theory of disease-spirits is well stated in its extreme form among the Mintira, a low race of the Malay peninsula. Their "hantu" or spirits have among their functions that of causing ailments; thus the "hantu kalumbahan" causes small-pox; the "hantu kamang" brings on inflammation and swellings in the hands and feet; when a person is wounded, the "hantu pari" fastens on the wound and sucks, and this is the cause of the blood flowing. And so on, as the describer says, "To enumerate the remainder of the hantus would be merely to convert the name of every species of disease known to the Mintira into a proper one. If any new disease appeared, it would be ascribed to a hantu bearing the same name."[2] It will help us to an idea of the distinct personality which the disease-demon has in the minds of the lower races, to notice the Orang Laut of this district placing thorns and brush in the paths leading to a part where small-pox had broken out, to keep the demons off; just as the Khonds of Orissa try with thorns, and ditches, and stinking oil poured on the ground, to barricade the paths to their hamlets against the goddess of small-pox, Jugah Pennu.[3] Among the Dayaks of Borneo, "to have been

---

[1] Oldfield in 'Tr. Eth. Soc.' vol. iii. p. 235; see Grey, 'Australia,' vol. ii. p. 337. Bonwick, 'Tasmanians,' pp. 183, 195.

[2] 'Journ. Ind. Archip.' vol. i. p. 307.

[3] Bastian, 'Psychologie,' p. 204; Mensch, vol. ii. p. 73, see p. 125 (Battas); Macpherson, 'India,' p. 370. See also Mason, 'Karens,' l. c. p. 201.

smitten by a spirit" is to be ill; sickness may be caused by invisible spirits inflicting invisible wounds with invisible spears, or entering men's bodies and driving out their souls, or lodging in their hearts and making them raving mad. In the Indian Archipelago, the personal semi-human nature of the disease-spirits is clearly acknowleged by appeasing them with feasts and dances and offerings of food set out for them away in the woods, to induce them to quit their victims, or by sending tiny proas to sea with offerings, that spirits which have taken up their abode in sick men's bowels may embark and not come back.[1] The animistic theory of disease is strongly marked in Polynesia, where every sickness is ascribed to spiritual action of deities, brought on by the offerings of enemies, or by the victim's violation of the laws of tapu. Thus in New Zealand each ailment is caused by a spirit, particularly an infant or undeveloped human spirit, which sent into the patient's body gnaws and feeds inside; and the exorcist, finding the path by which such a disease-spirit came from below to feed on the vitals of a sick relative, will persuade it by a charm to get upon a flax-stalk and set off home. We hear, too, of an idea of the parts of the body—forehead, breast, stomach, feet, etc.—being apportioned each to a deity who inflicts aches and pains and ailments there.[2] So in the Samoan group, when a man was near death, people were anxious to part on good terms with him, feeling assured that if he died with angry feelings towards any one, he would certainly return and bring calamity on that person or some one closely allied to him. This was considered a frequent source of disease and death, the spirit of a departed member of the family returning and taking up his abode in the head, chest, or stomach of a living man, and so causing sickness and

[1] 'Journ. Ind. Archip.' vol. iii. p. 110, vol. iv. p. 194; St. John, 'Far East,' vol. i. pp. 71, 87; Beeckman in Pinkerton, vol. ix. p. 133; Meiners, vol. i. p. 278. See also Doolittle, 'Chinese,' vol. i. p. 159.
[2] Shortland, 'Trads. of N. Z.' pp. 97, 114, 125; Taylor, 'New Zealand,' pp. 48, 137.

death. If a man died suddenly, it was thought that he was eaten by the spirit that took him; and though the soul of one thus devoured would go to the common spirit-land of the departed, yet it would have no power of speech there, and if questioned could but beat its breast. It completes this account to notice that the disease-inflicting souls of the departed were the same which possessed the living under more favourable circumstances, coming to talk through a certain member of the family, prophesying future events, and giving directions as to family affairs.[1] Farther east, in the Georgian and Society Islands, evil demons are sent to scratch and tear people into convulsions and hysterics, to torment poor wretches as with barbed hooks, or to twist and knot inside them till they die writhing in agony. But madmen are to be treated with great respect, as entered by a god, and idiots owe the kindness with which they are appeased and coaxed to the belief in their superhuman inspiration.[2] Here, and elsewhere in the lower culture, the old real belief has survived which has passed into a jest of civilized men in the famous phrase of the "inspired idiot."

American ethnography carries on the record of rude races ascribing disease to the action of evil spirits. Thus the Dacotas believe that the spirits punish them for misconduct, especially for neglecting to make feasts for the dead; these spirits have the power to send the spirit of something, as of a bear, deer, turtle, fish, tree, stone, worm, or deceased person, which entering the patient causes disease; the medicine-man's cure consists in reciting charms over him, singing "He-le-li-lah, etc.," to the accompaniment of a gourd-rattle with beads inside, ceremonially shooting a symbolic bark representation of the intruding creature, sucking over the seat of pain to get the spirit out, and

---

[1] Turner, 'Polynesia,' p. 236.
[2] Ellis, 'Polyn. Res.' vol. i. pp. 363, 395, etc., vol. ii. pp. 193, 274; Cook, '3rd Voy.' vol. iii. p. 131. Details of the superhuman character ascribed to weak or deranged persons among other races, in Schoolcraft, part iv. p. 49; Martius, vol. i. p. 633; Meiners, vol. i. p. 323; Waitz, vol. ii. p. 181.

firing guns at it as it is supposed to be escaping.[1] Such processes were in full vogue in the West Indies in the time of Columbus, when Friar Roman Pane put on record his quaint account of the native sorcerer pulling the disease off the patient's legs (as one pulls off a pair of trousers), going out of doors to blow it away, and bidding it begone to the mountain or the sea; the performance concluding with the regular sucking-cure and the pretended extraction of some stone or bit of flesh, or such thing, which the patient is assured that his patron-spirit or deity (cemi) put into him to cause the disease, in punishment for neglect to build him a temple or honour him with prayer or offerings of goods.[2] Patagonians considered sickness as caused by a spirit entering the patient's body; "they believe every sick person to be possessed of an evil demon; hence their physicians always carry a drum with figures of devils painted on it, which they strike at the beds of sick persons to drive out from the body the evil demon which causes the disorder."[3] In Africa, according to the philosophy of the Basutos and the Zulus, the causes of disease are the ghosts of the dead, come to draw the living to themselves, or to compel them to sacrifice meat-offerings. They are recognized by the diviners, or by the patient himself, who sees in dreams the departed spirit come to torment him. Congo tribes in like manner consider the souls of the dead, passed into the ranks of powerful spirits, to cause disease and death among mankind. Thus, in both these districts, medicine becomes an almost entirely religious matter of propitiatory sacrifice and prayer addressed to the disease-inflicting manes. The

[1] Schoolcraft, 'Indian Tribes,' part i. p. 250, part ii. pp. 179, 199, part iii. p. 498; M. Eastman, 'Dahcotah,' p. xxiii. 34, 41, 72. See also Gregg, 'Commerce of Prairies,' vol. ii. p. 297 (Comanches); Morgan, 'Iroquois,' p. 163; Sproat, p. 174 (Ahts); Egede, 'Greenland,' p. 186; Cranz, p. 269.
[2] Roman Pane, xix. in 'Life of Colon'; in Pinkerton, vol. xii. p. 87.
[3] D'Orbigny, 'L'Homme Américain,' vol. ii. pp. 73, 168; Musters, 'Patagonians,' p. 180. See also J G. Müller, pp. 207, 231 (Caribs); Spix and Martius, 'Brasilien,' vol. i. p. 70; Martius, 'Ethnog. Amer.' vol. i. p. 646 (Macusis).

Barolongs give a kind of worship to deranged persons, as under the direct influence of a deity; while in East Africa the explanation of madness and idiocy is simple and typical—"he has fiends."[1] Negroes of West Africa, on the supposition that an attack of illness has been caused by some spiritual being, can ascertain to their satisfaction what manner of spirit has done it, and why. The patient may have neglected his "wong" or fetish-spirit, who has therefore made him ill; or it may be his own "kla" or personal guardian-spirit, who on being summoned explains that he has not been treated respectfully enough, etc.; or it may be a "sisa" or ghost of some dead man, who has taken this means of making known that he wants perhaps a gold ornament that was left behind when he died.[2] Of course, the means of cure will then be to satisfy the demands of the spirit. Another aspect of the negro doctrine of disease-spirits is displayed in the following description from Guinea, by the Rev. J. L. Wilson, the missionary:—"Demoniacal possessions are common, and the feats performed by those who are supposed to be under such influence are certainly not unlike those described in the New Testament. Frantic gestures, convulsions, foaming at the mouth, feats of supernatural strength, furious ravings, bodily lacerations, gnashing of teeth, and other things of a similar character, may be witnessed in most of the cases which are supposed to be under diabolical influence."[3] The remark several times made by travellers is no doubt true, that the spiritualistic theory of disease has tended strongly to prevent progress in the medical art among the lower races. Thus among the Bodo and Dhimal of North-East India, who ascribe all diseases to a deity tormenting the patient for some impiety or neglect, the exorcists divine the offended

[1] Casalis, 'Basutos,' p. 247; Callaway, 'Rel. of Amazulu,' p. 147, etc.; Magyar, 'Süd-Afrika,' p. 21, etc.; Burton, 'Central Afr.' vol. ii. pp. 320, 354; Steere in 'Journ. Anthrop. Inst.' vol. i. 1871, p. cxlvii.

[2] Steinhauser, 'Religion des Negers,' in 'Magaz. der Evang. Missions und Bibel-Gesellschaften,' Basel, 1856, No. 2, p. 139.

[3] J. L. Wilson, 'W. Afr.' pp. 217, 388.

god and appease him with the promised sacrifice of a hog; these exorcists are a class of priests, and the people have no other doctors.[1] Where the world-wide doctrine of disease-demons has held sway, men's minds, full of spells and ceremonies, have scarce had room for thought of drugs and regimen.

The cases in which disease-possession passes into oracle-possession are especially connected with hysterical, convulsive, and epileptic affections. Mr. Backhouse describes a Tasmanian native sorcerer, "affected with fits of spasmodic contraction of the muscles of one breast, which he attributes, as they do all other diseases, to the devil"; this malady served to prove his inspiration to his people.[2] When Dr. Mason was preaching near a village of heathen Pwo, a man fell down in an epileptic fit, his familiar spirit having come over him to forbid the people to listen to the missionary, and he sang out his denunciations like one frantic. This man was afterwards converted, and told the missionary that "he could not account for his former exercises, but that it certainly appeared to him as though a spirit spoke, and he must tell what was communicated." In this Karen district flourishes the native "wee" or prophet, whose business is to work himself into the state in which he can see departed spirits, visit their distant home, and even recall them to the body, thus raising the dead; these wees are nervous excitable men, such as would become mediums, and in giving oracles they go into actual convulsions.[3] Dr. Callaway's details of the state of the Zulu diviners are singularly instructive. Their symptoms are ascribed to possession by "amatongo" or ancestral spirits; the disease is common, from some it departs of its own accord, others have the ghost laid which causes it, and others let the affection take its course and become professional diviners, whose powers of finding hidden things and giving apparently inaccessible

[1] Hodgson, 'Abor. of India,' pp. 163, 170.
[2] Backhouse, 'Australia,' p. 103.
[3] Mason in Bastian, 'Oestl. Asien,' vol. ii. p. 414. Cross, l. c. p. 305.

information are vouched for by native witnesses, who at the same time are not blind to their tricks and their failures. The most perfect description is that of a hysterical visionary, who had "the disease which precedes the power to divine." This man describes that well-known symptom of hysteria, the heavy weight creeping up within him to his shoulders, his vivid dreams, his waking visions of objects that are not there when he approaches, the songs that come to him without learning, the sensation of flying in the air. This man was " of a family who are very sensitive, and become doctors."[1] Persons whose constitutional unsoundness induces morbid manifestations are indeed marked out by nature to become seers and sorcerers. Among the Patagonians, patients seized with falling sickness or St. Vitus's dance were at once selected for magicians, as chosen by the demons themselves who possessed, distorted, and convulsed them.[2] Among Siberian tribes, the shamans select children liable to convulsions as suitable to be brought up to the profession, which is apt to become hereditary with the epileptic tendencies it belongs to.[3] Thus, even in the lower culture, a class of sickly brooding enthusiasts begin to have that power over the minds of their lustier fellows, which they have kept in so remarkable a way through the course of history.

Morbid oracular manifestations are habitually excited on purpose, and moreover the professional sorcerer commonly exaggerates or wholly feigns them. In the more genuine manifestations the medium may be so intensely wrought upon by the idea that a possessing spirit is speaking from within him, that he may not only give this spirit's name and speak in its character, but possibly may in good faith alter his voice to suit the spiritual utterance. This gift of spirit-utterance, which belongs to " ventriloquism " in the ancient and proper sense of the term, of course lapses into sheer

[1] Callaway, 'Religion of Amazulu,' pp. 183, etc., 259, etc.
[2] Falkner, 'Patagonia,' p. 116. See also Rochefort, ' Iles Antilles,' p. 418 (Caribs).
[3] Georgi, ' Reise im Russ. Reich,' vol. i. p. 280; Meiners, vol. ii. p. 488.

trickery. But that the phenomena should be thus artificially excited or dishonestly counterfeited, rather confirms than alters the present argument. Real or simulated, the details of oracle-possession alike illustrate popular belief. The Patagonian wizard begins his performance with drumming and rattling till the real or pretended epileptic fit comes on by the demon entering him, who then answers questions from within him with a faint and mournful voice.[1] Among the wild Veddas of Ceylon, the " devil-dancers " have to work themselves into paroxysms, to gain the inspiration whereby they profess to cure their patients.[2] So, with furious dancing to the music and chanting of the attendants, the Bodo priest brings on the fit of maniacal inspiration in which the deity fills him and gives oracles through him.[3] In Kamchatka the female shamans, when Billukai came down into them in a thunderstorm, would prophesy; or, receiving spirits with a cry of "hush!" their teeth chattered as in fever, and they were ready to divine.[4] Among the Singpho of South-East Asia, when the "natzo" or conjuror is sent for to a sick patient, he calls on his "nat" or demon, the soul of a deceased foreign prince, who descends into him and gives the required answers.[5] In the Pacific Islands, spirits of the dead would enter for a time the body of a living man, inspiring him to declare future events, or to execute some commission from the higher deities. The symptoms of oracular possession among savages have been especially well described in this region of the world. The Fijian priest sits looking steadfastly at a whale's tooth ornament, amid dead silence. In a few minutes he trembles, slight twitchings of face and limbs come on, which increase to strong convulsions, with swelling of the veins, murmurs and sobs. Now the god has entered

[1] Falkner, l. c.
[2] Tennent, 'Ceylon,' vol. ii. p. 441. See Latham, 'Descr. Eth.' vol. ii. p. 469.
[3] Hodgson, 'Abor. of India,' p. 172.
[4] Steller, 'Kamtschatka,' p. 278.
[5] Bastian, 'Oestl. Asien,' vol. ii. p. 328, see vol. iii. p. 201, 'Psychologie,' p. 139. See also Römer, 'Guinea,' p. 59.

him, and with eyes rolling and protruding, unnatural voice, pale face and livid lips, sweat streaming from every pore, and the whole aspect of a furious madman, he gives the divine answer, and then, the symptoms subsiding, he looks round with a vacant stare, and the deity returns to the land of spirits. In the Sandwich Islands, where the god Oro thus gave his oracles, his priest ceased to act or speak as a voluntary agent, but with his limbs convulsed, his features distorted and terrific, his eyes wild and strained, he would roll on the ground foaming at the mouth, and reveal the will of the possessing god in shrill cries and sounds violent and indistinct, which the attending priests duly interpreted to the people. In Tahiti, it was often noticed that men who in the natural state showed neither ability nor eloquence, would in such convulsive delirium burst forth into earnest lofty declamation, declaring the will and answers of the gods, and prophesying future events, in well-knit harangues full of the poetic figure and metaphor of the professional orator. But when the fit was over, and sober reason returned, the prophet's gifts were gone.[1] Lastly, the accounts of oracular possession in Africa show the primitive ventriloquist in perfect types of morbid knavery. In Sofala, after a king's funeral, his soul would enter into a sorcerer, and speaking in the familiar tones that all the bystanders recognized, would give counsel to the new monarch how to govern his people.[2] About a century ago, a negro fetish-woman of Guinea is thus described in the act of answering an enquirer who has come to consult her. She is crouching on the earth, with her head between her knees and her hands up to her face, till, becoming inspired by the fetish, she snorts and foams and gasps. Then the suppliant may put his question, "Will my friend or brother get well of this sickness?"—"What shall I give thee to set him free from his sickness?" and so

[1] Ellis, 'Polyn. Res.' vol. i. pp. 352, 373; Moerenhout, 'Voyage,' vol. i, p. 479; Mariner, 'Tonga Islands,' vol. i. p. 105; Williams, 'Fiji,' vol. i. p. 373.
[2] Dos Santos, 'Ethiopia,' in Pinkerton, vol. xvi. p. 686.

forth. Then the fetish-woman answers in a thin, whistling voice, and with the old-fashioned idioms of generations past; and thus the suppliant receives his command, perhaps to kill a white cock and put him at a four-cross way, or tie him up for the fetish to come and fetch him, or perhaps merely to drive a dozen wooden pegs into the ground, so to bury his friend's disease with them.[1]

The details of demoniacal possession among barbaric and civilized nations need no elaborate description, so simply do they continue the savage cases.[2] But the state of things we notice here agrees with the conclusion that the possession-theory belongs originally to the lower culture, and is gradually superseded by higher medical knowledge. Surveying its course through the middle and higher civilization, we shall notice first a tendency to limit it to certain peculiar and severe affections, especially connected with mental disorder, such as epilepsy, hysteria, delirium, idiocy, madness; and after this a tendency to abandon it altogether, in consequence of the persistent opposition of the medical faculty. Among the nations of South-East Asia, obsession and possession by demons is strong at least in popular belief. The Chinese attacked with dizziness, or loss of the use of his limbs, or other unaccountable disease, knows that he has been influenced by a malignant demon, or punished for some offence by a deity whose name he will mention, or affected by his wife of a former existence, whose spirit has after a long search discovered him. Exorcism of course exists, and when the evil spirit or influence is expelled, it is especially apt to enter some person standing near; hence the common saying, "idle spectators should not be present at an exorcism." Divination by possessed mediums is usual in China among such is the professional woman who sits at a table in contemplation, till the soul of a deceased person from whom

[1] Römer, 'Guinea,' p. 57. See also Steinhauser, l. c. pp. 132, 139; J. B. Schlegel, 'Ewe-Sprache,' p. xvi.
[2] Details from Tatar races in Castrén, 'Finn. Myth.' pp. 164, 173, etc. Bastian, 'Psychologie,' p. 90; from Abyssinia in Parkyns, 'Life in A.,' ch xxxiii.

communication is desired enters her body and talks through her to the living; also the man into whom a deity is brought by invocations and mesmeric passes, when, assuming the divine figure and attitude, he pronounces the oracle.[1] In Birma, the fever-demon of the jungle seizes trespassers on his domain, and shakes them in ague till he is exorcised, while falls and apoplectic fits are the work of other spirits. The dancing of women by demoniacal possession is treated by the doctor covering their heads with a garment, and thrashing them soundly with a stick, the demon and not the patient being considered to feel the blows; the possessing spirit may be prevented from escaping by a knotted and charmed cord hung round the bewitched person's neck, and when a sufficient beating has induced it to speak by the patient's voice and declare its name and business, it may either be allowed to depart, or the doctor tramples on the patient's stomach till the demon is stamped to death. For an example of invocation and offerings, one characteristic story told by Dr. Bastian will suffice. A Bengali cook was seized with an apoplectic fit, which his Birmese wife declared was but a just retribution, for the godless fellow had gone day after day to market to buy pounds and pounds of meat, yet in spite of her remonstrances would never give a morsel to the patron-spirit of the town; as a good wife, however, she now did her best for her suffering husband, placing near him little heaps of coloured rice for the "nat," and putting on his fingers rings with prayers addressed to the same offended being—"Oh ride him not!"—"Ah let him go!"—"Grip him not so hard!"—"Thou shalt have rice!"—"Ah, how good that tastes!" How explicitly Buddhism recognizes such ideas, may be judged from one of the questions officially put to candidates for admission as monks or talapoins—"Art thou afflicted by madness or the other ills caused by giants, witches, or evil demons of the forest and mountain?"[2] Within our own domain of British India,

[1] Doolittle, 'Chinese,' vol. i. p. 143, vol. ii. pp. 110, 320.
[2] Bastian, 'Oestl. Asien,' vol. ii. pp. 103, 152, 381, 418, vol. iii. p. 247,

the possession-theory and the rite of exorcism belonging to it may be perfectly studied to this day. There the doctrine of sudden ailment or nervous disease being due to a blast or possession by a "bhut," or being, that is, a demon, is recognized as of old; there the old witch who has possessed a man and made him sick or deranged, will answer spiritually out of his body and say who she is and where she lives; there the frenzied demoniac may be seen raving, writhing, tearing, bursting his bonds, till, subdued by the exorcist, his fury subsides, he stares and sighs, falls helpless to the ground, and comes to himself; and there the deities caused by excitement, singing, and incense to enter into men's bodies, manifest their presence with the usual hysterical or epileptic symptoms, and speaking in their own divine name and personality, deliver oracles by the vocal organs of the inspired medium.[1]

Opinions similar to these were current in ancient Greece and Rome, to whose languages indeed our own owes the technical terms of the subject, such as "demoniac" and "exorcist." Thus Homer's sick men racked with pain are tormented by a hateful demon (στυγερὸς δέ οἱ ἔχραε δαίμων). So to Pythagoras the causes of disease in men and beasts are demons pervading the air. "Epilepsy" (ἐπίληψις) was, as its name imports, the "seizure" of the patient by a superhuman agent: the agent being more exactly defined in "nympholepsy," the state of being seized or possessed by a nymph, i. e., rapt or entranced (νυμφόληπτος, lymphatus). The causation of mental derangement and delirious utterance by spiritual possession was an accepted tenet of Greek philosophy. To be insane was simply to have an evil spirit, as when Sokrates said of those who denied demoniac or spiritual

etc. See also Bowring, 'Siam,' vol. i. p. 139; 'Journ. Ind. Archip.' vol. iv. p. 507, vol. vi. p. 614; Turpin in Pinkerton, vol. ix. p. 761; Kempfer, 'Japan,' *ibid.* vol. vii. pp. 701, 730, etc.

[1] Ward, 'Hindoos,' vol. i. p. 155, vol. ii. p. 183; Roberts, 'Oriental Illustrations of the Scriptures,' p. 529; Bastian, 'Psychologie,' pp. 164, 184-7. Sanskrit paiçâcha-graha = demon-seizure, possession. Ancient evidence in Pictet, 'Origines Indo-Europ.' part ii. ch. v.; Spiegel, 'Avesta.'

knowledge, that they themselves were demoniac (δαιμονᾶν ἔφη), and Alexander ascribed to the influence of offended Dionysos the ungovernable drunken fury in which he killed his friend Kleitos; raving madness was obsession or possession by an evil demon (κακοδαιμονία). So the Romans called madmen " larvati," " larvarum pleni," full of ghosts. Patients possessed by demons stared and foamed, and the spirits spoke from within them by their voices. The craft of the exorcist was well known. As for oracular possession, its theory and practice remained in fullest vigour through the classic world, scarce altered from the times of lowest barbarism. Could a South Sea islander have gone to Delphi to watch the convulsive struggles of the Pythia, and listen to her raving, shrieking utterances, he would have needed no explanation whatever of a rite so absolutely in conformity with his own savage philosophy.[1]

The Jewish doctrine of possession[2] at no time in its long course exercised a direct influence on the opinion of the civilized world comparable to that produced by the mentions of demoniacal possession in the New Testament. It is needless to quote here even a selection from the familiar passages of the Gospels and Acts which display the manner in which certain described symptoms were currently accounted for in public opinion. Regarding these documents from an ethnographic point of view, it need only be said that they prove, incidentally but absolutely, that Jews and Christians at that time held the doctrine which had prevailed for ages before, and continued to prevail for ages after, referring to possession and obsession by spirits the symptoms of mania, epilepsy, dumbness, delirious and oracular utterance, and other morbid conditions, mental and bodily.[3] Modern missionary works, such as have been cited

[1] Homer. Odyss. v. 396, x. 64; Diog. Laert. viii. 1; Plat. Phædr. Tim. etc.; Pausan. iv. 27, 2; Xen. Mem. I. i. 9; Plutarch. Vit. Alex. De Orac. Def.; Lucian. Philopseudes; Petron. Arbiter, Sat.; etc. etc.
[2] Joseph. Ant. Jud. viii. 2, 5. Eisenmenger, 'Entdecktes Judenthum,' part ii. p. 454. See Maury, p. 290.
[3] Matth. ix. 32, xi. 18, xii. 22, xvii. 15; Mark, i. 23, ix. 17; Luke, iv.

here, give the most striking evidence of the correspondence of these demoniac symptoms with such as may still be observed among uncivilized races. During the early centuries of Christianity, demoniacal possession indeed becomes peculiarly conspicuous, perhaps not from unusual prevalence of the animistic theory of disease, but simply because a period of intense religious excitement brought it more than usually into requisition. Ancient ecclesiastical records describe, under the well-known names of " dæmoniacs " (δαιμονιζόμενοι), " possessed " (κατεχόμενοι), "energumens " (ἐνεργούμενοι), the class of persons whose bodies are seized or possessed with an evil spirit; such attacks being frequently attended with great commotions and vexations and disturbances of the body, occasioning sometimes frenzy and madness, sometimes epileptic fits, and other violent tossings and contortions. These energumens formed a recognized part of an early Christian congregation, a standing-place apart being assigned to them in the church. The church indeed seems to have been the principal habitation of these afflicted creatures, they were occupied in sweeping and the like out of times of worship, daily food was provided for them, and they were under the charge of a special order of clergy, the exorcists, whose religious function was to cast out devils by prayer and adjuration and laying on of hands. As to the usual symptoms of possession, Justin, Tertullian, Chrysostom, Cyril, Minucius, Cyprian, and other early Fathers, give copious descriptions of demons entering into the bodies of men, disordering their health and minds, driving them to wander among the tombs, forcing them to writhe and wallow and rave and foam, howling and declaring their own diabolical names by the patients' voices, but when overcome by conjuration or by blows administered to their victims, quitting the bodies they had entered, and acknowledging the pagan deities to be but devils.[1]

33, 39, vii. 33, viii. 27, ix. 39, xiii. 11 ; John, x. 20 ; Acts, xvi. 16, xix. 13 ; etc.

[1] For general evidence see Bingham, 'Antiquities of Christian Church,

On a subject so familiar to educated readers I may be excused from citing at length a vast mass of documents, barbaric in nature and only more or less civilized in circumstance, to illustrate the continuance of the doctrine of possession and the rite of exorcism through the middle ages and into modern times. A few salient examples will suffice. For a type of medical details, we may instance the recipes in the 'Early English Leechdoms,': a cake of the "thost" of a white hound baked with meal is to be taken against the attack by dwarves (*i. e.* convulsions); a drink of herbs worked up off clear ale with the aid of garlic, holy water, and singing of masses, is to be drunk by a fiend-sick patient out of a church-bell. Philosophical argument may be followed in the dissertations of the 'Malleus Maleficarum,' concerning demons substantially inhabiting men and causing illness in them, enquiries which may be pursued under the auspices of Glanvil in the 'Saducismus Triumphatus.' Historical anecdote bears record of the convulsive clairvoyant demon who possessed Nicola Aubry, and under the Bishop of Laon's exorcism testified in an edifying manner to the falsity of Calvinism; of Charles VI. of France, who was possessed, and whose demon a certain priest tried in vain to transfer into the bodies of twelve men who were chained up to receive it; of the German woman at Elbingerode who in a fit of toothache wished the devil might enter into her teeth, and who was possessed by six demons accordingly, which gave their names as Schalk der Wahrheit, Wirk, Widerkraut, Myrrha, Knip, Stüp; of George Lukins of Yatton, whom seven devils threw into fits and talked and sang and barked out of, and who was delivered by a solemn exorcism by seven clergymen at the Temple Church at Bristol in the year 1788.[1] A strong

book iii. ch. iv.; Calmet, 'Dissertation sur les Esprits;' Maury, 'Magie,' etc.; Lecky, 'Hist. of Rationalism.' Among particular passages are Tertull. Apolog. 23; De Spectaculis, 26; Chrysostom. Homil. xxviii. in Matth. iv.; Cyril. Hierosol. Catech. xvi. 16; Minuc. Fel. Octavius. xxi.; Concil. Carthag. iv.; etc., etc.

[1] Details in Cockayne, 'Leechdoms, &c., of Early England,' vol. i. p. 365,

sense of the permanence of the ancient doctrine may be gained from accounts of the state of public opinion in Europe, from Greece and Italy to France, where within the last century derangement and hysteria were still popularly ascribed to possession and treated by exorcism, just as in the dark ages.[1] In the year 1861, at Morzine, at the south of the Lake of Geneva, there might be seen in full fury an epidemic of diabolical possession worthy of a Red Indian settlement or a negro kingdom of West Africa, an outburst which the exorcisms of a superstitious priest had so aggravated that there were a hundred and ten raving demoniacs in that single village.[2] The following is from a letter written in 1862 by Mgr. Anouilh, a French missionary-bishop in China. "Le croiriez-vous ? dix villages se sont convertis. Le diable est furieux et fait les cent coups. Il y a eu, pendant les quinze jours que je viens de prêcher, cinq ou six possessions. Nos catéchumènes avec l'eau bénite chassent les diables, guérissent les malades. J'ai vu des choses merveilleuses. Le diable m'est d'un grand secours pour convertir les païens. Comme au temps de Notre-Seigneur, quoique père du mensonge, il ne peut s'empêcher de dire la vérité. Voyez ce pauvre possédé faisant mille contorsions et disant à grands cris : ' Pourquoi prêches-tu la vraie religion ? Je ne puis souffrir que tu m'enlèves mes disciples.'—' Comment t'appelles-tu ? ' lui demande le catéchiste. Après quelques refus : ' Je suis l'envoyé de Lucifer ' —' Combien êtes-vous ? '—' Nous sommes vingt-deux.' L'eau bénite et le signe de la croix ont délivré ce possédé."[3] To conclude the series with a modern spiritualistic instance,

vol. ii. p. 137, 355 ; Sprenger, 'Malleus Maleficarum,' part ii. ; Calmet, ' Dissertation,' vol. i. ch. xxiv.; Horst, ' Zauber-Bibliothek ;' Bastian, 'Mensch,' vol. ii. p. 557, &c. ; 'Psychologie,' p. 115, etc. ; Voltaire, 'Questions sur l'Encyclopédie,' art. 'Superstition' ; 'Encyclopædia Britannica,' art. 'Possession.'

[1] See Maury, 'Magie,' etc. part ii. ch. ii.

[2] A. Constans, ' Rel. sur une Epidémie d'Hystéro-Démonopathie, en 1861.' 2nd ed. Paris, 1863. For descriptions of such outbreaks, among the North American Indians, see Le Jeune in ' Rel. des Jés. dans la Nouvelle France,' 1639 ; Brinton. p. 275, and in Guinea, see J. L. Wilson, ' Western Africa,' p. 217.

[3] Gaume, 'L'Eau Bénite au Dix-Neuvième Siècle,' 3rd ed. Paris, 1866, p. 353.

one of those where the mediums feel themselves entered and acted through by a spirit other than their own soul. The Rev. Mr. West of Philadelphia describes how a certain possessed medium went through the sword exercise, and fell down senseless; when he came to himself again, the spirit within him declared itself to be the soul of a deceased ancestor of the minister's, who had fought and died in the American War.[1] We in England now hardly hear of demoniacal possession except as a historical doctrine of divines. We have discarded from religious services the solemn ceremony of casting out devils from the bodies of the possessed, a rite to this day officially retained in the Rituals of the Greek and Roman Churches. Cases of diabolical influence alleged from time to time among ourselves are little noticed except by newspaper paragraphs on superstition and imposture. If, however, we desire to understand the doctrine of possession, its origin and influence in the world, we must look beyond countries where public opinion has passed into this stage, and must study the demoniac theory as it still prevails in lower and lowest levels of culture.

It has to be thoroughly understood that the changed aspect of the subject in modern opinion is not due to disappearance of the actual manifestations which early philosophy attributed to demoniacal influence. Hysteria and epilepsy, delirium and mania, and such like bodily and mental derangement, still exist. Not only do they still exist, but among the lower races, and in superstitious districts among the higher, they are still explained and treated as of old. It is not too much to assert that the doctrine of demoniacal possession is kept up, substantially the same theory to account for substantially the same facts, by half the human race, who thus stand as consistent representatives of their forefathers back into primitive antiquity. It is in the civilized world, under the influence of the medical doctrines which have been developing since classic times, that the early animistic theory of these morbid phenomena has been

[1] West in 'Spiritual Telegraph,' cited by Bastian.

gradually superseded by views more in accordance with modern science, to the great gain of our health and happiness. The transition which has taken place in the famous insane colony of Gheel in Belgium is typical. In old days, the lunatics were carried there in crowds to be exorcised from the demons at the church of St. Dymphna; to Gheel they still go, but the physician reigns in the stead of the exorcist. Yet wherever, in times old or new, we find demoniacal influences brought forward to account for affections which scientific physicians now explain on a different principle, we must be careful not to misjudge the ancient doctrine and its place in history. As belonging to the lower culture it is a perfectly rational philosophical theory to account for certain pathological facts. But just as mechanical astronomy gradually superseded the animistic astronomy of the lower races, so biological pathology gradually supersedes animistic pathology, the immediate operation of personal spiritual beings in both cases giving place to the operation of natural processes.

We now pass to the consideration of another great branch of the lower religion of the world, a development of the same principles of spiritual operation with which we have become familiar in the study of the possession-theory. This is the doctrine of Fetishism. Centuries ago, the Portuguese in West Africa, noticing the veneration paid by the negroes to certain objects, such as trees, fish, plants, idols, pebbles, claws of beasts, sticks, and so forth, very fairly compared these objects to the amulets or talismans with which they were themselves familiar, and called them *feitiço* or "charm," a word derived from Latin *factitius*, in the sense of "magically artful." Modern French and English adopted this word from the Portuguese as *fétiche, fetish*, although curiously enough both languages had already possessed the word for ages in a different sense, Old French *faitis*, "well made, beautiful," which Old English adopted as *fetys*, "well made, neat." It occurs in the commonest of all quotations from Chaucer:

"And Frensch sche spak ful faire and *fetysly*,
Aftur the scole of Stratford atte Bowe,
For Frensch of Parys was to hire unknowe."

The President de Brosses, a most original thinker of the last century, struck by the descriptions of the African worship of material and terrestrial objects, introduced the word Fétichisme as a general descriptive term,[1] and since then it has obtained great currency by Comte's use of it to denote a general theory of primitive religion, in which external objects are regarded as animated by a life analogous to man's. It seems to me, however, more convenient to use the word Animism for the doctrine of spirits in general, and to confine the word Fetishism to that subordinate department which it properly belongs to, namely, the doctrine of spirits embodied in, or attached to, or conveying influence through, certain material objects. Fetishism will be taken as including the worship of "stocks and stones," and thence it passes by an imperceptible gradation into Idolatry.

Any object whatsoever may be a fetish. Of course, among the endless multitude of objects, not as we should say physically active, but to which ignorant men ascribe mysterious power, we are not to apply indiscriminately the idea of their being considered vessels or vehicles or instruments of spiritual beings. They may be mere signs or tokens set up to represent ideal notions or ideal beings, as fingers or sticks are set up to represent numbers. Or they may be symbolic charms working by imagined conveyance of their special properties, as an iron ring to give firmness, or a kite's foot to give swift flight. Or they may be merely regarded in some undefined way as wondrous ornaments or curiosities. The tendency runs through all human nature to collect and admire objects remarkable in beauty, form, quality, or scarceness. The shelves of ethnological museums show heaps of the objects which the lower races treasure up

[1] (C. de Brosses.) 'Du culte des dieux fétiches ou Parallèle de l'ancienne Religion de l'Egypte avec la religion actuelle de Nigritie.' 1760. [De Brosses supposed the word *fétiche* connected with *chose fée, fatum.*]

and hang about their persons—teeth and claws, roots and berries, shells and stones, and the like. Now fetishes are in great measure selected from among such things as these, and the principle of their attraction for savage minds is clearly the same which still guides the superstitious peasant in collecting curious trifles " for luck." The principle is one which retains its force in far higher ranges of culture than the peasant's. Compare the Ostyak's veneration for any peculiar little stone he has picked up, with the Chinese love of collecting curious varieties of tortoise-shell, or an old-fashioned English conchologist's delight in a reversed shell. The turn of mind which in a Gold-Coast negro would manifest itself in a museum of monstrous and most potent fetishes, might impel an Englishman to collect scarce postage-stamps or queer walking-sticks. In the love of abnormal curiosities there shows itself a craving for the marvellous, an endeavour to get free from the tedious sense of law and uniformity in nature. As to the lower races, were evidence more plentiful as to the exact meaning they attach to objects which they treat with mysterious respect, it would very likely appear more often and more certainly than it does now, that these objects seem to them connected with the action of spirits, so as to be, in the strict sense in which the word is here used, real fetishes. But this must not be taken for granted. To class an object as a fetish, demands explicit statement that a spirit is considered as embodied in it or acting through it or communicating by it, or at least that the people it belongs to do habitually think this of such objects; or it must be shown that the object is treated as having personal consciousness and power, is talked with, worshipped, prayed to, sacrificed to, petted or ill-treated with reference to its past or future behaviour to its votaries. In the instances now selected, it will be seen that in one way or another they more or less satisfy such conditions. In investigating the exact significance of fetishes in use among men, savage or more civilized, the peculiar difficulty is to know whether the effect of the object is thought due to a whole personal spirit embodied in or attached to it, or to some less definable influence exerted through it. In some cases this point is made clear, but in many it remains doubtful.

It will help us to a clearer conception of the nature of a fetish, to glance at a curious group of notions which connect a disease at once with spiritual influence, and with the presence of some material object. They are a set of illustrations of the savage principle, that a disease or an actual disease-spirit may exist embodied in a stick or stone or such like material object. Among the natives of Australia, we hear of the sorcerers extracting from their own bodies by passes and manipulations a magical essence called "boylya," which they can make to enter the patient's body like pieces of quartz, which causes pain there and consumes the flesh, and may be magically extracted either as invisible or in the form of a bit of quartz. Even the spirit of the waters, "nguk-wonga," which had caused an attack of erysipelas in a boy's leg (he had been bathing too long when heated) is declared to have been extracted by the conjurors from the affected part in the shape of a sharp stone.[1] The Caribs, who very distinctly referred diseases to the action of hostile demons or deities, had a similar sorcerer's process of extracting thorns or splinters from the affected part as the peccant causes, and it is said that in the Antilles morsels of stone and bone so extracted were wrapped up in cotton by the women, as protective fetishes in childbirth.[2] The Malagasy, considering all diseases as inflicted by an evil spirit, consult a diviner, whose method is often to remove the disease by means of a "faditra;" this is some object, such as a little grass, ashes, a sheep, a pumpkin, the water the patient has rinsed his mouth with, or what not, and when the priest has counted on it the evils

[1] Grey, 'Australia,' vol. ii. p. 337 ; Eyre, 'Australia,' vol. ii. p. 362; Oldfield in 'Tr. Eth. Soc.' vol. iii. p. 235, etc. ; G. F. Moore, 'Vocab. of S. W Austr.' pp. 18, 98, 103. See Bonwick, 'Tasmanians,' p. 195.
[2] Rochefort, 'Iles Antilles,' pp. 419, 508 ; J. G. Müller, pp. 173, 207, 217.

that may injure the patient, and charged the faditra to take them away for ever, it is thrown away, and the malady with it.[1] Among those strong believers in disease-spirits, the Dayaks of Borneo, the priest, waving and jingling charms over the affected part of the patient, pretends to extract stones, splinters, and bits of rag, which he declares are spirits; of such evil spirits he will occasionally bring half-a-dozen out of a man's stomach, and as he is paid a fee of six gallons of rice for each, he is probably disposed (like a chiropodist under similar circumstances) to extract a good many.[2] The most instructive accounts of this kind are those which reach us from Africa. Dr. Callaway has taken down at length a Zulu account of the method of stopping out disease caused by spirits of the dead. If a widow is troubled by her late husband's ghost coming and talking to her night after night as though still alive, till her health is affected and she begins to waste away, they find a "nyanga" or sorcerer who can bar out the disease. He bids her not lose the spittle collected in her mouth while she is dreaming, and gives her medicine to chew when she wakes. Then he goes with her to lay the "itongo," or ghost; perhaps he shuts it up in a bulb of the inkomfe plant, making a hole in the side of this, putting in the medicine and the dream-spittle, closing the hole with a stopper, and replanting the bulb. Leaving the place, he charges her not to look back till she gets home. Thus the dream is barred; it may still come occasionally, but no longer infests the woman; the doctor prevails over the dead man as regards that dream. In other cases the cure of a sick man attacked by the ancestral spirits may be effected with some of his blood put into a hole in an anthill by the doctor, who closes the hole with a stone, and departs without looking back; or the patient may be scarified over the painful place, and the blood put into the mouth of a frog, caught for the purpose and carried back. So the disease is barred out from the

[1] Ellis, 'Madagascar,' vol. i. pp. 221, 232, 422.
[2] St. John, 'Far East,' vol. i. p. 211, see 72.

man.[1] In West Africa, a case in point is the practice of transferring a sick man's ailment to a live fowl, which is set free with it, and if any one catches the fowl, the disease goes to him.[2] Captain Burton's account from Central Africa is as follows. Disease being possession by a spirit or ghost, the "mganga" or sorcerer has to expel it, the principal remedies being drumming, dancing, and drinking, till at last the spirit is enticed from the body of the patient into some inanimate article, technically called a "keti" or stool for it. This may be an ornament, such as a peculiar bead or a leopard's claw, or it may be a nail or rag, which by being driven into or hung to a "devil's tree" has the effect of laying the disease-spirit. Or disease-spirits may be extracted by chants, one departing at the end of each stave, when a little painted stick made for it is flung on the ground, and some patients may have as many as a dozen ghosts extracted, for here also the fee is so much apiece.[3] In Siam, the Laos sorcerer can send his "phi phob" or demon into a victim's body, where it turns into a fleshy or leathery lump, and causes disease ending in death.[4] Thus, on the one hand, the spirit-theory of disease is brought into connexion with that sorcerer's practice so extraordinarily prevalent among the lower races, of pretending to extract objects from the patient's body, such as stones, bones, balls of hair, &c., which are declared to be causes of disease conveyed by magical means into him; of this proceeding I have given some account elsewhere, under the name of the "sucking-cure."[5] On the other hand, we see among the lower races that well known conception of a disease or evil influence as an individual being, which may be not merely conveyed by an infected object (though this of course may have much to do with the idea), but may be

[1] Callaway, 'Religion of Amazulu,' p. 314.
[2] Steinhauser, l. c. p. 141. See also Steere, 'East Afr. Tribes,' in 'Journ. Anthrop. Soc.' vol. i. p. cxlviii.
[3] Burton, 'Central Africa,' vol. ii. p. 352. See 'Sindh,' p. 177.
[4] Bastian, 'Oestl. Asien,' vol. iii. p. 275.
[5] 'Early Hist. of Mankind,' ch. x. See Bastian, 'Mensch,' vol. ii. p. 116, etc.

removed by actual transfer from the patient into some other animal or object. Thus Pliny informs us how pains in the stomach may be cured by transmitting the ailment from the patient's body into a puppy or duck, which will probably die of it;[1] it is considered baneful to a Hindu woman to be a man's third wife, wherefore the precaution is taken of first betrothing him to a tree, which dies in her stead;[2] after the birth of a Chinese baby, its father's trousers are hung in the room wrong side up, that all evil influences may enter into them instead of into the child.[3] Modern folklore still cherishes such ideas. The ethnographer may still study in the " white witchcraft " of European peasants the arts of curing a man's fever or headache by transferring it to a crawfish or a bird, or of getting rid of ague or gout or warts by giving them to a willow, elder, fir, or ash-tree, with suitable charms, " Goe morgen, olde, ick geef oe de Kolde," "Goden Abend, Herr Fleder, hier bring ick mien Feber, ick bind em di an und gah davan," "Ash-tree, ashen tree, pray buy this wart of me," and so forth; or of nailing or plugging an ailment into a tree-trunk, or conveying it away by some of the patient's hair or nail-parings or some such thing, and so burying it. Looking at these proceedings from a moral point of view, the practice of transferring the ailment to a knot or a lock of hair and burying it is the most harmless, but another device is a very pattern of wicked selfishness. In England, warts may be touched each with a pebble, and the pebbles in a bag left on the road to church, to give up their ailments to the unlucky finder; in Germany, a plaister from a sore may be left at a cross-way to transfer the disease to a passer-by; I am told on medical authority that the bunches of flowers which children offer to travellers in Southern Europe are sometimes intended for the ungracious purpose of sending some disease away from their homes.[4] One case of this

[1] Plin. xxx. 14, 20. Cardan, ' De Var. Rerum,' cap. xliii.
[2] Ward, ' Hindoos,' vol. i. p. 134, vol. ii. p. 247.
[3] Doolittle, ' Chinese,' vol. i. p. 122.
[4] Grimm, 'D. M.' pp. 1118-23 ; Wuttke, ' Volksaberglaube,' pp. 155-70 ;

group, mentioned to me by Mr. Spottiswoode, is particularly interesting. In Thuringia it is considered that a string of rowan-berries, a rag, or any small article, touched by a sick person and then hung on a bush beside some forest path, imparts the malady to any person who may touch this article in passing, and frees the sick person from the disease. This gives great probability to Captain Burton's suggestion that the rags, locks of hair, and what not, hung on trees near sacred places by the superstitious from Mexico to India and from Ethiopia to Ireland, are deposited there as actual receptacles of disease; the African "devil's trees" and the sacred trees of Sindh, hung with rags through which votaries have transferred their complaints, being typical cases of a practice surviving in lands of higher culture.

The spirits which enter or otherwise attach themselves to objects may be human souls. Indeed one of the most natural cases of the fetish-theory is when a soul inhabits or haunts the relics of its former body. It is plain enough that by a simple association of ideas the dead person is imagined to keep up a connexion with his remains. Thus we read of the Mandan women going year after year to take food to the skulls of their dead kinsfolk, and sitting by the hour to chat and jest in their most endearing strain with the relics of a husband or child;[1] thus the Guinea negroes, who keep the bones of parents in chests, will go to talk with them in the little huts which serve for their tombs.[2] And thus, from the savage who keeps and carries with his household property the cleaned bones of his forefathers,[3] to

Brand, 'Pop. Ant.' vol. ii. p. 375, vol. iii. p. 286; Halliwell, 'Pop. Rhymes,' p. 208; R. Hunt, 'Pop. Romances,' 2nd Series, p. 211; Hylten-Cavallius, 'Wärend och Wirdarne,' vol. i. p. 173. It is said, however, that rags fastened on trees by Gypsies, which passers-by avoid with horror as having diseases thus banned into them, are only signs left for the information of fellow vagrants; Liebich, 'Die Zigeuner,' p. 96.
[1] Catlin, 'N. A. Indians,' vol. i. p. 90.
[2] J. L. Wilson, 'W Africa,' p. 394.
[3] Meiners, 'Gesch. der Rel.' vol. i. p. 305; J. G. Müller, p. 209.

the mourner among ourselves who goes to weep at the grave of one beloved, imagination keeps together the personality and the relics of the dead. Here, then, is a course of thought open to the animistic thinker, leading him on from fancied association to a belief in the real presence of a spiritual being in a material object. Thus there is no difficulty in understanding how the Karens thought the spirits of the dead might come back from the other world to re-animate their bodies;[1] nor how the Marian islanders should have kept the dried bodies of their dead ancestors in their huts as household gods, and even expected them to give oracles out of their skulls;[2] nor how the soul of a dead Carib might be thought to abide in one of his bones, taken from the grave and carefully wrapped in cotton, in which state it could answer questions, and even bewitch an enemy if a morsel of his property were wrapped up with it;[3] nor how the dead Santal should be sent to his fathers by the ceremony of committing to the sacred river morsels of his skull from the funeral-pile.[4] Such ideas are of great interest in studying the burial rites of mankind, especially the habit of keeping relics of the dead as vehicles of superhuman power, and of even preserving the whole body as a mummy, as in Peru and Egypt. The conception of such human relics becoming fetishes, inhabited or at least acted through by the souls which formerly belonged to them, would give a rational explanation of much relic-worship otherwise obscure.

A further stretch of imagination enables the lower races to associate the souls of the dead with mere objects, a practice which may have had its origin in the merest childish make-believe, but which would lead a thorough savage animist straight on to the conception of the soul entering

---

[1] Mason, Karens, l. c. p. 231.
[2] Meiners, vol. ii. pp. 721-3.
[3] Rochefort, 'Iles Antilles,' p. 418. See Martius, 'Ethnog. Amer.' vol. i. p. 485 (Yumanas swallow ashes of deceased with liquor, that he may live again in them.)
[4] Hunter, 'Rural Bengal,' p. 210. See Bastian, 'Psychologie,' p. 73; J. G. Müller, 'Amer. Urrel.' pp. 209, 262, 289, 401, 419.

the object as a body. Mr. Darwin saw two Malay women in Keeling Island who held a wooden spoon dressed in clothes like a doll; this spoon had been carried to the grave of a dead man, and becoming inspired at full moon, in fact lunatic, it danced about convulsively like a table or a hat at a modern spirit-séance.[1] Among the Salish Indians of Oregon, the conjurors bring back men's lost souls as little stones or bones or splinters, and pretend to pass them down through the tops of their heads into their hearts, but great care must be taken to remove the spirits of any dead people that may be in the lot, for the patient receiving one would die.[2] There are indigenous Kol tribes of India who work out this idea curiously in bringing back the soul of a deceased man into the house after the funeral, apparently to be worshipped as a household spirit; while some catch the spirit re-embodied in a fowl or fish, the Binjwar of Raepore bring it home in a pot of water, and the Bunjia in a pot of flour.[3] The Chinese hold such theories with extreme distinctness, considering one of a man's three spirits to take up its abode in the ancestral tablet, where it receives messages and worship from the survivors; while the long keeping of the dead man's gilt and lacquered coffin, and the reverence and offerings continued at the tomb, are connected with the thought of a spirit lingering about the corpse. Consistent with these quaint ideas are ceremonies in vogue in China, of bringing home in a cock (live or artificial) the spirit of a man deceased in a distant place, and of enticing into a sick man's coat the departing spirit which has already left his body, and so conveying it back.[4] Tatar folk-lore illustrates the idea of soul-embodiment in the quaint but intelligible story of the demon-giant who could not be slain, for he did not keep his soul in his body, but in a twelve-

[1] Darwin, 'Journal,' p. 458.
[2] Bastian, 'Mensch,' vol. ii. p. 320.
[3] Report of Jubbulpore Ethnological Committee, Nagpore, 1868, part i. p. 5.
[4] Doolittle, 'Chinese,' vol. i. pp. 151, 207, 214, vol. ii. p. 401; see Plath, 'Religion der alten Chinesen,' part i. p. 59, part ii. p. 101.

headed snake carried in a bag on his horse's back; the hero finds out the secret and kills the snake, and then the giant dies too. This tale is curious, as very likely indicating the original sense of a well-known group of stories in European folklore, the Scandinavian one, for instance, where the giant cannot be made an end of, because he keeps his heart not in his body, but in a duck's egg in a well far away; at last the young champion finds the egg and crushes it, and the giant bursts.[1] Following the notion of soul-embodiment into civilized times, we learn that " A ghost may be laid for any term less than an hundred years, and in any place or body, full or empty; as, a solid oak—the pommel of a sword —a barrel of beer, if a yeoman or simple gentleman—or a pipe of wine, if an esquire or a justice." This is from Grose's bantering description in the last century of the art of " laying " ghosts,[2] and it is one of the many good instances of articles of serious savage belief surviving as jests among civilized men.

Thus other spiritual beings, roaming free about the world, find fetish-objects to act through, to embody themselves in, to present them visibly to their votaries. It is extremely difficult to draw a distinct line of separation between the two prevailing sets of ideas relating to spiritual action through what we call inanimate objects. Theoretically we can distinguish the notion of the object acting as it were by the will and force of its own proper soul or spirit, from the notion of some foreign spirit entering its substance or acting on it from without, and so using it as a body or instrument. But in practice these conceptions blend almost inextricably. This state of things is again a confirmation of the theory of animism here advanced, which treats both sets of ideas as similar developments of the same original

---

[1] Castrén, 'Finn. Myth,' p. 187; Dasent, 'Norse Tales,' p. 69; Lane, 'Thousand and One Nights,' vol. iii. p. 316; Grimm, 'D. M.' p. 1033. See also Bastian, 'Psychologie,' p. 213. Eisenmenger, 'Judenthum,' part ii. p. 39.
[2] Brand, 'Pop. Ant.' vol. iii. p. 72.

idea, that of the human soul, so that they may well shade imperceptibly into one another. To depend on some typical descriptions of fetishism and its allied doctrines in different grades of culture, is a safer mode of treatment than to attempt too accurate a general definition.

There is a quaint story, dating from the time of Columbus, which shows what mysterious personality and power rude tribes could attach to lifeless matter. The cacique Hatuey, it is related, heard by his spies in Hispaniola that the Spaniards were coming to Cuba. So he called his people together, and talked to them of the Spaniards—how they persecuted the natives of the islands, and how they did such things for the sake of a great lord whom they much desired and loved. Then, taking out a basket with gold in it, he said, "Ye see here their lord whom they serve and go after; and, as ye have heard, they are coming hither to seek this lord. Therefore let us make him a feast, that when they come he may tell them not to do us harm." So they danced and sang from night to morning before the gold-basket, and then the cacique told them not to keep the Christian's lord anywhere, for if they kept him in their very bowels they would have to bring him out; so he bade them cast him to the bottom of the river, and this they did.[1] If this story be thought too good to be true, at any rate it does not exaggerate authentic savage ideas. The "maraca" or ceremonial rattle, used by certain rude Brazilian tribes, was an eminent fetish. It was a calabash with a handle and a hole for a mouth, and stones inside; yet to its votaries it seemed no mere rattle, but the receptacle of a spirit that spoke from it when shaken; therefore the Indians set up their maracas, talked to them, set food and drink and burned incense before them, held annual feasts in their honour, and would even go to war with their neighbours to satisfy the rattle-spirits' demand for human victims.[2] Among the North American Indians, the fetish-theory seems involved in that

[1] Herrera, 'Hist. de las Indias Occidentales,' Dec. i. ix. 3.
[2] Lery, Brésil, p. 249; J. G. Müller, pp. 210, 262.

remarkable and general proceeding known as getting "medicine." Each youth obtains in a vision or dream a sight of his medicine, and considering how thoroughly the idea prevails that the forms seen in visions and dreams are spirits, this of itself shows the animistic nature of the matter. The medicine thus seen may be an animal, or part of one, such as skin or claws, feather or shell, or such a thing as a plant, a stone, a knife, a pipe; this object he must obtain, and thenceforward through life it becomes his protector. Considered as a vehicle or receptacle of a spirit, its fetish-nature is shown in many ways; its owner will do homage to it, make feasts in its honour, sacrifice horses, dogs, and other valuable objects to it or its spirit, fast to appease it if offended, have it buried with him to conduct him as a guardian-spirit to the happy hunting-grounds. Beside these special protective objects, the Indians, especially the medicine-men (the word is French, "médecin," applied to these native doctors or conjurors, and since stretched to take in all that concerns their art), use multitudes of other fetishes as means of spiritual influence.[1] Among the Turanian tribes of Northern Asia, where Castrén describes the idea of spirits contained in material objects, to which they belong, and wherein they dwell in the same incomprehensible way as the souls in a man's body, we may notice the Ostyak's worship of objects of scarce or peculiar quality, and also the connexion of the shamans or sorcerers with fetish-objects, as where the Tatars consider the innumerable rags and tags, bells and bits of iron, that adorn the shaman's magic costume, to contain spirits helpful to their owner in his magic craft.[2] John Bell, in his journey across Asia in 1719, relates a story which well illustrates Mongol ideas as to the action of self-moving objects. A certain Russian merchant told him that once some pieces of damask

---

[1] Schoolcraft, 'Indian Tribes'; Waitz, vol. iii.; Catlin, 'N. A. Ind.' vol. i. p. 36; Keating, Narrative, vol. i. p. 421; J. G. Müller, p. 74, etc. See Cranz, Grönland, p. 274.

[2] Castrén, 'Finn. Myth.' pp. 162, 221, 230; Meiners, vol. i. p. 170.

were stolen out of his tent. He complained, and the Kutuchtu Lama ordered the proper steps to be taken to find out the thief. One of the Lamas took a bench with four feet, and after turning it several times in different directions, at last it pointed directly to the tent where the stolen goods lay concealed. The Lama now mounted astride the bench, and soon carried it, or, as was commonly believed, it carried him, to the very tent, where he ordered the damask to be produced. The demand was directly complied with: for it is vain in such cases to offer any excuse.[1]

A more recent account from Central Africa may be placed as a pendant to this Asiatic account of divination by a fetish-object. The Rev. H. Rowley says of the Manganja, that they believed the medicine-men could impart a power for good or evil to objects either animate or inanimate, which objects the people feared, though they did not worship them. This missionary once saw this art employed to detect the thief who had stolen some corn. The people assembled round a large fig-tree. The magician, a wild-looking man, produced two sticks, like our broomsticks, which after mysterious manipulation and gibberish he delivered to four young men, two holding each stick. A zebra-tail and a calabash-rattle were given to a young man and a boy. The medicine-man rolled himself about in hideous fashion, and chanted an unceasing incantation; the bearers of the tail and rattle went round the stick-holders, and shook these implements over their heads. After a while the men with the sticks had spasmodic twitchings of the arms and legs, these increased nearly to convulsions, they foamed at the mouth, their eyes seemed starting from their heads, they realized to the full the idea of demoniacal possession. According to the native notion, it was the sticks which were possessed primarily, and through them the men, who could hardly hold them. The sticks whirled and dragged the men round and round like mad, through bush and thorny shrub, and over every obstacle, nothing stopped them, their bodies

[1] Bell in Pinkerton, vol. vii. p. 357.

were torn and bleeding; at last they came back to the assembly, whirled round again, and rushed down the path to fall panting and exhausted in the hut of one of a chief's wives, the sticks rolling to her very feet, denouncing her as the thief. She denied it, but the medicine-man answered, "The spirit has declared her guilty, the spirit never lies." However, the "muavi" or ordeal-poison was administered to a cock, as deputy for the woman; the bird threw it up, and she was acquitted.[1]

Fetishism in the lower civilization is thus by no means confined to the West African negro with whom we specially associate the term. Yet, what with its being in fact extremely prevalent there, and what with the attention of foreign observers have been particularly drawn to it, the accounts from West Africa are certainly the fullest and most minute on record. The late Professor Waitz's generalization of the principle involved in these is much to the purpose. He thus describes the negro's conception of his fetish. "According to his view, a spirit dwells or can dwell in every sensible object, and often a very great and mighty one in an insignificant thing. This spirit he does not consider as bound fast and unchangeably to the corporeal thing it dwells in, but it has only its usual or principal abode in it. The negro indeed in his conception not uncommonly separates the spirit from the sensible object which it inhabits, he even sometimes contrasts the one with the other, but most usually combines the two as forming a whole, and this whole is (as the Europeans call it) the "fetish," the object of his religious worship." Some further particulars will show how this principle is worked out. Fetishes (native names for them are "grigri," "juju," etc.), may be mere curious mysterious objects that strike a negro's fancy, or they may be consecrated or affected by a priest or fetish-man; the theory of their influence is that they belong to or are made effectual by a spirit or demon, yet they have to stand the test of experience, and if they

[1] H. Rowley, 'Universities' Mission to Central Africa,' p. 217.

fail to bring their owner luck and safety, he discards them for some more powerful medium. The fetish can see and hear and understand and act, its possessor worships it, talks familiarly with it as a dear and faithful friend, pours libations of rum over it, and in times of danger calls loudly and earnestly on it as if to wake up its spirit and energy. To give an idea of the sort of things which are chosen as fetishes, and of the manner in which they are associated with spiritual influences, Römer's account from Guinea about a century ago may serve. In the fetish-house, he says, there hang or lie thousands of rubbishy trifles, a pot with red earth and a cock's feather stuck in it, pegs wound over with yarn, red parrots' feathers, men's hair, and so forth. The principal thing in the hut is the stool for the fetish to sit on, and the mattress for him to rest on, the mattress being no bigger than a man's hand and the stool in proportion, and there is a little bottle of brandy always ready for him. Here the word fetish is used as it often is, to denote the spirit which dwells in this rudimentary temple, but we see that the innumerable quaint trifles which we call fetishes were associated with the deity in his house. Römer once peeped in at an open door, and found an old negro caboceer sitting amid twenty thousand fetishes in his private fetish-museum, thus performing his devotions. The old man told him he did not know the hundredth part of the use they had been to him; his ancestors and he had collected them, each had done some service. The visitor took up a stone about as big as a hen's egg, and its owner told its history. He was once going out on important business, but crossing the threshold he trod on this stone and hurt himself. Ha ha! thought he, art thou here? So he took the stone, and it helped him through his undertaking for days. In our own time, West Africa is still a world of fetishes. The traveller finds them on every path, at every ford, on every house-door, they hang as amulets round every man's neck, they guard against sickness or inflict it if neglected, they bring rain, they fill the sea with fishes willing to swim into the fisherman's net, they catch and punish thieves, they give their owner a bold heart and confound his enemies, there is nothing that the fetish cannot do or undo, if it be but the right fetish. Thus the one-sided logic of the barbarian, making the most of all that fits and glossing over all that fails, has shaped a universal fetish-philosophy of the events of life. So strong is the pervading influence, that the European in Africa is apt to catch it from the negro, and himself, as the saying is, "become black." Thus even yet some traveller, watching a white companion asleep, may catch a glimpse of some claw or bone or such-like sorcerer's trash secretly fastened round his neck.[1]

European life, lastly, shows well-marked traces of the ancient doctrine of spirits or mysterious influences inhabiting objects. Thus a mediæval devil might go into an old sow, a straw, a barleycorn, or a willow-tree. A spirit might be carried about in a solid receptacle for use :—

"Besides in glistering glasses fayre, or else in christall cleare,
They sprightes enclose."

Modern peasant folklore knows that spirits must have some animal body or other object to dwell in, a feather, a bag, a bush, for instance. The Tyrolese object to using grass for toothpicks because of the demons that may have taken up their abode in the straws. The Bulgarians hold it a great sin not to fumigate the flour when it is brought from the mill (particularly if the mill be kept by a Turk) in order to prevent the devil from entering into it.[2] Amulets are still carried in the most civilized countries of the world, by the

[1] Waitz, 'Anthropologie,' vol. ii. p. 174; Römer, 'Guinea,' p. 56, etc., J. L. Wilson, 'West Africa,' pp. 135, 211-6, 275, 338; Burton, 'Wit and Wisdom from W. Afr.' pp. 174, 455, Steinhauser, l. c. p. 134; Bosman, Guinea, in Pinkerton, vol. xvi. p. 397; Meiners, 'Gesch. der Relig.' vol. i. p. 173. See also Ellis, 'Madagascar,' vol. i. p. 396; Flacourt, 'Madag.' p. 191.

[2] Brand, 'Popular Antiquities,' vol. iii. p. 255, etc. Bastian, 'Psychologie,' p. 171. Wuttke, 'Deutsche Volksaberglaube,' pp. 75-95, 225, etc. St. Clair and Brophy, 'Bulgaria,' p. 46.

ignorant and superstitious with real savage faith in their mysterious virtues, by the more enlightened in quaint survival from the past. The mental and physical phenomena of what is now called "table-turning" belong to a class of proceedings which we have seen to be familiar to the lower races, and accounted for by them on a theory of extra-human influence which is in the most extreme sense spiritualistic.

In giving its place in the history of mental development to the doctrine of the lower races as to embodiment in or penetration of an object by a spirit or an influence, there is no slight interest in comparing it with theories familiar to the philosophy of cultured nations. Thus Bishop Berkeley remarks on the obscure expressions of those who have described the relation of power to the objects which exert it. He cites Torricelli as likening matter to an enchanted vase of Circe serving as a receptacle of force, and declaring that power and impulse are such subtle abstracts and refined quintessences, that they cannot be enclosed in any other vessels but the inmost materiality of natural solids; also Leibnitz as comparing active primitive power to soul or substantial form. Thus, says Berkeley, must even the greatest men, when they give way to abstraction, have recourse to words having no certain signification, and indeed mere scholastic shadows.[1] We may fairly add that such passages show the civilized metaphysician falling back on such primitive conceptions as still occupy the minds of the rude natives of Siberia and Guinea. To go yet farther, I will venture to assert that the scientific conceptions current in my own schoolboy days, of heat and electricity as invisible fluids passing in and out of solid bodies, are ideas which reproduce with extreme closeness the special doctrine of Fetishism.

Under the general heading of Fetishism, but for convenience sake separately, may be considered the worship of "stocks and stones." Such objects, if merely used as

[1] Berkeley, 'Concerning Motion,' in 'Works,' vol. ii. p. 86.

altars, are not of the nature of fetishes, and it is first necessary to ascertain that worship is actually addressed to them. Then arises the difficult question, are the stocks and stones set up as mere ideal representatives of deities, or are these deities considered as physically connected with them, embodied in them, hovering about them, acting through them? In other words, are they only symbols, or have they passed in the minds of their votaries into real fetishes? The conceptions of the worshippers are sometimes in this respect explicitly stated, may sometimes be fairly inferred from the circumstances, and are often doubtful.

Among the lower races of America, the Dacotas would pick up a round boulder, paint it, and then, addressing it as grandfather, make offerings to it and pray to it to deliver them from danger:[1] in the West India Islands, mention is made of three stones to which the natives paid great devotion—one was profitable for the crops, another for women to be delivered without pain, the third for sunshine and rain when they were wanted;[2] and we hear of Brazilian tribes setting up stakes in the ground, and making offerings before them to appease their deities or demons.[3] Stone-worship held an important place in the midst of the comparatively high culture of Peru, where not only was reverence given to especial curious pebbles and the like, but stones were placed to represent the penates of households and the patron-deities of villages. It is related by Montesinos that when the worship of a certain sacred stone was given up, a parrot flew from it into another stone, to which adoration was paid: and though this author is not of good credit, he can hardly have invented a story which, as we shall see, so curiously coincides with the Polynesian idea of a bird conveying to and from an idol the spirit which embodies itself in it.[4]

[1] Schoolcraft, 'Indian Tribes,' part ii. p. 196, part iii. p. 229
[2] Herrera, 'Indias Occidentales,' dec. i. iii. 3.
[3] De Laet, Novus Orbis, xv. 2.
[4] Garcilaso de la Vega, 'Commentarios Reales,' i. 9; J. G. Müller, pp. 263, 311, 371, 387; Waitz, vol. iv p. 454; see below, p. 175.

In Africa, stock-and-stone worship is found among the Damaras of the South, whose ancestors are represented at the sacrificial feasts by stakes cut from trees or bushes consecrated to them, to which stakes the meat is first offered;[1] among the Dinkas of the White Nile, where the missionaries saw an old woman in her hut offering the first of her food and drink before a short thick staff planted in the ground, that the demon might not hurt her;[2] among the Gallas of Abyssinia, a people with a well-marked doctrine of deities, and who are known to worship stones and logs, but not idols.[3] In the island of Sambawa, the Orang Dongo attribute all supernatural or incomprehensible force to the sun, moon, trees, &c., but especially to stones, and when troubled by accident or disease, they carry offerings to certain stones to implore the favour of their genius or dewa.[4] Similar ideas are to be traced through the Pacific islands, both among the lighter and the darker races. Thus in the Society Islands, rude logs or fragments of basalt columns, clothed in native cloth and anointed with oil, received adoration and sacrifice as divinely powerful by virtue of the atua or deity which had filled them.[5] So in the New Hebrides worship was given to water-worn pebbles,[6] while Fijian gods and goddesses had their abodes or shrines in black stones like smooth round milestones, and there received their offerings of food.[7] The curiously anthropomorphic idea of stones being husbands and wives, and even having children, is familiar to the Fijians as it is to the Peruvians and the Lapps.

The Turanian tribes of North Asia display stock-and-stone worship in full sense and vigour. Not only were

[1] Hahn, 'Gramm. des Hereró,' s. v. 'omu-makisina.'
[2] Kaufmann, 'Central-Afrika,' (White Nile), p. 131.
[3] Waitz, vol. ii. pp. 518, 523.
 Zollinger in 'Journ. Ind. Archip.' vol. ii. p. 692.
[5] Ellis, 'Polyn. Res.' vol. i. p. 337. See also Ellis, 'Madagascar,' vol. i. p. 399.
[6] Turner, 'Polynesia,' pp. 347, 526.
[7] Williams, 'Fiji,' vol. i. p. 220; Seemann, Viti, pp. 66, 89.

stones, especially curious ones and such as were like men or animals, objects of veneration, but we learn that they were venerated because mighty spirits dwelt in them. The Samoyed travelling ark-sledge, with its two deities, one with a stone head, the other a mere black stone, both dressed in green robes with red lappets, and both smeared with sacrificial blood, may serve as a type of stone-worship. And as for the Ostyaks, had the famous King Log presented himself among them, they would without more ado have wrapped his sacred person in rags, and set him up for worship on a mountain-top or in the forest.[1] The frequent stock-and-stone worship of modern India belongs especially to races non-Hindu or part-Hindu in race and culture. Among such may serve as examples the bamboo which stands for the Bodo goddess Mainou, and for her receives the annual hog, and the monthly eggs offered by the women;[2] the stone under the great cotton-tree of every Khond village, shrine of Nadzu Pennu the village deity;[3] the clod or stone under a tree, which in Behar will represent the deified soul of some dead personage who receives worship and inspires oracles there;[4] the stone kept in every house by the Bakadâra and Betadâra, which represents their god Bûta, whom they induce by sacrifice to restrain the demon-souls of the dead from troubling them;[5] the two rude stones placed under a shed among the Shanars of Tinnevelly, by the medium of which the great god and goddess receive sacrifice, but which are thrown away or neglected when done with.[6] The remarkable groups of standing-stones in India

---

[1] Castrén, 'Finn. Myth.' p. 193, etc., 204, etc.; 'Voyages au Nord,' vol. viii. pp. 103, 410; Klemm, 'C. G.' vol. iii. p. 120. See also Steller, 'Kamtschatka,' pp. 265, 276.
[2] Hodgson, 'Abor. of India,' p. 174. See also Macrae in 'As. Res.' vol. vii. p. 196; Dalton, Kols, in 'Tr. Eth. Soc.' vol. vi. p. 33.
[3] Macpherson, India, pp. 103, 358.
[4] Bastian, 'Psychologie,' p. 177. See also Shortt, 'Tribes of Neilgherries' in 'Tr. Eth. Soc.' vol. vii. p. 281.
[5] Elliot in 'Journ. Eth. Soc.' vol. i. 1869, p. 115.
[6] Buchanan, 'Mysore,' in Pinkerton, vol. vii. p. 739.

are in many cases at least set up for each stone to represent or embody a deity. Mr. Hislop remarks that in every part of Southern India, four or five stones may often be seen in the ryot's field, placed in a row and daubed with red paint, which they consider as guardians of the field and call the five Pândus; he reasonably takes these Hindu names to have superseded more ancient native appellations. In the Indian groups it is a usual practice to daub each stone with red paint, forming as it were a great blood-spot where the face would be if it were a shaped idol.[1] In India, moreover, the rites of stone-worship are not unexampled among the Hindus proper. Shashtî, protectress of children, receives worship, vows, and offerings, especially from women; yet they provide her with no idol or temple, but her proper representative is a rough stone as big as a man's head, smeared with red paint and set at the foot of the sacred vata-tree. Even Siva is worshipped as a stone, especially that Siva who will afflict a child with epileptic fits, and then, speaking by its voice, will announce that he is Panchânana the Five-faced, and is punishing the child for insulting his image; to this Siva, in the form of a clay idol or of a stone beneath a sacred tree, there are offered not only flowers and fruits, but also bloody sacrifices.[2]

This stone-worship among the Hindus seems a survival of a rite belonging originally to a low civilization, probably a rite of the rude indigenes of the land, whose religion, largely incorporated into the religion of the Aryan invaders, has contributed so much to form the Hinduism of to-day. It is especially interesting to survey the stock-and-stone worship of the lower culture, for it enables us to explain by the theory of survival the appearance in the Old World, in the very midst of classic doctrine and classic art, of the

[1] Elliot in 'Journ. Eth. Soc.' vol. i. pp. 96, 115, 125. Lubbock, 'Origin of Civilization,' p. 222. Forbes Leslie, 'Early Races of Scotland,' vol. ii. p. 462, etc. Prof. Liebrecht, in 'Ztschr. für Ethnologie,' vol. v., p. 100, compares the field-protecting Priapos-hermes of ancient Italy, daubed with minium.

[2] Ward, 'Hindoos,' vol. ii. pp. 142, 182, etc., see 221. See also Latham, 'Descr. Eth.' vol. ii. p. 239. (Siah-push, stone offered to the representative of deity)

worship of the same rude objects, whose veneration no doubt dated from remote barbaric antiquity. As Mr. Grote says, speaking of Greek worship, "The primitive memorial erected to a god did not even pretend to be an image, but was often nothing more than a pillar, a board, a shapeless stone or a post, receiving care and decoration from the neighbourhood, as well as worship." Such were the log that stood for Artemis in Eubœa, the stake that represented Pallas Athene, "sine effigie rudis palus, et informe lignum," the unwrought stone (λίθος ἀργὸς) at Hyettos which "after the ancient manner" represented Herakles, the thirty such stones which the Pharæans in like archaic fashion worshipped for the gods, and that one which received such honour in Bœotian festivals as representing the Thespian Eros. Theophrastus, in the 4th century B.C., depicts the superstitious Greek passing the anointed stones in the streets, taking out his phial and pouring oil on them, falling on his knees to adore, and going his way. Six centuries later, Arnobius could describe from his own heathen life the state of mind of the stock-and-stone worshipper, telling how when he saw one of the stones anointed with oil, he accosted it in flattering words, and asked benefits from the senseless thing as though it contained a present power.[1] The ancient and graphic passage in the book of Isaiah well marks stone-worship within the range of the Semitic race:

"Among the smooth stones of the valley is thy portion:
They, they are thy lot:
Even to them hast thou poured a drink-offering,
Hast thou offered a meat-offering."[2]

---

[1] Grote, 'Hist. of Greece,' vol. iv. p. 132; Welcker, 'Griechische Götterlehre,' vol. i. p. 220. Meiners, vol. i. p. 150, etc. Details esp. in Pausanias; Theophrast. Charact. xvi.; Tacit. Hist. ii. 3; Arnobius, Adv. Gent.; Tertullianus; Clemens Alexandr.

[2] Is. lvii. 6. The first line, "behhalkey-nahhal hhêlkech," turns on the pun

Long afterwards, among the local deities which Mohammed found in Arabia, and which Dr. Sprenger thinks he even acknowledged as divine during a moment when he well nigh broke down in his career, were Manah and Lât, the one a rock, the other a stone or a stone idol; while the veneration of the black stone of the Kaaba, which Captain Burton thinks an aërolite, was undoubtedly a local rite which the Prophet transplanted into his new religion, where it flourishes to this day.[1] The curious passage in Sanchoniathon which speaks of the Heaven-god forming the "bætyls, animated stones" (θεὸς Οὐρανὸς Βαιτύλια, λίθους ἐμψύχους, μηχανησάμενος) perhaps refers to meteorites or supposed thunderbolts fallen from the clouds. To the old Phœnician religion, which made so deep a contact with the Jewish world on the one side and the Greek and Roman on the other, there belonged the stone pillars of Baal and the wooden Ashera-cones, but how far these objects were of the character of altars, symbols, or fetishes, is a riddle.[2] We may still say with Tacitus, describing the conical pillar which stood instead of an image to represent the Paphian Venus—" et ratio in obscuro."

There are accounts of formal Christian prohibitions of stone-worship in France and England, reaching on into the early middle ages,[3] which show this barbaric cultus as then distinctly lingering in popular religion. Coupling this fact with the accounts of the groups of standing-stones set up to represent deities in South India, a plausible solution is suggested for an interesting problem of Prehistoric Archæology in Europe. Are the menhirs, cromlechs, etc., idols, and circles

---

on hhlk=smooth (stone), and also lot or portion; a double sense probably connected with the use of smooth pebbles for casting lots.

[1] Sprenger, 'Mohammad,' vol. ii. p. 7, etc. Burton, El Medinah, etc. vol. ii. p. 157.

[2] Euseb. Præp. Evang. i. 10. Movers, Phönizier, vol. i. pp. 105, 569, and see index, 'Säule,' etc. See De Brosses, 'Dieux Fétiches,' p. 135 (considers bætyl=beth-el, etc.).

[3] Lubbock, 'Origin of Civ.' p. 225. Leslie, 'Early Races of Scotland, vol. i. p. 256.

and lines of idols, worshipped by remotely ancient dwellers in the land as representatives or embodiments of their gods ? It may well be so: yet the ideas with which stone-worship is carried on by different races are multifarious, and the analogy may be misleading. It is remarkable to what late times full and genuine stone-worship has survived in Europe. In certain mountain districts of Norway, up to the end of the last century, the peasants used to preserve round stones, washed them every Thursday evening (which seems to show some connection with Thor), smeared them with butter before the fire, laid them in the seat of honour on fresh straw, and at certain times of the year steeped them in ale, that they might bring luck and comfort to the house.[1] In an account dating from 1851, the islanders of Inniskea, off Mayo, are declared to have a stone carefully wrapped in flannel, which is brought out and worshipped at certain periods, and when a storm arises it is supplicated to send a wreck on the coast.[2] No savage ever showed more clearly by his treatment of a fetish that he considered it a personal being, than did these Norwegians and Irishmen. The ethnographic argument from the existence of stock-and-stone worship among so many nations of comparatively high culture seems to me of great weight as bearing on religious development among mankind. To imagine that peoples skilled in carving wood and stone, and using these arts habitually in making idols, should have gone out of their way to invent a practice of worshipping logs and pebbles, is not a likely theory. But on the other hand, when it is considered how such a rude object serves to uncultured men as a divine image or receptacle, there is nothing strange in its being a relic of early barbarism holding its place against more artistic models through ages of advancing

[1] Nilsson, 'Primitive Inhabitants of Scandinavia,' p. 241. See also Meiners, vol. ii. p. 671 (speaking stones in Norway, etc.).
[2] Earl of Roden, 'Progress of Reformation in Ireland,' London, 1851, p. 51. Sir J. E. Tennent in 'Notes and Queries,' Feb. 7, 1852. See Borlase, 'Antiquities of Cornwall,' Oxford, 1754, book iii. ch. 2.

civilization, by virtue of the traditional sanctity which belongs to survival from remote antiquity.

By a scarcely perceptible transition, we pass to Idolatry. A few chips or scratches or daubs of paint suffice to convert the rude post or stone into an idol. Difficulties which complicate the study of stock-and-stone worship disappear in the worship of even the rudest of unequivocal images, which can no longer be mere altars, and if symbols must at least be symbols of a personal being. Idolatry occupies a remarkable district in the history of religion. It hardly belongs to the lowest savagery, which simply seems not to have attained to it, and it hardly belongs to the highest civilization, which has discarded it. Its place is intermediate, ranging from the higher savagery where it first clearly appears, to the middle civilization where it reaches its extreme development, and thenceforward its continuance is in dwindling survival and sometimes expanding revival. The position thus outlined is, however, very difficult to map exactly. Idolatry does not seem to come in uniformly among the higher savages; it belongs, for instance, fully to the Society Islanders, but not to the Tongans and Fijians. Among higher nations, its presence or absence does not necessarily agree with particular national affinities or levels of culture—compare the idol-worshipping Hindu with his ethnic kinsman the idol-hating Parsi, or the idolatrous Phœnician with his ethnic kinsman the Israelite, among whose people the incidental relapse into the proscribed image-worship was a memory of disgrace. Moreover, its tendency to revive is ethnographically embarrassing. The ancient Vedic religion seems not to recognize idolatry, yet the modern Brahmans, professed followers of Vedic doctrine, are among the greatest idolators of the world. Early Christianity by no means abrogated the Jewish law against image-worship, yet image-worship became and still remains widely spread and deeply rooted in Christendom.

Of Idolatry, so far as its nature is symbolic or representative, I have given some account elsewhere.[1] The old and greatest difficulty in investigating the general subject is this, that an image may be, even to two votaries kneeling side by side before it, two utterly different things; to the one it may be only a symbol, a portrait, a memento; while to the other it is an intelligent and active being, by virtue of a life or spirit dwelling in it or acting through it. In both cases Image-worship is connected with the belief in spiritual beings, and is in fact a subordinate development of animism. But it is only so far as the image approximates to the nature of a material body provided for a spirit, that Idolatry comes properly into connection with Fetishism. It is from this point of view that it is proposed to examine here its purpose and its place in history. An idol, so far as it belongs to the theory of spirit-embodiment, must combine the characters of portrait and fetish. Bearing this in mind, and noticing how far the idol is looked on as in some way itself an energetic object, or as the very receptacle enshrining a spiritual god, let us proceed to judge how far, along the course of civilization, the idea of the image itself exerting power or being actually animate has prevailed in the mind of the idolator.

As to the actual origin of idolatry, it need not be supposed that the earliest idols made by man seemed to their maker living or even active things. It is quite likely that the primary intention of the image was simply to serve as a sign or representative of some divine personage, and certainly this original character is more or less maintained in the world through the long history of image-worship. At a stage succeeding this original condition, it may be argued, the tendency to identify the symbol and the symbolized, a tendency so strong among children and the ignorant everywhere, led to the idol being treated as a living powerful being, and thence even to explicit doctrines as to the manner of its energy or animation. It is, then, in this secondary stage, where the once merely representative image is passing

[1] 'Early Hist. of Mankind,' chap. vi.

into the active image-fetish, that we are particularly concerned to understand it. Here it is reasonable to judge the idolator by his distinct actions and beliefs. A line of illustrative examples will carry the personality of the idol through grade after grade of civilization. Among the lower races, such thoughts are displayed by the Kurile islander throwing his idol into the sea to calm the storm; by the negro who feeds ancestral images and brings them a share of his trade profits, but will beat an idol or fling it into the fire if it cannot give him luck or preserve him from sickness; by famous idols of Madagascar, of which one goes about of himself or guides his bearers, and another answers when spoken to—at least, they did this till they were ignominiously found out a few years ago. Among Tatar peoples of North Asia and Europe, conceptions of this class are illustrated by the Ostyak, who clothes his puppet and feeds it with broth, but if it brings him no sport will try the effect of a good thrashing on it, after which he will clothe and feed it again; by the Lapps, who fancied their uncouth images could go about at will; or the Esths, who wondered that their idols did not bleed when Dieterich the Christian priest hewed them down. Among high Asiatic nations, what could be more anthropomorphic than the rites of modern Hinduism, the dances of the nautch-girls before the idols, the taking out of Jagannath in procession to pay visits, the spinning of tops before Krishna to amuse him? Buddhism is a religion in its principles little favourable to idolatry. Yet, from setting up portrait-statues of Gautama and other saints, there developed itself the full worship of images, and even of images with hidden joints and cavities, which moved and spoke as in our own middle ages. In China, we read stories of worshippers abusing some idol that has failed in its duty. "How now," they say, "you dog of a spirit; we have given you an abode in a splendid temple, we gild you and feed you and fumigate you with incense, and yet you are so ungrateful that you won't listen to our prayers!" So they drag him in the dirt, and then, if they get what they want, it is but to clean him and set him up again, with apologies and promises of a new coat of gilding. There is what appears a genuine story of a Chinaman who had paid an idol priest to cure his daughter, but she died; whereupon the swindled worshipper brought an action of law against the god, who for his fraud was banished from the province. The classic instances, again, are perfect—the dressing and anointing of statues, feeding them with delicacies and diverting them with raree-shows, summoning them as witnesses; the story of the Arkadian youths coming back from a bad day's hunting and revenging themselves by scourging and pricking Pan's statue, with its companion tale of the image which fell upon the man who ill-treated it; the Tyrians chaining the statue of the Sun-god that he might not abandon their city; Augustus chastising in effigy the ill-behaved Neptune; Apollo's statue that moved when it would give an oracle; and the rest of the images which brandished weapons, or wept, or sweated, to prove their supernatural powers. Such ideas continued to hold their place in Christendom, as was natural, considering how directly the holy image or picture took the place of the household god or the mightier idol of the temple. The Russian boor covering up the saint's picture that it may not see him do wrong; the Mingrelian borrowing a successful neighbour's saint when his own crop fails, or when about to perjure himself, choosing for the witness of his deceitful oath a saint of mild countenance and merciful repute; the peasant of Southern Europe, alternately coaxing and trampling on his special saint-fetish, and ducking the Virgin or St. Peter for rain; the winking and weeping images that are worked, even at this day, to the greater glory of God, or rather to the greater shame of Man—these are but the extreme instances of the worshipper's endowment of the sacred image with a life and personality modelled on his own.[1]

[1] For general collections of evidence, see especially Meiners, 'Geschichte der Religionem,' vol. i. books i. and v.; Bastian, 'Mensch,' vol. ii.; Waitz,

The appearance of idolatry at a grade above the lowest of known human culture, and its development in extent and elaborateness under higher conditions of civilization, are well displayed among the native races of America. "Conspicuous by its absence" among many of the lower tribes, image-worship comes plainly into view toward the upper levels of savagery, as where, for instance, Brazilian native tribes set up in their huts, or in the recesses of the forest, their pigmy heaven-descended figures of wax or wood;[1] or where the Mandans, howling and whining, made their prayers before puppets of grass and skins; or where the spiritual beings of the Algonquins ("manitu" or "oki") were represented by, and in language identified with, the carved wooden heads or more complete images to which worship and sacrifice were offered. Among the Virginians and other of the more cultured Southern tribes, these idols even had temples to dwell in.[2] The discoverers of the New World found idolatry an accepted institution among the islanders of the West Indies. These strong animists are recorded to have carved their little images in the shapes in which they believed the spirits themselves to have appeared to them; and some human figures bore the names of ancestors in memory of them. The images of such "cemi" or spirits, some animal, but most of human type, were found by thousands; and it is even declared that an island near Hayti had a population of idol-makers, who especially made images of nocturnal spectres. The spirit could be conveyed with the image, both were called "cemi," and in the local accounts of sacrifices, oracles, and miracles, the deity and the idol are mixed together in a way which at least shows the extreme closeness of their connexion in the native

'Anthropologie;' De Brosses, 'Dieux Fétiches,' etc. Particular details in J. L. Wilson, 'W. Afr.' p. 393; Ellis, 'Madagascar,' vol. i. p. 395; Castrén, 'Finnische Mythologie,' p. 193, etc.; Ward, 'Hindoos,' vol. ii.; Köppen, 'Rel. des Buddha,' vol i. p. 493, etc.; Grote, 'Hist. of Greece.'

[1] J. G. Müller, 'Amer. Urrelig.' p. 263; Meiners, vol. i. p. 163.

[2] Loskiel, 'Ind. of N. A.' vol. i. p. 39. Smith, 'Virginia,' in Pinkerton, vol. xiii. p. 14. Waitz, vol. iii. p. 203; J. G. Müller, pp. 95-8, 128.

mind.[1] If we pass to the far higher culture of Peru, we find idols in full reverence, some of them complete figures, but the great deities of Sun and Moon figured by discs with human countenances, like those which to this day represent them in symbol among ourselves. As for the conquered neighbouring tribes brought under the dominion of the Incas, their idols were carried, half trophies and half hostages, to Cuzco, to rank among the inferior deities of the Peruvian Pantheon.[2] In Mexico, idolatry had attained to its full barbaric development. As in the Aztec mind the world swarmed with spiritual deities, so their material representatives the idols stood in the houses, at the corners of the streets, on every hill and rock, to receive from passers-by some little offering—a nosegay, a whiff of incense, a drop or two of blood; while in the temples more huge and elaborate images enjoyed the dances and processions in their honour, were fed by the bloody sacrifice of men and beasts, and received the tribute and reverence paid to the great national gods.[3] Up to a certain point, such evidence bears upon the present question. We learn that the native races of the New World had idols, that those idols in some sort represented ancestral souls and other deities, and for them received adoration and sacrifice. But whether the native ideas of the connexion of spirit and image were obscure, or whether the foreign observers did not get at these ideas, or partly for both reasons, there is a general want of express statement how far the idols of America remained mere symbols or portraits, or how far they had come to be considered the animated bodies of the gods.

It is not always thus, however. In the island regions of

[1] Fernando Colombo, 'Vita del Amm. Cristoforo Colombo,' Venice, 1571, p. 127, etc.; and 'Life of Colon,' in Pinkerton, vol. xii. p. 84. Herrera, dec. i. iii. 3. Rochefort, 'Iles Antilles,' pp. 421–4. Waitz, vol. iii. p. 384; J. G. Müller, pp. 171–6, 182, 210, 232.

[2] Prescott, 'Peru,' vol. i. pp. 71, 89; Waitz, vol. iv. p. 458; J. G. Müller, pp. 322, 371.

[3] Brasseur, 'Mexique,' vol. iii. p. 486; Waitz, vol. iv. p. 148; J. G. Müller, p. 642.

the Southern Hemisphere, while image-worship scarcely appears among the Andaman islanders, Tasmanians, or Australians, and is absent or rare in various Papuan and Polynesian districts, it prevails among the majority of the island tribes who have attained to middle and high savage levels. In Polynesian islands, where the meaning of the native idolatry has been carefully examined, it is found to rest on the most absolute theory of spirit-embodiment. Thus, New Zealanders set up memorial idols of deceased persons near the burial-place, talking affectionately to them as if still alive, and casting garments to them when they passed by, and preserve in their houses small carved wooden images, each dedicated to the spirit of an ancestor. It is distinctly held that such an atua or ancestral deity enters into the substance of an image in order to hold converse with the living. A priest can by repeating charms cause the spirit to enter into the idol, which he will even jerk by a string round its neck to arrest its attention; it is the same atua or spirit which will at times enter not the image but the priest himself, throw him into convulsions, and deliver oracles through him; while it is quite understood that the images themselves are not objects of worship, nor do they possess in themselves any virtue, but derive their sacredness from being the temporary abodes of spirits.[1] In the Society Islands, it was noticed in Captain Cook's exploration that the carved wooden images at burial-places were not considered mere memorials, but abodes into which the souls of the departed retired. In Mr. Ellis's account of the Polynesian idolatry, relating as it seems especially to this group, the sacred objects might be either mere stocks and stones, or carved wooden images, from six to eight feet long down to as many inches. Some of these were to represent "tii," divine manes or spirits of the dead, while others were to represent "tu," or deities of higher rank and power. At certain seasons, or in answer to the prayers of the priests, these spiritual beings entered into the idols,

[1] Shortland, 'Trads. of N. Z.' etc. p. 83; Taylor, pp. 171, 183, 212.

which then became very powerful, but when the spirit departed, the idol remained only a sacred object. A god often came to and passed from an image in the body of a bird, and spiritual influence could be transmitted from an idol by imparting it by contact to certain valued kinds of feathers, which could be carried away in this "inhabited" state, and thus exert power elsewhere, and transfer it to new idols. Here then we have the similarity of souls to other spirits shown by the similar way in which both become embodied in images, just as these same people consider both to enter into human bodies. And we have the pure fetish, which here is a feather or a log or stone, brought together with the more elaborate carved idol, all under one common principle of spirit embodiment.[1] In Borneo, notwithstanding the Moslem prohibition of idolatry, not only do images remain in use, but the doctrine of spirit-embodiment is distinctly applied to them. Among the tribes of Western Sarawak the priestesses have made for them rude figures of birds, which none but they may touch. These are supposed to become inhabited by spirits, and at the great harvest feasts are hung up in bunches of ten or twenty in the long common room, carefully veiled with coloured handkerchiefs. Again, among some Dayak tribes, they will make rude figures of a naked man and woman, and place these opposite to one another on the path to the farms. On their heads are head-dresses of bark, by their sides is the betel-nut basket, and in their hands a short wooden spear. These figures are said to be inhabited each by a spirit who prevents inimical influences from passing on to the farms, and likewise from the farms to the village, and evil betide the profane wretch who lifts his hand against them—violent fever and sickness would be sure to follow.[2]

West Africa naturally applies its familiar fetish-doctrine

[1] J. R. Forster, 'Obs. during Voyage,' London, 1778, p. 534, etc. Ellis, 'Polyn. Res.' vol. i. p. 281, etc., 323, etc. See also Earl, 'Papuans,' p. 84; Bastian, 'Psychologie,' p. 78 (Nias).
[2] St. John, 'Far East,' vol. i. p. 198.

of spirit-embodiment to images or idols. How an image may be considered a receptacle for a spirit, is well shown here by the straw and rag figures of men and beasts made in Calabar at the great triennial purification, for the expelled spirits to take refuge in, whereupon they are got rid of over the border.[1] As to positive idols, nothing could be more explicit than the Gold-Coast account of certain wooden figures called "amagai," which are specially treated by a "wong-man" or priest, and have a "wong" or deity in connexion with them; so close is the connexion conceived between spirit and image, that the idol is itself called "wong."[2] So in the Ewe district, the same "edro" or deity who inspires the priest is also present in the idol, and "edro" signifies both god and idol.[3] Waitz sums up the principles of West African idolatry in a distinct theory of embodiment, as follows: "The god himself is invisible, but the devotional feeling and especially the lively fancy of the negro demands a visible object to which worship may be directed. He wishes really and sensibly to behold the god, and seeks to shape in wood or clay the conception he has formed of him. Now if the priest, whom the god himself at times inspires and takes possession of, consecrates this figure to him, the idea has only to follow that the god may in consequence be pleased to take up his abode in the figure, to which he may be specially invited by the consecration, and thus image-worship is seen to be comprehensible enough. Denham found that even to take a man's portrait was dangerous and caused mistrust, from the fear that a part of the living man's soul might be conveyed by magic into the artificial figure. The idols are not, as Bosman thinks, deputies of the gods, but merely objects in which the god loves to place himself, and which at the same time display him in sensible presence to his adorers. The

[1] Hutchinson in 'Tr. Eth. Soc.' vol. i. p. 336; see Bastian, 'Psychologie,' p. 172.
[2] Steinhauser, in 'Magaz. der Evang. Missionen,' Basel, 1856, No. 2, p. 131.
[3] Schlegel, Ewe-Sprache, p. xvi.

god is also by no means bound fast to his dwelling in the image, he goes out and in, or rather is present in it sometimes with more and sometimes with less intensity."[1]

Castrén's wide and careful researches among the rude Turanian tribes of North Asia, led him to form a similar conception of the origin and nature of their idolatry. The idols of these people are uncouth objects, often mere stones or logs with some sort of human countenance, or sometimes more finished images, even of metal; some are large, some mere dolls; they belong to individuals, or families, or tribes; they may be kept in the yurts for private use, or set up in sacred groves or on the steppes or near the hunting and fishing places they preside over, or they may even have special temple-houses; some open-air gods are left naked, not to spoil good clothes, but others under cover are decked out with all an Ostyak's or Samoyed's wealth of scarlet cloths and costly furs, necklaces and trinkets; and lastly, to the idols are made rich offerings of food, clothes, furs, kettles, pipes, and the rest of the inventory of Siberian nomade wealth. Now these idols are not to be taken as mere symbols or portraits of deities, but the worshippers mostly imagine that the deity dwells in the image or, so to speak, is embodied in it, whereby the idol becomes a real god capable of giving health and prosperity to man. On the one hand, the deity becomes serviceable to the worshipper by being thus contained and kept for his use, and on the other hand, the god profits by receiving richer offerings, failing which it would depart from its receptacle. We even hear of numerous spirits being contained in one image, and flying off at the death of the shaman who owned it. In Buddhist Tibet, as in West Africa, the practice of conjuring into puppets the demons which molest men is a recognized rite; while in Siam the making of clay puppets to be exposed on trees or by the roadside, or set adrift with food-

[1] Waitz, 'Anthropologie,' vol. ii. p. 183; Denham, 'Travels,' vol. i. p. 113; Römer, 'Guinea;' Bosman, 'Guinea,' in Pinkerton, vol. xvi. See also Livingstone, 'S. Afr.' p. 282 (Balonda.)

offerings in baskets, is a recognized manner of expelling disease-spirits.[1] In the image-worship of modern India, there crop up traces of the embodiment-theory. It is possible for the intelligent Hindu to attach as little real personality to a divine image, as to the man of straw which he makes in order to celebrate the funeral rites of a relative whose body cannot be recovered. He can even protest against being treated as an idolater at all, declaring the images of his gods to be but symbols, bringing to his mind thoughts of the real deities, as a portrait reminds one of a friend no longer to be seen in the body. Yet in the popular religion of his country, what could be more in conformity with the fetish-theory than the practice of making temporary hollow clay idols by tens of thousands, which receive no veneration for themselves, and only become objects of worship when the officiating brahman has invited the deity to dwell in the image, performing the ceremony of the "adhivâsa" or inhabitation, after which he puts in the eyes and the "prâna," *i. e.*, breath, life, or soul.[2]

Nowhere, perhaps, in the wide history of religion, can we find definitions more full and absolute of the theory of deities actually animating their images, than in those passages from early Christian writers which describe the nature and operation of the heathen idols. Arnobius introduces the heathen as declaring that it is not the bronze or gold and silver material they consider to be gods, but they worship in them those beings which sacred dedication introduces, and causes to inhabit the artificial images.[3] Augustine cites as follows the opinions attributed to Hermes Trismegistus. This Egyptian, he tells us, considers some gods as made by the highest Deity, and some by men; "he asserts the visible and tangible images to be as it were bodies of

[1] Castrén, 'Finn. Myth.' p. 193, etc.; Bastian, 'Psych.' p. 34, 208, 'Oestl. Asien,' vol. iii. p. 293, 486. See 'Journ. Ind. Archip.' vol. ii. p. 350 (Chinese.)
[2] Max Müller, 'Chips,' vol. i. p. xvii.; Ward, 'Hindoos,' vol. i. p. 198, vol. ii. p. xxxv. 164, 234, 292, 485.
[3] Arnobius Adversus Gentes, vi. 17–19.

gods, for there are within them certain invited spirits, of some avail for doing harm, or for fulfilling certain desires of those who pay them divine honours and rites of worship. By a certain art to connect these invisible spirits with visible objects of corporeal matter, that such may be as it were animated bodies, effigies dedicate and subservient to the spirits—this is what he calls making gods, and men have received this great and wondrous power." And further, this Trismegistus is made to speak of "statues animated with sense and full of spirit, doing so great things ; statues prescient of the future, and predicting it by lots, by priests, by dreams, and by many other ways."[1] This idea, as accepted by the early Christians themselves, with the qualification that the spiritual beings inhabiting the idols were not beneficent deities but devils, is explicitly stated by Minucius Felix, in a passage in the 'Octavius,' which gives an instructive account of the animistic philosophy of Christianity towards the beginning of the third century : " Thus these impure spirits or demons, as shown by the magi, by the philosophers, and by Plato, are concealed by consecration in statues and images, and by their afflatus obtain the authority as of a present deity when at times they inspire priests, inhabit temples, occasionally animate the filaments of the entrails, govern the flight of birds, guide the falling of lots, give oracles enveloped in many falsehoods . . . also secretly creeping into (men's) bodies as thin spirits, they feign diseases, terrify minds, distort limbs, in order to compel men to their worship ; that fattening on the steam of altars or their offered victims from the flocks, they may seem to have cured the ailments which they had constrained. And these are the madmen whom ye see rush forth into

---

[1] Augustinus 'De Civ. Dei,' viii. 23 : "at ille visibilia et contrectabilia simulacra, velut corpora deorum esse asserit ; inesse autem his quosdam spiritus invitatos, etc. . . . . Hos ergo spiritus invisibiles per artem quandam visibilibus rebus corporalis materiæ copulare, ut sint quasi animata corpora, illis spiritibus dicata et subdita simulacra, etc." See also Tertullianus De Spectaculis xii. : " In mortuorum autem idolis dæmonia consistunt, etc."

public places; and the very priests without the temple thus go mad, thus rave, thus whirl about. . . . All these things most of you know, how the very demons confess of themselves, so often as they are expelled by us from the patient's bodies with torments of word and fires of prayer. Saturn himself, and Serapis, and Jupiter, and whatsoever demons ye worship, overcome by pain declare what they are; nor surely do they lie concerning their iniquity, above all when several of you are present. Believe these witnesses, confessing the truth of themselves, that they are demons. For adjured by the true and only God, they shudder reluctant in the wretched bodies; and either they issue forth at once, or vanish gradually, according as the faith of the patient aids, or the grace of the curer favours."[1]

The strangeness with which such words now fall upon our ears is full of significance. It is one symptom of that vast quiet change which has come over animistic philosophy in the modern educated world. Whole orders of spiritual beings, worshipped in polytheistic religion, and degraded in early Christendom to real but evil demons, have since passed from objective to subjective existence, have faded from the Spiritual into the Ideal. By the operation of similar intellectual changes, the general theory of spirit-embodiment, having fulfilled the great work it had for ages to do in religion and philosophy, has now dwindled within the limits of the educated world to near its vanishing-point. The doctrines of Disease-possession and Oracle-possession, once integral parts of the higher philosophy, and still maintaining a vigorous existence in the lower culture, seem to be dying out within the influence of the higher into dogmatic survival, conscious metaphor, and popular superstition. The doctrine of spirit-embodiment in objects, Fetishism, now scarcely appears outside barbaric regions,

---

[1] Marcus Minucius Felix, Octavius, cap. xxvii. : "Isti igitur impuri spiritus, dæmones, ut ostensum a magis, a philosophis, et a Platone sub statuis et imaginibus consecrati delitescunt, etc."

save in the peasant folklore which keeps it up amongst us with so many other remnants of barbaric thought. And the like theory of spiritual influence as applied to Idolatry, though still to be studied among savages and barbarians, and on record in past ages of the civilized world, has perished so utterly amongst ourselves, that few but students are aware of its ever having existed.

To bring home to our minds the vastness of the intellectual tract which separates modern from savage philosophy, and to enable us to look back along the path where step by step the mind's journey was made, it will serve us to glance over the landmarks which language to this day keeps standing. Our modern languages reach back through the middle ages to classic and barbaric times, where in this matter the transition from the crudest primæval animism is quite manifest. We keep in daily use, and turn to modern meaning, old words and idioms which carry us home to the philosophy of ancient days. We talk of "genius" still, but with thought how changed. The genius of Augustus was a tutelary demon, to be sworn by and to receive offerings on an altar as a deity. In modern English, Shakspere, Newton, or Wellington, is said to be led and prompted by his genius, but that genius is a shrivelled philosophic metaphor. So the word "spirit" and its kindred terms keep up with wondrous pertinacity the traces which connect the thought of the savage with its hereditary successor, the thought of the philosopher. Barbaric philosophy retains as real what civilized language has reduced to simile. The Siamese is made drunk with the demon of the arrack that possesses the drinker, while we with so different sense still extract the "spirit of wine."[1] Look at the saying ascribed to Pythagoras, and mentioned by Porphyry. "The sound indeed which is given by striking brass, is the voice of a certain demon contained in that brass." These might have been the representative words of some savage animistic philo-

[1] Bastian, 'Oestl. Asien,' vol. ii. p. 455. See Spiegel, 'Avesta,' vol. ii. p. 54.

sopher; but with the changed meaning brought by centuries of philosophizing, Oken hit upon a definition almost identical in form, that "What sounds, announces its spirit" ("Was tönt, gibt seinen Geist kund.")[1] What the savage would have meant, or Porphyry after him did mean, was that the brass was actually animated by a spirit of the brass apart from its matter, but when a modern philosopher takes up the old phrase, all he means is the qualities of the brass. As for our own selves and our feelings, we still talk of "animal spirits," of being in "good and bad spirits," only recalling with an effort the long past metaphysics which such words once expressed. The modern theory of the mind considers it capable of performing even exalted and unusual functions without the intervention of prompting or exciting demons; yet the old recognition of such beings crops up here and there in phrases which adapt animistic ideas to commonplaces of human disposition, as when a man is still said to be animated by a patriotic spirit, or possessed by a spirit of disobedience. In old times the ἐγγαστρίμυθος, or "ventriloquus" was really held to have a spirit rumbling or talking from inside his body, as when Eurykles the soothsayer was inspired by such a familiar; or when a certain Patriarch mentioning a demon heard to speak out of a man's belly, remarks on the worthy place it had chosen to dwell in. In the time of Hippokrates, the giving of oracular responses by such ventriloquism was practised by certain women as a profession. To this day in China one may get an oracular response from a spirit apparently talking out of a medium's stomach, for a fee of about twopence-halfpenny. How changed a philosophy it marks, that among ourselves the word "ventriloquist" should have sunk to its present meaning.[2] Nor is that

---

[1] Porphyr. de Vita Pythagoræ. Oken, 'Lehrbuch der Naturphilosophie,' 2753.
[2] Suidas, s. v. ἐγγαστρίμυθος; Isidor. Gloss s. v. 'præcantatores;' Bastian, 'Mensch.' vol. ii. p. 578. Maury, 'Magie,' etc. p. 269. Doolittle, 'Chinese,' vol. ii. p. 115.

change less significant which, starting with the conception of a man being really ἔνθεος, possessed by a deity within him, carries on a metamorphosed relic of this thorough animistic thought, from ἐνθουσιασμός to "enthusiasm." With all this, let it not be supposed that such change of opinion in the educated world has come about through wanton incredulity or decay of the religious temperament. Its source is the alteration in natural science, assigning new causes for the operations of nature and the events of life. The theory of the immediate action of personal spirits has here, as so widely elsewhere, given place to ideas of force and law. No indwelling deity now regulates the life of the burning sun, no guardian angels drive the stars across the arching firmament, the divine Ganges is water flowing down into the sea to evaporate into cloud and descend again in rain. No deity simmers in the boiling pot, no presiding spirits dwell in the volcano, no howling demon shrieks from the mouth of the lunatic. There was a period of human thought when the whole universe seemed actuated by spiritual life. For our knowledge of our own history, it is deeply interesting that there should remain rude races yet living under the philosophy which we have so far passed from, since Physics, Chemistry, Biology, have seized whole provinces of the ancient Animism, setting force for life and law for will.

CHAPTER XV.

ANIMISM (continued).

Spirits regarded as personal causes of Phenomena of the World—Pervading Spirits as good and evil Demons affecting man—Spirits manifest in Dreams and Visions: Nightmares; Incubi and Succubi; Vampires; Visionary Demons—Demons of darkness repelled by fire—Demons otherwise manifest: seen by animals; detected by footprints—Spirits conceived and treated as material—Guardian and Familiar Spirits—Nature-Spirits; historical course of the doctrine—Spirits of Volcanos, Whirlpools, Rocks —Water-Worship: Spirits of Wells, Streams, Lakes, &c.—Tree-Worship: Spirits embodied in or inhabiting Trees; Spirits of Groves and Forests—Animal-Worship: Animals worshipped, directly, or as incarnations or representatives of Deities; Totem-Worship; Serpent-Worship— Species-Deities; their relation to Archetypal Ideas.

WE have now to enter on the final topic of the investigation of Animism, by completing the classified survey of spiritual beings in general, from the myriad souls, elves, fairies, genii, conceived as filling their multifarious offices in man's life and the world's, up to the deities who reign, few and mighty, over the spiritual hierarchy. In spite of endless diversity of detail, the general principles of this investigation seem comparatively easy of access to the enquirer, if he will use the two keys which the foregoing studies supply: first, that spiritual beings are modelled by man on his primary conception of his own human soul, and second, that their purpose is to explain nature on the primitive childlike theory that it is truly and throughout " Animated Nature." If, as the poet says, " Felix qui potuit rerum cognoscere causas," then rude tribes of ancient men had within them this source of happiness, that they could explain to their own content the causes of things. For to them spiritual beings, elves and gnomes, ghosts and manes, demons and deities, were the living personal causes of universal life. "The first men found everything easy, the mysteries of nature were not so hidden from them as from us," said Jacob Böhme the mystic. True, we may well answer, if these primitive men believed in that animistic philosophy of nature which even now survives in the savage mind. They could ascribe to kind or hostile spirits all good and evil of their own lives, and all striking operations of nature; they lived in familiar intercourse with the living and powerful souls of their dead ancestors, with the spirits of the stream and grove, plain and mountain, they knew well the living mighty Sun pouring his beams of light and heat upon them, the living mighty Sea dashing her fierce billows on the shore, the great personal Heaven and Earth protecting and producing all things. For as the human body was held to live and act by virtue of its own inhabiting spirit-soul, so the operations of the world seemed to be carried on by the influence of other spirits. And thus Animism, starting as a philosophy of human life, extended and expanded itself till it became a philosophy of nature at large.

To the minds of the lower races it seems that all nature is possessed, pervaded, crowded, with spiritual beings. In seeking by a few types to give an idea of this conception of pervading Spirits in its savage and barbaric stage, it is not indeed possible to draw an absolute line of separation between spirits occupied in affecting for good and ill the life of Man, and spirits specially concerned in carrying on the operations of Nature. In fact these two classes of spiritual beings blend into one another as inextricably as do the original animistic doctrines they are based on. As, however, the spirits considered directly to affect the life and fortune of Man lie closest to the centre of the animistic scheme, it is well to give them precedence. The description and function of these beings extend upwards from among the rudest human tribes. Milligan writes of the natives of Tasmania: "They were polytheists, that is, they believed in guardian angels or spirits, and in a plurality of powerful, but generally evil-disposed beings, inhabiting crevices and caverns of rocky mountains, and making temporary abode in hollow trees and solitary valleys : of these a few were supposed to be of great power, while the majority were much of the nature and attributes of the goblins and elves of our native land."[1] Oldfield writes of the aborigines of Australia, " The number of supernatural beings, feared if not loved, that they acknowledge, is exceedingly great; for not only are the heavens peopled with such, but the whole face of the country swarms with them; every thicket, most watering-places, and all rocky places abound with evil spirits. In like manner, every natural phenomenon is believed to be the work of demons, none of which seem of a benign nature, one and all apparently striving to do all imaginable mischief to the poor black fellow."[2] It must be indeed an unhappy race among whom such a demonology could shape itself, and it is a relief to find that other people of low culture, while recognizing the same spiritual world swarming about them, do not find its main attribute to be spite against themselves. Among the Algonquin Indians of North America, Schoolcraft finds the very groundwork of their religion in the belief " that the whole visible and invisible creation is animated with various orders of malignant or benign spirits, who preside over the daily affairs and over the final destinies of men."[3] Among the Khonds of Orissa, Macpherson describes the greater gods and tribal manes, and below these the order of minor and local deities : " They are the tutelary gods of every spot on earth, having power over the functions of nature which operate there, and over everything relating to human life in it. Their number is

[1] Bonwick, 'Tasmanians,' p. 182.
[2] Oldfield in 'Tr Eth. Soc.' vol iii. p. 228.
[3] Schoolcraft, 'Algic Res.' vol. i. p. 41. 'Indian Tribes,' vol. iii. p. 327. Waitz, vol. iii. p. 191. See also J. G. Müller, p. 175. (Antilles Islanders); Brasseur, ' Mexique,' vol. iii. p. 482.

unlimited. They fill all nature, in which no power or object, from the sea to the clods of the field, is without its deity. They are the guardians of hills, groves, streams, fountains, paths, and hamlets, and are cognizant of every human action, want, and interest in the locality, where they preside."[1] Describing the animistic mythology of the Turanian tribes of Asia and Europe, Castrén has said that every land, mountain, rock, river, brook, spring, tree, or whatsoever it may be, has a spirit for an inhabitant; the spirits of the trees and stones, of the lakes and brooks, hear with pleasure the wild man's pious prayers and accept his offerings.[2] Such are the conceptions of the Guinea negro, who finds the abodes of his good and evil spirits in great rocks, hollow trees, mountains, deep rivers, dense groves, echoing caverns, and who passing silently by these sacred places leaves some offering, if it be but a leaf or a shell picked up on the beach.[3] Such are examples which not unfairly picture the belief of the lower races in a world of spirits on earth, and such descriptions apply to the state of men's minds along the course of civilization.

The doctrine of ancient philosophers such as Pythagoras[4] and Iamblichus,[5] of spiritual beings swarming through the atmosphere we breathe, was carried on and developed in special directions in the discussions concerning the nature and functions of the world-pervading host of angels and devils, in the writings of the early Christian Fathers.[6] Theologians of modern centuries have for the most part seen reason to reduce within comparatively narrow limits the action ascribed to external spiritual beings on mankind;

---

[1] Macpherson, 'India,' p. 90. See also Cross, Karens, in 'Journ. Amer. Or. Soc.' vol. iv. p. 315 ; Williams, ' Fiji,' vol. i. p. 239.
[2] Castrén, ' Finn. Myth,' p. 114, 182, etc.
[3] J. L. Wilson, ' W. Afr.' p. 218, 388 ; Waitz, vol. ii. p. 171.
[4] Diog. Laert. Vita Pythagor. 32.
[5] Iamblichus, ii.
[6] Collected passages in Calmet, ' Diss. sur les Esprits ;' Horst, ' Zauber-Bibliothek,' vol. ii. p. 263, etc. ; vol. vi. p. 49, etc. ; see Migne's Dictionaries.

yet there are some who retain to the full the angelology and demonology of Origen and Tertullian. These two views may be well contrasted by setting side by side the judgments of two ecclesiastics of the Roman Church, as to the belief in pervading demons prevalent in uncivilized countries. The celebrated commentator, Dom Calmet, lays down in the most explicit terms the doctrine of angels and demons, as a matter of dogmatic theology. But he is less inclined to receive unquestioned the narratives of particular manifestations in the mediæval and modern world. He mentions indeed the testimony of Louis Vivez, that in the newly discovered countries of America, nothing is more common than to see spirits which appear at noonday, not only in the country but in towns and villages, speaking, commanding, sometimes even striking men; and the account by Olaus Magnus of the spectres or spirits seen in Sweden and Norway, Finland and Lapland, which do wonderful things, some even serving men as domestics and driving the cattle out to pasture. But what Calmet remarks on these stories, is that the greater ignorance prevails in a country, the more superstition reigns there.[1] It seems that in our own day, however, the tendency is to encourage less sceptical views. Monsignor Gaume's book on 'Holy Water,' which not long since received the special and formal approval of Pius IX., appears "at an epoch when the millions of evil angels which surround us are more enterprising than ever;" and here Olaus Magnus' story of the demons infesting Northern Europe is not only cited but corroborated.[2] On the whole, the survey of the doctrine of pervading spirits through all the grades of culture is a remarkable display of intellectual continuity. Most justly does Ellis the missionary, depicting the South Sea Islanders' world crowded with its innumerable pervading spirits, point out the closeness of correspondence here between doctrines of the savage and the

[1] Calmet, 'Dissertation sur les Esprits,' vol. i. ch. xlviii.
[2] Gaume, 'L'Eau Bénite au XIXme Siècle,' p. 295, 341.

civilized animist, expressed as both may be in Milton's familiar lines :—

"Millions of spiritual creatures walk the earth,
Unseen, both when we wake, and when we sleep."[1]

As with souls, so with other spirits, man's most distinct and direct intercourse is had where they become actually present to his senses in dreams and visions. The belief that such phantoms are real and personal spirits, suggested and maintained as it is by the direct evidence of the senses of sight, touch, and hearing, is naturally an opinion usual in savage philosophy, and indeed elsewhere, long and obstinately resisting the attacks of the later scientific doctrine. The demon Koin strives to throttle the dreaming Australian;[2] the evil "na" crouches on the stomach of the Karen;[3] the North American Indian, gorged with feasting, is visited by nocturnal spirits;[4] the Caribs, subject to hideous dreams, often woke declaring that the demon Maboya had beaten them in their sleep, and they could still feel the pain.[5] These demons are the very elves and nightmares that to this day in benighted districts of Europe ride and throttle the snoring peasant, and whose names, not forgotten among the educated, have only made the transition from belief to jest.[6] A not less distinct product of the savage animistic theory of dreams as real visits from personal spiritual beings, lasted on without a shift or break into the belief of mediæval Christendom. This is the doctrine of the incubi and succubi, those male and female nocturnal demons which

[1] Ellis, 'Polyn. Res.' vol. i. p. 331.
[2] Backhouse, 'Australia,' p. 555; Grey, 'Australia,' vol. ii. p. 337.
[3] Mason, 'Karens,' l. c. p. 211.
[4] Schoolcraft, 'Indian Tribes,' part iii. p. 226.
[5] Rochefort, 'Antilles,' p. 419.
[6] Grimm 'D. M.' p. 1193; Hanusch, 'Slaw. Myth.' p. 332; St. Clair & Brophy, 'Bulgaria,' p. 59; Wuttke, 'Volksaberglaube,' p. 122; Bastian, 'Psychologie,' p. 103; Brand, vol. iii. p 279; The *mare* in *nightmare* means spirit, elf, or nymph; compare Anglo-Sax. *wudumære* (wood-mare) = echo.

consort sexually with men and women. We may set out with their descriptions among the islanders of the Antilles, where they are the ghosts of the dead, vanishing when clutched;[1] in New Zealand, where ancestral deities "form attachments with females and pay them repeated visits," while in the Samoan Islands such intercourse of mischievous inferior gods caused "many supernatural conceptions;"[2] and in Lapland, where details of this last extreme class have also been placed on record.[3] From these lower grades of culture we may follow the idea onward. Formal rites are specified in the Hindu Tantra, which enable a man to obtain a companion-nymph by worshipping her and repeating her name by night in a cemetery.[4] Augustine, in an instructive passage, states the popular notions of the visits of incubi, vouched for, he tells us, by testimony of such quantity and quality that it may seem impudence to deny it; yet he is careful not to commit himself to a positive belief in such spirits.[5] Later theologians were less cautious, and grave argumentation on nocturnal intercourse with incubi and succubi was carried on till, at the height of mediæval civilization, we find it accepted in full belief by ecclesiastics and lawyers. Nor are we to count it as an ugly but harmless superstition, when for example we find it set forth in the Bull of Pope Innocent VIII. in 1484, as an

[1] 'Vita del Amm. Christoforo Colombo,' ch. xiii.; and Life of Colon in Pinkerton, vol. xii. p. 84.
[2] Taylor, 'New Zealand,' p. 149, 389. Mariner, 'Tonga Is.' vol. ii. p. 119.
[3] Högström, 'Lapmark,' ch. xi.
[4] Ward, 'Hindoos,' vol. ii. p. 151. See also Borri, 'Cochin-China,' in Pinkerton, vol. ix. p. 823.
[5] Augustin. 'De Civ. Dei,' xv. 23 : "Et quoniam creberrima fama est, multique se expertos, vel ab eis qui experti essent, de quorum fide dubitandum non esset, audisse confirmant, Silvanos et Faunos, quos vulgo incubos vocant, improbos sæpe extitisse mulieribus, et earum appetisse ac peregisse concubitum; et quosdam dæmones, quos Dusios Galli nuncupant, hanc assidue immunditiam et tentare et efficere ; plures talesque asseverant, ut hoc negare impudentiæ videatur ; non hinc aliquid audeo definire, utrum aliqui spiritus . . . possint etiam hanc pati libidinem; ut . . . . sentientibus feminibus misceantur." See also Grimm, 'D. M.' p. 449, 479; Hanusch, 'Slaw. Myth.' p. 332; Cockayne, 'Leechdoms of Early England,' vol. i. p. xxxviii. vol. ii. p. 345.

accepted accusation against "many persons of both sexes, forgetful of their own salvation, and falling away from the Catholic faith." The practical outcome of this belief is known to students who have traced the consequence of the Papal Bull in the legal manual of the witchcraft tribunals, drawn up by the three appointed Inquisitors, the infamous Malleus Maleficarum ; and have followed the results of this again into those dreadful records which relate in their bald matter-of-fact phraseology the confessions of the crime of diabolic intercourse, wrung from the wretched victims worked on by threat and persuasion in the intervals of the rack, till enough evidence was accumulated for clear judgment, and sentence of the stake.[1] I need not dwell on the mingled obscenity and horror of these details, which here only have their bearing on the history of animism. But it will aid the ethnographer to understand the relation of modern to savage philosophy, if he will read Richard Burton's seriously believing account in the 'Anatomy of Melancholy,' where he concludes with acquiescence in a declaration lately made by Lipsius, that on the showing of daily narratives and judicial sentences, in no age had these lecherous demons appeared in such numbers as in his own time—and this was about A.D. 1600.[2]

In connexion with the nightmare and the incubus, another variety of nocturnal demon requires notice, the vampire. Inasmuch as certain patients are seen becoming day by day, without apparent cause, thin, weak, and bloodless, savage animism is called upon to produce a satisfactory explanation, and does so in the doctrine that there exist certain demons which eat out the souls or hearts or suck the blood of their victims. The Polynesians said that it was the

[1] The 'Malleus Maleficarum' was published about 1489. See on the general subject, Horst, 'Zauber-Bibliothek,' vol. vi. ; Ennemoser, 'Magic,' vol. ii. : Maury, 'Magie,' etc. p. 256 ; Lecky, 'Hist. of Rationalism,' vol. i.
[2] Burton, 'Anatomy of Melancholy,' iii. 2. "Unum dixero, non opinari me ullo retro ævo tantam copiam Satyrorum, et salacium istorum Geniorum se ostendisse, quantum nunc quotidianæ narrationes, et judiciales sententiæ proferunt."

departed souls (tii) which quitted the graves and grave-idols to creep by night into the houses, and devour the heart and entrails of the sleepers, and these died.[1] The Karens tell of the "kephu," which is a wizard's stomach going forth in the shape of a head and entrails, to devour the souls of men, and they die.[2] The Mintira of the Malay Peninsula have their "hantu penyadin;" he is a water-demon, with a dog's head and an alligator's mouth, who sucks blood from men's thumbs and great toes, and they die.[3] It is in Slavonia and Hungary that the demon blood-suckers have their principal abode, and to this district belongs their special name of *vampire*, Polish *upior*, Russian *upir*. There is a whole literature of hideous vampire-stories, which the student will find elaborately discussed in Calmet. The shortest way of treating the belief is to refer it directly to the principles of savage animism. We shall see that most of its details fall into their places at once, and that vampires are not mere creations of groundless fancy, but causes conceived in spiritual form to account for specific facts of wasting disease. As to their nature and physical action, there are two principal theories, but both keep close to the original animistic idea of spiritual beings, and consider these demons to be human souls. The first theory is that the soul of a living man, often a sorcerer, leaves its proper body asleep and goes forth, perhaps in the visible form of a straw or fluff of down, slips through keyholes and attacks its sleeping victim. If the sleeper should wake in time to clutch this tiny soul-embodiment, he may through it have his revenge by maltreating or destroying its bodily owner. Some say these "mury" come by night to men, sit upon their breasts and suck their blood, while others think it is only children's blood they suck, they being to grown people mere nightmares. Here we have the actual phenomenon of nightmare, adapted to a particular purpose. The second

[1] J. R. Forster, 'Observations during Voyage round World,' p. 548.
[2] Cross, 'Karens,' l. c. p. 312.
[3] 'Journ. Ind. Archip.' vol. i. p. 307.

theory is that the soul of a dead man goes out from its buried corpse and sucks the blood of living men. The victim becomes thin, languid, and bloodless, falls into a rapid decline and dies. Here again is actual experience, but a new fancy is developed to complete the idea. The corpse thus supplied by its returning soul with blood, is imagined to remain unnaturally fresh and supple and ruddy; and accordingly the means of detecting a vampire is to open his grave, where the re-animated corpse may be found to bleed when cut, and even to move and shriek. One way to lay a vampire is to stake down the corpse (as with suicides and with the same intention); but the more effectual plan is to behead and burn it. This is the substance of the doctrine of vampires. Still, as one order of demons is apt to blend into others, the vampire-legends are much mixed with other animistic folk-lore. Vampires appear in the character of the poltergeist or knocker, as causing those disturbances in houses which modern spiritualism refers in like manner to souls of the departed. Such was the ghost of a certain surly peasant who came out of his grave in the island of Mycone in 1700, after he had been buried but two days; he came into the houses, upset the furniture, put the lamps out, and carried on his tricks till the whole population went wild with terror. Tournefort happened to be there and was present at the exhumation; his account is curious evidence of the way an excited mob could persuade themselves, without the least foundation of fact, that the body was warm and its blood red. Again, the blood-sucker is very generally described under the Slavonic names of werewolf (wilkodlak, brukolaka, &c.); the descriptions of the two creatures are inextricably mixed up, and a man whose eyebrows meet, as if his soul were taking flight like a butterfly, to enter some other body, may be marked by this sign either as a werewolf or a vampire. A modern account of vampirism in Bulgaria well illustrates the nature of spirits as conceived in such beliefs as these. A sorcerer armed with a saint's picture will hunt a vampire into a bottle containing some of the filthy food that the demon loves; as soon as he is fairly inside he is corked down, the bottle is thrown into the fire, and the vampire disappears for ever.[1]

As to the savage visionary and the phantoms he beholds, the Greenlander preparing for the profession of sorcerer may stand as type, when, rapt in contemplation in his desert solitude, emaciated by fasting and disordered by fits, he sees before him scenes with figures of men and animals, which he believes to be spirits. Thus it is interesting to read the descriptions by Zulu converts of the dreadful creatures which they see in moments of intense religious exaltation, the snake with great eyes and very fearful, the leopard creeping stealthily, the enemy approaching with his long assagai in his hand—these coming one after another to the place where the man has gone to pray in secret, and striving to frighten him from his knees.[2] Thus the visionary temptations of the Hindu ascetic and the mediæval saint are happening in our own day, though their place is now rather in the medical handbook than in the record of miracle. Like the disease-demons and the oracle-demons, these spiritual groups have their origin not in fancy, but in real phenomena interpreted on animistic principles.

In the dark especially, harmful spirits swarm. Round native Australian encampments, Sir George Grey used to see the bush dotted with little moving points of fire; these were the firesticks carried by the old women sent to look after the young ones, but who dared not quit the firelight without a brand to protect them from the evil spirits.[3] So South American Indians would carry brands or torches for fear of evil demons when they ventured into the dark.[4]

[1] J. V. Grohmann, 'Aberglauben aus Böhmen,' etc. p. 24; Calmet, 'Diss. sur les Esprits.' vol. ii.; Grimm, 'D. M.' p. 1048, etc.; St. Clair and Brophy, 'Bulgaria,' p. 49; see Ralston, 'Songs of Russian People,' p. 409.
[2] Cranz, 'Grönland,' p. 268. Callaway, 'Rel. of Amazulu,' p. 246, etc.
[3] Grey, 'Australia,' vol. ii. p. 302. See also Bonwick, 'Tasmanians,' p. 180.
[4] Southey, 'Brazil,' part i. p. 238. See also Rochefort, p. 418; J. G.

Tribes of the Malay Peninsula light fires near a mother at childbirth, to scare away the evil spirits.[1] Such notions extend to higher levels of civilization. In Southern India, where for fear of pervading spirits only pressing need will induce a man to go abroad after sundown, the unlucky wight who has to venture into the dark will carry a firebrand to keep off the spectral foes. Even in broad daylight, the Hindu lights lamps to keep off the demons,[2] a ceremony which is to be noticed again at a Chinese wedding.[3] In Europe, the details of the use of fire to drive off demons and witches are minute and explicit. The ancient Norse colonists in Iceland carried fire round the lands they intended to occupy, to expel the evil spirits. Such ideas have brought into existence a whole group of Scandinavian customs, still remembered in the country, but dying out in practice. Till a child is baptized, the fire must never be let out, lest the trolls should be able to steal the infant; a live coal must be cast after the mother as she goes to be churched, to prevent the trolls from carrying her off bodily or bewitching her; a live coal is to be thrown after a trollwife or witch as she quits a house, and so forth.[4] Into modern times, the people of the Hebrides continued to protect the mother and child from evil spirits, by carrying fire round them.[5] In modern Bulgaria, on the Feast of St. Demetrius, lighted candles are placed in the stables and the wood-shed, to prevent evil spirits from entering into

Müller, p. 273 (Caribs); Cranz, 'Grönland,' p. 301; Schoolcraft, 'Indian Tribes,' part iii. p. 140.
[1] 'Journ. Ind. Archip.' vol. i. pp. 270, 298; vol. ii. 'N. S.' p. 117.
[2] Roberts, 'Oriental Illustrations,' p. 531; Colebrooke in 'As. Res.' vol. vii. p. 274.
[3] Doolittle, 'Chinese,' vol. i. p. 77.
[4] Hylten-Cavallius, 'Wärend och Wirdarne,' vol. i. p. 191; Atkinson, 'Glossary of Cleveland Dial.' p. 597. [Prof. Liebrecht, in 'Zeitschrift für Ethnologie,' vol. v. 1873, p. 99, adds a comparison of the still usual German custom of keeping a light burning in the lying-in room till the child is baptized (Wuttke, 2nd ed. No. 583), and the similar ancient Roman practice whence the goddess Candelifera had her name (note to 2nd ed.).]
[5] Martin, 'Western Islands,' in Pinkerton, vol. iii. p. 612.

the domestic animals.[1] Nor did this ancient idea remain a mere lingering notion of peasant folklore. Its adoption by the Church is obvious in the ceremonial benediction of candles in the Roman Ritual: "Ut quibuscumque locis accensæ, sive positæ fuerint, discedant principes tenebrarum, et contremiscant, et fugiant pavidi cum omnibus ministris suis ab habitationibus illis, etc." The metrical translation of Naogeorgus shows perfectly the retention of primitive animistic ideas in the middle ages:—

". . . . a wondrous force and might
Doth in these candels lie, which if at any time they light,
They sure beleve that neyther storm or tempest dare abide,
Nor thunder in the skies be heard, nor any devil's spide,
Nor fearefull sprightes that walke by night, nor hurts of frost
or haile."[2]

Animals stare and startle when we see no cause; is it that they see spirits invisible to man? Thus the Greenlander says that the seals and wildfowl are scared by spectres, which no human eye but the sorcerer's can behold;[3] and thus the Khonds hold that their flitting ethereal gods, invisible to man, are seen by beasts.[4] The thought holds no small place in the folklore of the world. Telemachos could not discern Athene standing near him, for not to all do the gods visibly appear; but Odysseus saw her, and the dogs, and they did not bark, but with low whine slunk across the dwelling to the further side.[5] So in old Scandinavia, the dogs could see Hela the death-goddess move unseen by men;[6] so Jew and Moslem, hearing the dogs howl, know that they have seen the Angel of Death come on his awful errand;[7] while the

[1] St. Clair and Brophy, 'Bulgaria,' p. 44.
[2] Rituale Romanum; Benedictio Candelarum. Brand, 'Popular Antiquities,' vol. i. p. 46.
[3] Cranz, 'Grönland,' p. 267, see 296.
[4] Macpherson, 'India,' p. 100.
[5] Homer. Odyss. xvi. 160.
[6] Grimm, 'D. M.' p. 632.
[7] Eisenmenger, 'Judenthum,' part i. p. 872. Lane, Thousand and One Nights,' vol. ii. p. 56.

beliefs that animals see spirits, and that a dog's melancholy howl means death somewhere near, are still familiar to our own popular superstition.

Another means by which men may detect the presence of invisible spirits, is to adopt the thief-catcher's well-known device of strewing ashes. According to the ideas of a certain stage of animism, a spirit is considered substantial enough to leave a footprint. The following instances relate sometimes to souls, sometimes to other beings. The Philippine islanders expected the dead to return on the third day to his dwelling, wherefore they set a vessel of water for him to wash himself clean from the grave-mould, and strewed ashes to see footprints.[1] A more elaborate rite forms part of the funeral customs of the Hos of North-East India. On the evening of a death, the near relatives perform the ceremony of calling the dead. Boiled rice and a pot of water are placed in an inner room, and ashes sprinkled from thence to the threshold. Two relatives go to the place where the body was burnt, and walk round it beating ploughshares and chanting a plaintive dirge to call the spirit home; while two others watch the rice and water to see if they are disturbed, and look for the spirit-footsteps in the ashes. If a sign appears, it is received with shivering horror and weeping, the mourners outside coming in to join. Till the survivors are thus satisfied of the spirit's return, the rite must be repeated.[2] In Yucatan there is mention of the custom of leaving a child alone at night in a place strewn with ashes; if the footprint of an animal were found next morning, this animal was the guardian deity of the child.[3] Beside this may be placed the Aztec ceremony at the second festival of the Sun-god Tezcatlipoca, when they sprinkled maize-flour before his sanctuary, and his

[1] Bastian, 'Psychologie,' p. 162. Other localities in 'Journ. Ind. Archip.' vol. iv. p. 333.
[2] Tickell in 'Journ. As. Soc. Bengal,' vol. ix. p. 795. The dirge is given above, p. 32.
[3] De Brosses, 'Dieux Fétiches,' p. 46.

high-priest watched till he beheld the divine footprints, and then shouted to announce. " Our great god is come."[1] Among such rites in the Old World, the Talmud contains a salient instance ; there are a great multitude of devils, it is said ; and he who will be aware of them let him take sifted ashes and strew them by his bed, and in the early morning he shall see as it were marks of cocks' feet.[2] This is an idea that has widely spread in the modern world, as where in German folklore the little " earthmen " make footprints like a duck's or goose's in the strewn ashes. Other marks, too, betoken the passage of spirit-visitors ;[3] and as for ghosts, our own superstition is among the most striking of the series. On St. Mark's Eve, ashes are to be sifted over the hearth, and the footprint will be seen of any one who is to die within the year; many a mischievous wight has made a superstitious family miserable by slily coming down stairs and marking the print of some one's shoe.[4] Such details as these may justify us in thinking that the lower races are apt to ascribe to spirits in general that kind of ethereal materiality which we have seen they attribute to souls. Explicit statements on the subject are scarce till we reach the level of early Christian theology. The ideas of Tertullian and Origen, as to the thin yet not immaterial substance of angels and demons, probably represent the conceptions of primitive animism far more clearly than the doctrine which Calmet lays down with the weight of theological dogma, that angels, demons, and disembodied souls are pure immaterial spirit; but that when spirits appear, act, speak, walk, eat, and so forth, they must produce tangible bodies by either condensing the air, or substituting

[1] Clavigero, 'Messico,' vol. ii. p. 79.
[2] Tractat. Berachoth.
[3] Grimm, 'D. M.' pp. 420, 1117 ; St. Clair and Brophy, 'Bulgaria, p. 54. See also Bastian, 'Mensch.' vol. ii. p. 325 ; Tschudi, 'Peru,' vol. ii. p. 355.
[4] Brand, 'Popular Antiquities,' vol. i. p. 193. See Boecler, 'Ehsten Abergl.' p. 73.

other terrestrial solid bodies capable of performing these functions.[1]

No wonder that men should attack such material beings by material means, and even sometimes try to rid themselves by a general clearance from the legion of ethereal beings hovering around them. As the Australians annually drive from their midst the accumulated ghosts of the last year's dead, so the Gold Coast negroes from time to time turn out with clubs and torches to drive the evil spirits from their towns; rushing about and beating the air with frantic howling, they drive the demons into the woods, and then come home and sleep more easily, and for a while afterwards enjoy better health.[2] When a baby was born in a Kalmuk horde, the neighbours would rush about crying and brandishing cudgels about the tents, to drive off the harmful spirits who might hurt mother and child.[3] Keeping up a closely allied idea in modern Europe, the Bohemians at Pentecost, and the Tyrolese on Walpurgisnacht, hunt the witches, invisible and imaginary, out of house and stall.[4]

Closely allied to the doctrine of souls, and almost rivalling it in the permanence with which it has held its place through all the grades of animism, is the doctrine of patron, guardian, or familiar spirits. These are beings specially attached to individual men, soul-like in their nature, and sometimes considered as actually being human souls. These beings have, like all others of the spiritual world as originally conceived, their reason and purpose. The special functions which they perform are twofold. First, while man's own proper soul serves him for the ordinary purposes of life and thought, there are times when powers

[1] Tertullian. De Carne Christi, vi.; Adv. Marcion. ii.; Origen. De Princip. i. 7. See Horst, l. c. Calmet, 'Dissertation,' vol. i. ch. xlvi.
[2] J. L. Wilson, 'W. Afr.' p. 217. See Bosman, 'Guinea,' in Pinkerton, vol. xvi. p. 402.
[3] Pallas, 'Reisen,' vol. i. p. 360.
[4] Grimm, 'D. M.' p. 1212; Wuttke, 'Volksaberglaube,' p. 119; see Hyltén-Cavallius, part i. p. 178 (Sweden).

and impressions out of the course of the mind's normal action, and words that seem spoken to him by a voice from without, messages of mysterious knowledge, of counsel or warning, seem to indicate the intervention of as it were a second superior soul, a familiar demon. And as enthusiasts, seers, sorcerers, are the men whose minds most often show such conditions, so to these classes more than to others the informing and controlling patron-spirits are attached. Second, while the common expected events of daily life pass unnoticed as in the regular course of things, such events as seem to fall out with especial reference to an individual, demand an intervening agent; and thus the decisions, discoveries, and deliverances, which civilized men variously ascribe to their own judgment, to luck, and to special interposition of Providence, are accounted for in the lower culture by the action of the patron-spirit or guardian-genius. Not to crowd examples from all the districts of animism to which this doctrine belongs, let us follow it by a few illustrations from the lower grades of savagery upward. Among the Watchandis of Australia, it is held that when a warrior slays his first man, the spirit of the dead enters the slayer's body and becomes his "woorie" or warning spirit; taking up its abode near his liver, it informs him by a scratching or tickling sensation of the approach of danger.[1] In Tasmania, a native has been heard to ascribe his deliverance to the preserving care of his deceased father's spirit, now become his guardian angel.[2] That the most important act of the North American Indian's religion is to obtain his individual patron genius or deity, is well known. Among the Esquimaux, the sorcerer qualifies for his profession by getting a "torngak" or spirit which will henceforth be his familiar demon, and this spirit may be the soul of a deceased parent.[3] In Chili, as to guardian spirits, it has been re-

[1] Oldfield, 'Abor. of Australia,' in 'Tr. Eth. Soc.' vol. iii. p. 240.
[2] Bonwick, 'Tasmanians,' p. 182.
[3] Cranz, 'Grönland,' p. 268; Egede, p. 187.

marked that every Araucanian imagines he has one in his service; "I keep my amchi-malghen (guardian nymph) still," being a common expression when they succeed in any undertaking.[1] The Caribs display the doctrine well in both its general and special forms. On the one hand, there is a guardian deity for each man, which accompanies his soul to the next life; on the other hand, each sorcerer has his familiar demon, which he evokes in mysterious darkness by chants and tobacco-smoke; and when several sorcerers call up their familiars together, the consequence is apt to be a quarrel among the demons, and a fight.[2] In Africa, the negro has his guardian spirit—how far identified with what Europeans call soul or conscience, it may be hard to determine; but he certainly looks upon it as a being separate from himself, for he summons it by sorcery, builds a little fetish-hut for it by the wayside, rewards and propitiates it by libations of liquor and bits of food.[3] In Asia, the Mongols, each with his patron genius,[4] and the Laos sorcerers who can send their familiar spirits into others' bodies to cause disease,[5] are examples equally to the purpose.

Among the Aryan nations of Northern Europe,[6] the old doctrine of man's guardian spirit may be traced, and in classic Greece and Rome it renews with philosophic eloquence and cultured custom the ideas of the Australian and the African. The thought of the spiritual guide and protector of the individual man is happily defined by Menander, who calls the attendant genius, which each man has from the hour of birth, the good mystagogue (*i.e.*, the novice's guide to the mysteries) of this life.

[1] Molina, 'Chili,' vol. ii. p. 86.
[2] Rochefort, 'Iles Antilles,' p. 418; J. G. Müller, 'Amer. Urrel.' p. 171, 217.
[3] Waitz, vol. ii. p. 182; J. L. Wilson, 'W. Afr.' p. 387; Steinhauser, l. c. p. 134. Compare Callaway, p. 327, etc.
[4] Bastian, 'Psychologie,' p. 75.
[5] Bastian, 'Oestl. Asien,' vol. iii. p. 275.
[6] Grimm, 'D. M.' p. 829; Rochholz, 'Deutscher Glaube,' part i. p. 92; Hanusch, 'Slaw. Mythus,' p. 247.

"«Ἅπαντι δαίμων ἀνδρὶ συμπαρίσταται
Εὐθὺς γενομένῳ μυσταγωγὸς τοῦ βίου.
'Αγαθός· κακὸν γὰρ δαίμον' οὐ νομιστέον
Εἶναι τὸν βίον βλάπτοντα χρηστόν. Πάντα γὰρ
Δεῖ ἀγαθὸν εἶναι τὸν Θεόν."

Ingrained in the Platonic system, the doctrine has its salient example in the warning spirit which Sokrates felt within him dissuading from wrong.[1] In the Roman world, the doctrine came to be accepted as a philosophy of human life. Each man had his " genius natalis," associated with him from birth to death, influencing his action and his fate, standing represented by its proper image as a lar among the household gods; and at weddings and joyous times, and especially on the anniversary of the birthday when genius and man began their united career, worship was paid with song and dance to the divine image, adorned with garlands, and propitiated with incense and libations of wine. The demon or genius was, as it were, the man's companion soul, a second spiritual ego. The Egyptian astrologer warned Antonius to keep far from the young Octavius, " for thy demon," said he, " is in fear of his; " and truly in after years that genius of Augustus had become an imperial deity, by whom Romans swore solemn oaths, not to be broken.[2] The doctrine which could thus personify the character and fate of the individual man, proved capable of a yet further development. Converting into animistic entities the inmost operations of the human mind, a dualistic philosophy conceived as attached to every mortal a good and an evil genius, whose efforts through life drew him backward and forward toward virtue and vice, happiness and misery. It was the kakodæmon of Brutus

[1] Menander, 205, ap. Clement. Stromat. v. Xenophon, Memorab. See Plotin. Ennead. iii. 4; Porphyr. Plotin.
[2] Paulus Diaconus : "*Genium* appellant Deum, qui vim obtineret rerum omnium *generandarum.*" Censorin. de Die Natali, 3 : "Eundem esse genium et larem, multi veteres memoriæ prodiderunt." Tibull. Eleg. i. 2, 7 ; Ovid. Trist. iii. 13, 18, v. 5, 10 ; Horat. Epist. ii. 1, 140, Od. iv. 11, 7. Appian. de Bellis Parth. p. 156. Tertullian. Apol. xxiii.

which appeared to him by night in his tent: "I am thy evil genius," it said, "we meet again at Philippi."[1] As we study the shapes which the attendant spirits of the individual man assumed in early and mediæval Christendom, it is plain that the good and evil angels contending for man from birth to death, the guardian angel watching and protecting him, the familiar spirit giving occult knowledge or serving with magic art, continue in principle, and even in detail, the philosophy of earlier culture. Such beings even take visible form. St. Frances had a familiar angel, not merely that domestic one that is given as a guardian to every man, but this was as it were a boy of nine years old, with a face more splendid than the sun, clad in a little white tunic; it was in after years that there came to her a second angel, with a column of splendour rising to the sky, and three golden palm-branches in his hands. Or such attendant beings, though invisible, make their presence evident by their actions, as in Calmet's account of that Cistercian monk whose familiar genius waited on him, and used to get his chamber ready when he was coming back from the country, so that people knew when to expect him home.[2] There is a pleasant quaintness in Luther's remark concerning guardian angels, that a prince must have a greater, stronger, wiser angel than a count, and a count than a common man.[3] Bishop Bull, in one of his vigorous sermons, thus sums up a learned argument: "I cannot but judge it highly probable, that every faithful person at least hath his particular good *Genius* or *Angel*, appointed by God over him, as the Guardian and Guide of his Life." But he

[1] Serv. in Virg. Æn. vi. 743: "Cum nascimur, duos genios sortimur: unus hortatur ad bona, alter depravat ad mala, quibus assistentibus post mortem aut asserimur in meliorem vitam, aut condemnamur in deteriorem." Horat. Epist. ii. 187; Valer. Max. i. 7; Plutarch. Brutus. See Pauly, 'Real-Encyclop.;' Smith's 'Dic. of Biog. & Myth.' s. v. 'genius.'
[2] Acta Sanctorum Bolland: S. Francisca Romana ix. Mart. Calmet, 'Dissertation,' ch. iv. xxx.; Bastian, 'Mensch,' vol. ii. pp. 140, 347, vol. ii. p. 10; Wright, 'St. Patrick's Purgatory,' p. 33.
[3] Rochholz, p. 93.

will not insist on the belief, provided that the general ministry of angels be accepted.[1] Swedenborg will go beyond this. "Every man," he says, "is attended by an associate spirit; for without such an associate, a man would be incapable of thinking analytically, rationally, and spiritually."[2] Yet in the modern educated world at large, this group of beliefs has passed into the stage of survival. The conception of the good and evil genius contending for man through life, indeed, perhaps never had much beyond the idealistic meaning which art and poetry still give it. The traveller in France may hear in our own day the peasant's salutation, "Bonjour à vous et à votre compagnie!" (*i.e.*, your guardian angel).[3] But at the birthday festivals of English children, how few are even aware of the historical sequence, plain as it is, from the rites of the classic natal genius and the mediæval natal saint! Among us, the doctrine of guardian angels is to be found in commentaries, and may be sometimes mentioned in the pulpit; but the once distinct conception of a present guardian spirit, acting on each individual man and interfering with circumstances on his behalf, has all but lost its old reality. The familiar demon which gave occult knowledge and did wicked work for the magician, and sucked blood from miserable hags by witch-teats, was two centuries ago as real to the popular mind as the alembic or the black cat with which it was associated. Now, it has been cast down to the limbo of unhallowed superstitions.

To turn from Man to Nature. General mention has been made already of the local spirits which belong to mountain and rock and valley, to well and stream and lake, in brief to those natural objects and places which in early ages aroused the savage mind to mythological ideas, such as modern poets in their altered intellectual atmosphere strive

---

[1] Bull, 'Sermons,' 2nd Ed. London, 1714, vol. ii. p. 506.
[2] Swedenborg, 'True Christian Religion,' 380. See also A. J. Davis, 'Philosophy of Spiritual Intercourse,' p. 38.
[3] D. Monnier, 'Traditions Populaires,' p. 7.

to reproduce. In discussing these imaginary beings, it is above all things needful to bring our minds into sympathy with the lower philosophy. Here we must seek to realize to the utmost the definition of the Nature-Spirits, to understand with what distinct and full conviction savage philosophy believes in their reality, to discern how, as living causes, they filled their places and did their daily work in the natural philosophy of primæval man. Seeing how the Iroquois at their festivals could thank the invisible aids or good spirits, and with them the trees, shrubs, and plants, the springs and streams, the fire and wind, the sun, moon, and stars—in a word, every object that ministered to their wants—we may judge what real personality they attached to the myriad spirits which gave animated life to the world around them.[1] The Gold Coast negro's generic name for a fetish-spirit is "wong;" these aerial beings dwell in temple-huts and consume sacrifices, enter into and inspire their priests, cause health and sickness among men, and execute the behests of the mighty Heaven-god. But part or all of them are connected with material objects, and the negro can say, " In this river, or tree, or amulet, there is a wong." But he more usually says, " This river, or tree, or amulet is a wong." Thus among the wongs of the land are rivers, lakes, and springs, districts of land, termite-hills, trees, crocodiles, apes, snakes, birds, and so on.[2] In a word, his conceptions of animating souls and presiding spirits as efficient causes of all nature are two groups of ideas which we may well find it hard to distinguish, for the sufficient reason that they are but varying developments of the same fundamental animism.

In the doctrine of nature-spirits among nations which have reached a higher grade of culture, we find at once traces of such primitive thought, and of its change under

[1] L. H. Morgan, 'Iroquois,' p. 64. Brebeuf in 'Rel. des Jes.' 1636, p. 107. See Schoolcraft, 'Tribes,' vol. iii. p. 337.
[2] Steinhauser, ' Religion des Negers,' in 'Magazin der Evang. Missionen,' Basel, 1856; No. 2, p. 127, etc.

new intellectual conditions. Knowing the thoughts of rude Turanian tribes of Siberia as to pervading spirits of nature, we are prepared to look for re-modelled ideas of the same class among a nation whose religion shows plain traces of evolution from the low Turanian stage. The archaic system of manes-worship and nature-worship, which survives as the state religion of China, fully recognizes the worship of the numberless spirits which pervade the universe. The belief in their personality is vouched for by the sacrifices offered to them. "One must sacrifice to the spirits," says Confucius, "as though they were present at the sacrifice." At the same time, spirits were conceived as embodied in material objects. Confucius says, again: "The action of the spirits, how perfect is it! Thou perceivest it, and yet seest it not! Incorporated or immembered in things, they cannot quit them. They cause men, clean and pure and better clothed, to bring them sacrifice. Many, many, are there of them, as the broad sea, as though they were above and right and left." Here are traces of such a primitive doctrine of personal and embodied nature-spirits, as is still at home in the religion of rude Siberian hordes. But it was natural that Chinese philosophers should find means of refining into mere ideality these ruder animistic creations. Spirit (shin), they tell us, is the fine or tender part in all the ten thousand things; all that is extraordinary or supernatural is called spirit; the unsearchable of the male and female principles is called spirit; he who knows the way of passing away and coming to be, he knows the working of spirit.[1]

The classic Greeks had inherited from their barbaric ancestors a doctrine of the universe essentially similar to that of the North American Indian, the West African, and the Siberian. We know, more intimately than the heathen religion of our own land, the ancient Greek scheme of nature-spirits impelling and directing by their personal power and will the functions of the universe; the ancient

[1] Plath, 'Religion der Alten Chinesen,' part i. p. 44.

Greek religion of nature, developed by imagination, adorned by poetry, and consecrated by faith. History records for our instruction, how out of the midst of this splendid and honoured creed there were evolved the germs of the new philosophy. Led by minuter insight and stricter reason, thoughtful Greeks began the piecemeal supersession of the archaic scheme, and set in movement the transformation of animistic into physical science, which thence pervaded the whole cultured world. Such, in brief, is the history of the doctrine of nature-spirits from first to last. Let us endeavour, by classifying some of its principal special groups, to understand its place in the history of the human intellect.

What causes volcanos? The Australians account for volcanic rocks by the tradition that the sulky underground "ingna" or demons made great fires and threw up red-hot stones.[1] The Kamchadals say that just as they themselves warm up their winter-houses, so the "kamuli" or mountain-spirits heat up the mountains in which they dwell, and fling the brands out of the chimney.[2] The Nicaraguans offered human sacrifices to Masaya or Popogatepec (Smoking-Mountain), by throwing the bodies into the crater. It seems as though it were a controlling deity, not the mountain itself, that they worshipped; for we read of the chiefs going to the crater, whence a hideous old naked woman came out and gave them counsel and oracle; at the edge were placed earthen vessels of food to please her, or to appease her when there was a storm or earthquake.[3] Thus animism provided a theory of volcanos, and so it was likewise with whirlpools and rocks. In the Vei country in West Africa, there is a dangerous rock on the Mafa river, which is never passed without offering a tribute to the spirit of the flood—a leaf of tobacco, a handful of rice, or

---

[1] Oldfield, 'Abor. of Austr.' in 'Tr. Eth. Soc.' vol. iii. p. 232.
[2] Steller, 'Kamtschatka,' pp. 47, 265.
[3] Oviedo, 'Nicaragua,' in Ternaux-Compans, part xiv. pp. 132, 160. Compare Catlin, 'N. A. Ind.' vol. ii. p. 169.

a drink of rum. An early missionary account of a rock-demon worshipped by the Huron Indians will show with what absolute personality savages can conceive such a being. In the hollow of a certain sacred rock, it is related, dwells an "oki" or spirit who can give success to travellers, wherefore they put tobacco into one of the cracks, and pray thus : "Demon who dwellest in this place, behold tobacco I present to thee; help us, keep us from shipwreck, defend us against our enemies, and vouchsafe that when we have made a good trade, we may return safe and sound to our village." Father Marquette relates how, travelling on a river in the then little known region of Western America, he was told of a dreadful place to which the canoe was just drawing near, where dwells a demon waiting to devour such as dare to approach; this terrific manitu proved on arrival to be some high rocks in the bend of the river, against which the current runs violently.[2] Thus the missionary found in living belief among the savage Indians the very thought which had so long before passed into the classic tale of Skylla and Charybdis.

In those moments of the civilized man's life when he casts off hard dull science, and returns to childhood's fancy, the world-old book of animated nature is open to him anew. Then the well-worn thoughts come back fresh to him, of the stream's life that is so like his own; once more he can see the rill leap down the hillside like a child, to wander playing among the flowers; or can follow it as, grown to a river, it rushes through a mountain gorge, henceforth in sluggish strength to carry heavy burdens across the plain. In all that water does, the poet's fancy can discern its personality of life. It gives fish to the fisher, and crops to the husbandman; it swells in fury and lays waste the land; it grips the bather with chill

[1] Creswick, 'Veys,' in 'Tr. Eth. Soc.' vol. vi. p. 359. See Du Chaillu, 'Ashango-land,' p. 106.
[2] Brebeuf in 'Rel. des Jes.' 1636, p. 108. Long's Exp. vol. i. p. 46. See Loskiel, 'Indians of N. A.' part i. p. 45.

and cramp, and holds with inexorable grasp its drowning victim :[1]

"Tweed said to Till,
'What gars ye rin sae still?'
Till said to Tweed,
'Though ye rin wi' speed,
And I rin slaw,
Yet, where ye drown ae man,
I drown twa.'"

What ethnography has to teach of that great element of the religion of mankind, the worship of well and lake, brook and river, is simply this—that what is poetry to us was philosophy to early man; that to his mind water acted not by laws of force, but by life and will; that the water-spirits of primæval mythology are as souls which cause the water's rush and rest, its kindness and its cruelty; that lastly man finds, in the beings which with such power can work him weal and woe, deities with a wider influence over his life, deities to be feared and loved, to be prayed to and praised and propitiated with sacrificial gifts.

In Australia, special water-demons infest pools and watering-places. In the native theory of disease and death, no personage is more prominent than the water-spirit, which afflicts those who go into unlawful pools or bathe at unlawful times, the creature which causes women to pine and die, and whose very presence is death to the beholder, save to the native doctors, who may visit the water-spirit's subaqueous abode and return with bleared eyes and wet clothes to tell the wonders of their stay.[2] It would seem that creatures with such attributes come naturally into the category of spiritual beings, but already among the rude natives of Australia and Van Diemen's

[1] For details of the belief in water-spirits as the cause of drowning, see *ante*, vol. i. p. 109.
[2] Oldfield in 'Tr. Eth. Soc.' vol. iii. p. 328; Eyre, vol. ii. p. 362; Grey, vol. ii. p. 339; Bastian, 'Vorstellungen von Wasser und Feuer,' in 'Zeitschrift ür Ethnologie,' vol. i. (contains a general collection of details as to water-worship).

Land, in such stories as that of the bunyip which carries off the native women to his retreat below the waters, there appears that confusion between the spiritual water-demon and the material water-monster, which runs on into the midst of European mythology in such conceptions as that of the water-kelpie and the sea-serpent.[1] America gives cases of other principal animistic ideas concerning water. The water has its own spirits, writes Cranz, among the Greenlanders, so when they come to an untried spring, an angekok or the oldest man must drink first, to free it from a harmful spirit.[2] "Who makes this river flow?" asks the Algonquin hunter in a medicine-song, and his answer is, "The spirit, he makes this river flow." In any great river, or lake, or cascade, there dwell such spirits, looked upon as mighty manitus. Thus Carver mentions the habit of the Red Indians, when they reached the shores of Lake Superior or the banks of the Mississippi, or any other great body of water, to present to the spirit who resides there some kind of offering; this he saw done by a Winnebago chief who went with him to the Falls of St. Anthony. Franklin saw a similar sacrifice made by an Indian, whose wife had been afflicted with sickness by the water-spirits, and who accordingly to appease them tied up in a small bundle a knife and a piece of tobacco and some other trifling articles, and committed them to the rapids.[3] On the river-bank, the Peruvians would scoop up a handful of water and drink it, praying the river-deity to let them cross or to give them fish, and they threw maize into the stream as a propitiatory offering; even to this day the Indians of the Cordilleras perform the ceremonial sip before they will pass a river on foot or horseback.[4] Africa displays well the

[1] Compare Bonwick, 'Tasmanians,' p. 203, and Taylor, 'New Zealand,' p. 48, with Forbes Leslie, Brand, &c.
[2] Cranz, 'Grönland,' p. 267.
[3] Tanner, 'Narr.' p 341; Carver, 'Travels,' p. 383; Franklin, 'Journey to Polar Sea,' vol. ii. p. 245; Lubbock, 'Origin of Civilization,' pp. 213–20 (contains details as to water-worship); see Brinton, p. 124.
[4] Rivero and Tschudi, 'Peruvian Ant.' p. 161; Garcilaso de la Vega,

rites of water-worship. In the East, among the Wanika, every spring has its spirit, to which oblations are made; in the West, in the Akra district, lakes, ponds, and rivers received worship as local deities. In the South, among the Kafirs, streams are venerated as personal beings, or the abodes of personal deities, as when a man crossing a river will ask leave of its spirit, or having crossed will throw in a stone; or when the dwellers by a stream will sacrifice a beast to it in time of drought, or, warned by illness in the tribe that their river is angry, will cast into it a few handfuls of millet or the entrails of a slaughtered ox.[1] Not less strongly marked are such ideas among the Tatar races of the North. Thus the Ostyaks venerate the river Ob, and when fish is scanty will hang a stone about a rein-deer's neck and cast it in for a sacrifice. Among the Buraets, who are professing Buddhists, the old worship may still be seen at the picturesque little mountain lake of Ikeougoun, where they come to the wooden temple on the shore to offer sacrifices of milk and butter and the fat of the animals which they burn on the altars. So across in Northern Europe, almost every Esthonian village has its sacred sacrificial spring. The Esths could at times even see the churl with blue and yellow stockings rise from the holy brook Wöhhanda, no doubt that same spirit of the brook to whom in older days there were sacrificed beasts and little children; in newer times, when a German landowner dared to build a mill and dishonour the sacred water, there came bad seasons that lasted year after year, and the country people burned down the abominable thing.[2] As for the water-worship prevailing among non-Aryan indigenes of British India, it

'Comm. Real.' i. 10. See also J. G. Müller, 'Amer. Urrelig.' pp. 258, 260, 282.
[1] Krapf, 'E. Afr.' p. 198; Steinhauser, l. c. p. 131; Villault in Astley, vol. i. p. 668; Backhouse, 'Afr.' p. 230; Callaway, 'Zulu Tales, vol. i. p. 90; Bastian, l. c.
[2] Castrén, 'Vorlesungen über die Altaischen Völker,' p. 114. 'Finn. Myth.' p. 70. Atkinson, 'Siberia,' p. 444. Boecler, 'Ehsten Abergläub. Gebräuche,' ed. Kreutzwald, p. 6.

seems to reach its climax among the Bodo and Dhimal of the North-East, tribes to whom the local rivers are the local deities,[1] so that men worship according to their water-sheds, and the map is a pantheon.

Nor is such reverence strange to Aryan nations. To the modern Hindu, looking as he still does on a river as a living personal being to be adored and sworn by, the Ganges is no solitary water deity, but only the first and most familiar of the long list of sacred streams.[2] Turn to the classic world, and we but find the beliefs and and rites of a lower barbaric culture holding their place, consecrated by venerable antiquity and glorified by new poetry and art. To the great Olympian assembly in the halls of cloud-compelling Zeus, came the Rivers, all save Ocean, and thither came the nymphs who dwell in lovely groves and at the springs of streams, and in the grassy meads; and they sate upon the polished seats:—

"Οὔτε τις οὖν Ποταμῶν ἀπέην, νόσφ' Ὠκεανοῖο,
Οὔτ' ἄρα Νυμφάων ταί τ' ἄλσεα καλὰ νέμονται,
Καὶ πηγὰς ποταμῶν, καὶ πίσεα ποιήεντα.
Ἐλθόντες δ' ἐς δῶμα Διὸς νεφεληγερέταο,
Ξεστῆς αἰθούσησιν ἐφίζανον, ἃς Διῒ πατρὶ
Ἥφαιστος ποίησεν ἰδυίῃσι πραπίδεσσιν."

Even against Hephaistos the Fire-god, a River-god dared to stand opposed, deep-eddying Xanthos, called of men Skamandros. He rushed down to overwhelm Achilles and bury him in sand and slime, and though Hephaistos prevailed against him with his flames, and forced him, with the fish skurrying hither and thither in his boiling waves and the willows scorched upon his banks, to rush on no more but stand, yet at the word of white-armed Here, that it was not fit for mortals' sake to handle so roughly an immortal god, Hephaistos quenched his furious fire, and the returning flood sped again along his channel:—

[1] Hodgson, 'Abor. of India,' p. 164; Hunter, 'Rural Bengal,' p. 184. See also Lubbock, l. c.; Forbes Leslie, 'Early Races of Scotland,' vol. i. p. 163, vol. ii. p. 497.
[2] Ward, 'Hindoos,' vol. ii. p. 206, etc.

"'Ἥφαιστε, σχέο, τέκνον ἀγακλέες· οὐ γὰρ ἔοικεν
ἀθάνατον θεὸν ὧδε βροτῶν ἕνεκα στυφελίζειν.
Ὣς ἔφαθ'· Ἥφαιστος δὲ κατέσβεσε θεσπιδαὲς πῦρ·
ἄψορρον δ' ἄρα κῦμα κατέσσυτο καλὰ ῥέεθρα."

To beings thus conceived in personal divinity, full worship was given. Odysseus invokes the river of Scheria; Skamandros had his priest and Spercheios his grove; and sacrifice was done to the rival of Herakles, the river-god Acheloos, eldest of the three thousand river-children of old Okeanos.[1] Through the ages of the classic world, the river-gods and the water-nymphs held their places, till within the bounds of Christendom they came to be classed with ideal beings like them in the mythology of the northern nations, the kindly sprites to whom offerings were given at springs and lakes, and the treacherous nixes who entice men to a watery death. In times of transition, the new Christian authorities made protest against the old worship, passing laws to forbid adoration and sacrifice to fountains—as when Duke Bretislav forbade the still halfpagan country folk of Bohemia to offer libations and sacrifice victims at springs,[2] and in England Ecgbert's Poenitentiale proscribes the like rites, "if any man vow or bring his offerings to any well" "if one hold his vigils at any well."[3] But the old veneration was too strong to be put down, and with a varnish of Christianity and sometimes the substitution of a saint's name, water-worship has held its own to our day. The Bohemians will go to pray on the river-bank where a man has been drowned, and there they will cast in an offering, a loaf of new bread and a pair of

---

[1] Homer. ll. xx. xxi. See Gladstone, 'Juventus Mundi,' pp. 190, 345, etc. etc.

[2] Cosmas, book iii. p. 197, "superstitiosas institutiones, quas villani adhuc semipagani in Pentecosten tertia sive quarta feria observabant offerentes libamina super fontes mactabant victimas et dæmonibus immolabant."

[3] Poenitentiale Ecgberti, ii. 22, "gif hwilc man his ælmessan gehâte oththe bringe to hwilcon wylle;" iv. 19, "gif hwâ his wæccan æt ænigum wylle hæbbe." Grimm, 'D. M.' p. 549, etc. See Hyltén-Cavallius, 'Wärend och Wirdarne,' part i. pp. 131, 171 (Sweden).

wax-candles. On Christmas Eve they will put a spoonful of each dish on a plate, and after supper throw the food into the well, with an appointed formula, somewhat thus:—

> "House-father gives thee greeting,
> Thee by me entreating:
> Springlet, share our feast of Yule,
> But give us water to the full;
> When the land is plagued with drought,
> Drive it with thy well-spring out." [1]

It well shows the unchanged survival of savage thought in modern peasants' minds, to find still in Slavonic lands the very same fear of drinking a harmful spirit in the water, that has been noticed among the Esquimaux. It is a sin for a Bulgarian not to throw some water out of every bucket brought from the fountain; some elemental spirit might be floating on the surface, and if not thrown out, might take up his abode in the house, or enter into the body of some one drinking from the vessel.[2] Elsewhere in Europe, the list of still existing water-rites may be extended. The ancient lake-offerings of the South of France seem not yet forgotten in La Lozère, the Bretons venerate as of old their sacred springs, and Scotland and Ireland can show in parish after parish the sites and even the actual survivals of such observance at the holy wells. Perhaps Welshmen no longer offer cocks and hens to St. Tecla at her sacred well and church of Llandegla, but Cornish folk still drop into the old holy-wells offerings of pins, nails, and rags, expecting from their waters cure for disease, and omens from their bubbles as to health and marriage.[3]

The spirits of the tree and grove no less deserve our

[1] Grohmann, 'Aberglauben aus Böhmen and Mähren,' p. 43, etc. Hanusch, 'Slaw. Myth.' p. 291, etc. Ralston, 'Songs of Russian People,' p. 139, etc.
[2] St. Clair and Brophy, 'Bulgaria,' p. 46. Similar ideas in Grohmann, p. 44. Eisenmenger, 'Entd. Judenthum,' part i. p. 426.
[3] Maury, 'Magie,' etc. p. 158. Brand, 'Pop. Ant.' vol. ii. p. 366, etc. Hunt, 'Pop. Rom. 2nd Series,' p. 40, etc. Forbes Leslie, 'Early Races of Scotland,' vol. i. p. 156, etc.

study for their illustrations of man's primitive animistic theory of nature. This is remarkably displayed in that stage of thought where the individual tree is regarded as a conscious personal being, and as such receives adoration and sacrifice. Whether such a tree is looked on as inhabited, like a man, by its own proper life or soul, or as possessed, like a fetish, by some other spirit which has entered it and uses it for a body, is often hard to determine. Shelley's lines well express a doubting conception familiar to old barbaric thought—

"Whether the sensitive plant, or that
Which within its boughs like a spirit sat
Ere its outward form had known decay,
Now felt this change, I cannot say."

But this vagueness is yet again a proof of the principle which I have confidently put forward here, that the conceptions of the inherent soul and of the embodied spirit are but modifications of one and the same deep-lying animistic thought. The Mintira of the Malay Peninsula believe in "hantu kayu," *i. e.* "tree-spirits," or "tree-demons," which frequent every species of tree, and afflict men with diseases; some trees are noted for the malignity of their demons.[1] Among the Dayaks of Borneo, certain trees possessed by spirits must not be cut down; if a missionary ventured to fell one, any death that happened afterwards would naturally be set down to this crime.[2] The belief of certain Malays of Sumatra is expressly stated, that certain venerable trees are the residence, or rather the material frame, of spirits of the woods.[3] In the Tonga Islands, we hear of natives laying offerings at the foot of particular trees, with the idea of their being inhabited by spirits.[4] So in America, the Ojibwa medicine-man has heard the tree utter its complaint

[1] 'Journ. Ind. Archip.' vol. i. p. 307.
[2] Beeker, Dyaks, in 'Journ. Ind. Archip.' vol. iii. p. 111.
[3] Marsden, 'Sumatra,' p. 301.
[4] S. S. Farmer, 'Tonga,' p. 127.

when wantonly cut down.[1] A curious and suggestive description bearing on this point is given in Friar Roman Pane's account of the religion of the Antilles islanders, drawn up by order of Columbus. Certain trees, he declares, were believed to send for sorcerers, to whom they gave orders how to shape their trunks into idols, and these "cemi" being then installed in temple-huts, received prayer and inspired their priests with oracles.[2] Africa shows as well-defined examples. The negro woodman cuts down certain trees in fear of the anger of their inhabiting demons, but he finds his way out of the difficulty by a sacrifice to his own good genius, or, when he is giving the first cuts to the great asorin-tree, and its indwelling spirit comes out to chase him, he cunningly drops palm-oil on the ground, and makes his escape while the spirit is licking it up.[3] A negro was once worshipping a tree with an offering of food, when some one pointed out to him that the tree did not eat; the negro answered, "O the tree is not fetish, the fetish is a spirit and invisible, but he has descended into this tree. Certainly he cannot devour our bodily food, but he enjoys its spiritual part and leaves behind the bodily which we see."[4] Tree-worship is largely prevalent in Africa, and much of it may be of this fully animistic kind; as where in Whidah Bosman says that "the trees, which are the gods of the second rank of this country, are only prayed to and presented with offerings in time of sickness, more especially fevers, in order to restore the patients to health;"[5] or where in Abyssinia the Gallas made pilgrimage from all quarters to their sacred tree Wodanabe on the banks of the Hawash, worshipping it and praying to it for riches, health, life, and every blessing.[6]

[1] Bastian, 'Der Baum in vergleichender Ethnologie,' in Lazarus and Steinthal's 'Zeitschrift für Völkerpsychologie,' etc. vol. v. 1868.
[2] Chr. Colombo, ch. xix.; and in Pinkerton, vol. xii. p. 87.
[3] Burton, 'W. & W. fr. W. Afr.' pp. 205, 243.
[4] Waitz, vol. ii. p. 188.
[5] Bosman, letter 19, and in Pinkerton, vol. xvi. p. 500.
[6] Krapf, 'E. Afr.' p. 77; Prichard, 'N. H. of Man,' p. 290; Waitz, vol. ii. p. 518. See also Merolla, 'Congo,' in Pinkerton, vol. xvi. p. 236.

The position of tree-worship in Southern Asia in relation to Buddhism is of particular interest. To this day there are districts of this region, Buddhist or under strong Buddhist influence, where tree-worship is still displayed with absolute clearness of theory and practice. Here in legend a dryad is a being capable of marriage with a human hero, while in actual fact a tree-deity is considered human enough to be pleased with dolls set up to swing in the branches. The Talein of Birmah, before they cut down a tree, offer prayers to its "kaluk" (*i.q.*, "kelah"), its inhabiting spirit or soul. The Siamese offer cakes and rice to the takhien-tree before they fell it, and believe the inhabiting nymphs or mothers of trees to pass into guardian-spirits of the boats built of their wood, so that they actually go on offering sacrifice to them in this their new condition.[1] These people have indeed little to learn from any other race, however savage, of the principles of the lower animism. The question now arises, did such tree-worship belong to the local religions among which Buddhism established itself? There is strong evidence that this was the case. Philosophic Buddhism, as known to us by its theological books, does not include trees among sentient beings possessing mind, but it goes so far as to acknowledge the existence of the "dewa" or genius of a tree. Buddha, it is related, told a story of a tree crying out to the brahman carpenter who was going to cut it down, "I have a word to say, hear my word!" but then the teacher goes on to explain that it was not really the tree that spoke, but a dewa dwelling in it. Buddha himself was a tree-genius forty-three times in the course of his transmigrations. Legend says that during one such existence, a certain brahman used to pray for protection to the tree which Buddha was attached to; but the transformed teacher reproved the tree-worshipper for thus

[1] Bastian, 'Oestl. Asien,' vol. ii. pp. 457, 461, vol. iii. pp. 187, 251, 289, 497. For details of tree-worship from other Asiatic districts, see Ainsworth, 'Yezidis,' in 'Tr. Eth. Soc.' vol. i. p. 23 ; Jno. Wilson, 'Parsi Religion,' p. 262.

addressing himself to a senseless thing, which hears and knows nothing.[1] As for the famous Bo tree, its miraculous glories are not confined to the ancient Buddhist annals; for its surviving descendant, grown from the branch of the parent tree sent by King Asoka from India to Ceylon in the 3rd century B.C., to this day receives the worship of the pilgrims who come by thousands to do it honour, and offer prayer before it. Beyond these hints and relics of the old worship, however, Mr. Fergusson's recent investigations, published in his "Tree and Serpent Worship," have brought to light an ancient state of things which the orthodox Buddhist literature gives little idea of. It appears from the sculptures of the Sanchi tope in Central India, that in the Buddhism of about the 1st century A.D., sacred trees had no small place as objects of authorized worship. It is especially notable that the representatives of indigenous race and religion in India, the Nagas, characterized by their tutelary snakes issuing from their backs between their shoulders and curving over their heads, and other tribes actually drawn as human apes, are seen adoring the divine tree in the midst of unquestionable Buddhist surroundings.[2] Tree-worship, even now well marked among the indigenous tribes of India, was obviously not abolished on the Buddhist conversion. The new philosophic religion seems to have amalgamated, as new religions ever do, with older native thoughts and rites. And it is quite consistent with the habits of the Buddhist theologians and hagiologists, that when tree-worship was suppressed, they should have slurred over the fact of its former prevalence, and should even have used the recollection of it as a gibe against the hostile Brahmans.

Conceptions like those of the lower races in character, and rivalling them in vivacity, belong to the mythology of Greece and Rome. The classic thought of the tree inhabited by a deity and uttering oracles, is like that of other

[1] Hardy, 'Manual of Budhism,' pp. 100, 443.
[2] Fergusson, 'Tree and Serpent Worship,' pl. xxiv. xxvi. etc.

regions. Thus the sacred palm of Negra in Yemen, whose demon was propitiated by prayer and sacrifice to give oracular response,[1] or the tall oaks inhabited by the gods, where old Slavonic people used to ask questions and hear the answers,[2] have their analogue in the prophetic oak of Dodona, wherein dwelt the deity, "ναῖεν δ' ἐνὶ πυθμένι φηγοῦ."[3] The Homeric hymn to Aphrodite tells of the tree-nymphs, long-lived yet not immortal—they grow with their high-topped leafy pines and oaks upon the mountains, but when the lot of death draws nigh, and the lovely trees are sapless, and the bark rots away and the branches fall, then their spirits depart from the light of the sun:—

"Νύμφαι μιν θρέψουσιν ὀρεσκῷοι βαθύκολποι,
αἱ τόδε ναιετάουσιν ὄρος μέγα τε ζάθεόν τε·
αἵ ῥ' οὔτε θνητοῖς οὔτ' ἀθανάτοισιν ἔπονται·
δηρὸν μὲν ζώουσι καὶ ἄμβροτον εἶδαρ ἔδουσι,
καί τε μετ' ἀθανάτοισι καλὸν χορὸν ἐρρώσαντο.
τῇσι δὲ Σειληνοί τε καὶ εὔσκοπος Ἀργειφόντης
μίσγοντ' ἐν φιλότητι μυχῷ σπείων ἐροέντων.
τῇσι δ' ἅμ' ἢ ἐλάται ἠὲ δρύες ὑψικάρηνοι
γεινομένῃσιν ἔφυσαν ἐπὶ χθονὶ βωτιανείρῃ,
καλαί, τηλεθάουσαι, ἐν οὔρεσιν ὑψηλοῖσιν·

. . . . .

ἀλλ' ὅτε κεν δὴ μοῖρα παρεστήκῃ θανάτοιο,
ἀζάνεται μὲν πρῶτον ἐπὶ χθονὶ δένδρεα καλά,
φλοιὸς δ' ἀμφιπεριφθινύθει, πίπτουσι δ' ἀπ' ὄζοι,
τῶν δέ θ' ὁμοῦ ψυχὴ λείπει φάος ἠελίοιο."[4]

The hamadryad's life is bound to her tree, she is hurt when it is wounded, she cries when the axe threatens, she dies with the fallen trunk:—

"Non sine hamadryadis fato cadit arborea trabs."[5]

How personal a creature the tree-nympth was to the classic mind, is shown in legends like that of Paraibios,

[1] Tabary in Bastian, l. c. p. 295.
[2] Hartknoch, 'Alt-und Neues Preussen,' part i. ch. v.
[3] See Pauly, 'Real-Encyclopedie.' Homer. Odyss. xiv. 327, xix. 296.
[4] Hymn. Homer. Aphrod. 257.
[5] Ausonii Idyll. De Histor. 7.

whose father, regardless of the hamadryad's entreaties, cut down her ancient trunk, and in himself and in his offspring suffered her dire vengeance.[1] The ethnographic student finds a curious interest in transformation-myths like Ovid's, keeping up as they do vestiges of philosophy of archaic type—Daphne turned into the laurel that Apollo honours for her sake, the sorrowing sisters of Phaethon changing into trees, yet still dropping blood and crying for mercy when their shoots are torn.[2] Such episodes mediæval poetry could still adapt, as in the pathless infernal forest whose knotted dusk-leaved trees revealed their human animation to the Florentine when he plucked a twig,

> "Allor porsi la mano un poco avante,
> E colsi un ramoscel da un gran pruno:
> E' l tronco suo gridò: Perchè mi schiante?"[3]

or the myrtle to which Ruggiero tied his hippogriff, who tugged at the poor trunk till it murmured and oped its mouth, and with doleful voice told that it was Astolfo, enchanted by the wicked Alcina among her other lovers,

> "D' entrar o in fera o in fonte o in legno o in sasso."[4]

If these seem to us now conceits over quaint for beauty, we need not scruple to say so. They are not of Dante and Ariosto, they are sham antiques from classic models. And if even the classic originals have become unpleasing, we need not perhaps reproach ourselves with decline of poetic taste. We have lost something, and the loss has spoiled our appreciation of many an old poetic theme, yet it is not always our sense of the beautiful that has dwindled, but the old animistic philosophy of nature that is gone from us, dissipating from such fancies their meaning, and with

---

[1] Apollon. Rhod. Argonautica, ii. 476. See Welcker, 'Griech. Götterl. vol. iii. p. 57.
[2] Ovid. Metamm. i. 452, ii. 345, xi. 67.
[3] Dante, 'Divina Commedia,' 'Inferno,' canto xiii.
[4] Ariosto, 'Orlando Furioso,' canto vi.

their meaning their loveliness. Still, if we look for living men to whom trees are, as they were to our distant forefathers, the habitations and embodiments of spirits, we shall not look in vain. The peasant folklore of Europe still knows of willows that bleed and weep and speak when hewn, of the fairy maiden that sits within the fir-tree, of that old tree in Rugaard forest that must not be felled, for an elf dwells within, of that old tree on the Heinzenberg near Zell, which uttered its complaint when the woodman cut it down, for in it was Our Lady, whose chapel now stands upon the spot.[1] One may still look on where Franconian damsels go to a tree on St. Thomas's day, knock thrice solemnly, and listen for the indwelling spirit to give answer by raps from within, what manner of husbands they are to have.[2]

In the remarkable document of mythic cosmogony, preserved by Eusebius under the alleged authorship of the Phœnician Sanchoniathon, is the following passage : " But these first men consecrated the plants of the earth, and judged them gods, and worshipped the things upon which they themselves lived and their posterity, and all before them, and (to these) they made libations and sacrifices."[3] From examples such as have been here reviewed, it seems that direct and absolute tree-worship of this kind may indeed lie very wide and deep in the early history of religion. But the whole tree-cultus of the world must by no means be thrown indiscriminately into this one category. It is only on such distinct evidence as has been here put forward, that a sacred tree may be taken as having a spirit embodied in or attached to it. Beyond this limit, there is a wider range of animistic conceptions connected with tree and forest worship. The tree may be the spirit's perch or shelter or favourite haunt. Under this definition come the

[1] Grimm, 'D. M.' p. 615, etc. Bastian, 'Der Baum,' l. c. p. 297 ; Hanusch, 'Slaw. Myth.' p. 313.
[2] Wuttke, 'Volksaberglaube,' p. 57, see 183.
[3] Euseb. 'Præp. Evang.' i. 10.

trees hung with objects which are the receptacles of disease-spirits. As places of spiritual resort, there is no real distinction between the sacred tree and the sacred grove. The tree may serve as a scaffold or altar, at once convenient and conspicuous, where offerings can be set out for some spiritual being, who may be a tree-spirit, or perhaps the local deity, living there just as a man might do who had his hut and owned his plot of land around. The shelter of some single tree, or the solemn seclusion of a forest grove, is a place of worship set apart by nature, of some tribes the only temple, of many tribes perhaps the earliest. Lastly, the tree may be merely a sacred object patronized by or associated with or symbolizing some divinity, often one of those which we shall presently notice as presiding over a whole species of trees or other things. How all these conceptions, from actual embodiment or local residence or visit of a demon or deity, down to mere ideal association, can blend together, how hard it often is to distinguish them, and yet how in spite of this confusion they conform to the animistic theology in which all have their essential principles, a few examples will show better than any theoretical comment.[1] Take the groups of malicious wood-fiends so obviously devised to account for the mysterious influences that beset the forest wanderer. In the Australian bush, demons whistle in the branches, and stooping with outstretched arms sneak among the trunks to seize the wayfarer; the lame demon leads astray the hunter in the Brazilian forest; the Karen crossing a fever-haunted jungle shudders in the grip of the spiteful "phi," and runs to lay an offering by the tree he rested under last, from whose boughs the malaria-fiend came down upon him; the negro of Senegambia seeks to pacify the long-haired tree-demons that send diseases; the terrific cry of the wood-demon is heard in the Finland

---

[1] Further details as to tree-worship in Bastian, 'Der Baum,' etc. here cited; Lubbock, 'Origin of Civilization,' p. 206, etc.; Fergusson, 'Tree and Serpent Worship,' etc.

forest; the baleful shapes of terror that glide at night through our own woodland are familiar still to peasant and poet.[1] The North American Indians of the Far West, entering the defiles of the Black Mountains of Nebraska, will often hang offerings on the trees or place them on the rocks, to propitiate the spirits and procure good weather and hunting.[2] In South America, Mr. Darwin describes the Indians offering their adorations by loud shouts when they came in sight of the sacred tree standing solitary on a high part of the Pampas, a landmark visible from afar. To this tree were hanging by threads numberless offerings such as cigars, bread, meat, pieces of cloth, &c., down to the mere thread pulled from his poncho by the poor wayfarer who had nothing better to give. Men would pour libations of spirits and maté into a certain hole, and smoke upwards to gratify Walleechu, and all around lay the bleached bones of the horses slaughtered as sacrifices. All Indians made their offerings here, that their horses might not tire, and that they themselves might prosper. Mr. Darwin reasonably judges on this evidence that it was to the deity Walleechu that the worship was paid, the sacred tree being only his altar; but he mentions that the Gauchos think the Indians consider the tree as the god itself, a good example of the misunderstanding possible in such cases.[3] The New Zealanders would hang an offering of food or a lock of hair on a branch at a landing place, or near remarkable rocks or trees would throw a bunch of rushes as an offering to the spirit dwelling there.[4] The Dayaks fasten rags of their clothes on trees at cross roads, fearing for their health if they neglect the custom;[5] the Macassar man halting to eat in the forest will put a morsel of rice or fish on a leaf, and lay it on a stone or stump.[6] The divinities of African tribes

[1] Bastian, 'Der Baum,' l. c. etc.
[2] Irving, 'Astoria,' vol. ii. ch. viii.
[3] Darwin, 'Journal,' p. 68.
[4] Polack, 'New Z.' vol. ii. p. 6; Taylor, p. 171, see 99.
[5] St. John, 'Far East,' vol. i. p. 89.
[6] Wallace, 'Eastern Archipelago,' vol. i. p. 338.

may dwell in trees remarkable for size and age, or inhabit sacred groves where the priest alone may enter.[1] Trees treated as idols by the Congo people, who put calabashes of palm wine at their feet in case they should be thirsty;[2] and among West African negro tribes father north, trees hung with rags by the passers-by, and the great baobabs pegged to hang offerings to, and serving as shrines before which sheep are sacrificed,[3] display well the rites of tree sacrifice, though leaving undefined the precise relation conceived between deity and tree.

The forest theology that befits a race of hunters is dominant still among Turanian tribes of Siberia, as of old it was across to Lapland. Full well these tribes know the gods of the forest. The Yakuts hang on any remarkably fine tree iron, brass, and other trinkets; they choose a green spot shaded by a tree for their spring sacrifice of horses and oxen, whose heads are set up in the boughs; they chant their extemporised songs to the Spirit of the Forest, and hang for him on the branches of the trees along the roadside offerings of horsehair, emblems of their most valued possession. A clump of larches on a Siberian steppe, a grove in the recesses of a forest, is the sanctuary of a Turanian tribe. Gaily-decked idols in their warm fur-coats, each set up beneath its great tree swathed with cloth or tinplate, endless reindeer-hides and peltry hanging to the trees around, kettles and spoons and snuff-horns and household valuables strewn as offerings before the gods—such is the description of a Siberian holy grove, at the stage when the contact of foreign civilization has begun by ornamenting the rude old ceremonial it must end by abolishing.[4] A race ethnologically allied to these tribes, though risen to higher culture, kept up remarkable relics of tree-worship in Northern Europe. In Esthonian districts, within the pre-

[1] Prichard, 'Nat. Hist. of Man,' p. 531.
[2] Merolla in Pinkerton, vol. xvi. p. 236.
[3] Lubbock, p. 193; Bastian, l. c.; Park, 'Travels,' vol. i. pp. 64, 106.
[4] Castrén, 'Finn. Myth.' p. 86, etc. 191, etc.; Latham, 'Descr. Eth.' vol. i. p. 368; Simpson, 'Journey,' vol. ii. p. 261.

sent century, the traveller might often see the sacred tree, generally an ancient lime, oak, or ash, standing inviolate in a sheltered spot near the dwelling-house, and old memories are handed down of the time when the first blood of a slaughtered beast was sprinkled on its roots, that the cattle might prosper, or when an offering was laid beneath the holy linden, on the stone where the worshipper knelt on his bare knees moving from east to west and back, which stone he kissed thrice when he had said, " Receive the food as an offering!" It may well have been an indwelling tree-deity for whom this worship was intended, for folklore shows that the Esths recognized such a conception with the utmost distinctness; they have a tale of the tree-elf who appeared in personal shape outside his crooked birch-tree, whence he could be summoned by three knocks on the trunk and the inquiry, " Is the crooked one at home?" But also it may have been the Wood-Father or Tree-King, or some other deity, who received sacrifice and answered prayer beneath his sacred tree, as in a temple.[1] If, again, we glance at the tree-and-grove worship of the non-Aryan indigenous tribes of British India, we shall gather clear and instructive hints of its inner significance. In the courtyard of a Bodo house is planted the sacred " sij " or euphorbia of Batho, the national god, to whom under this representation the " deoshi " or priest offers prayer and kills a pig.[2] When the Khonds settle a new village, the sacred cotton-tree must be planted with solemn rites, and beneath it is placed the stone which enshrines the village deity.[3] Nowhere, perhaps, in the world in these modern days is the original meaning of the sacred grove more picturesquely shown than among the Mundas of Chota-Nagpur, in whose settlements a sacred grove of sal-trees, a remnant of the primæval forest spared by the woodman's axe, is left as a home for the

[1] Boecler, 'Ehsten Abergläubische Gebräuche,' etc. ed. Kreutzwald, pp. 2, 112, 146.
[2] Hodgson, 'Abor. of India,' pp. 165, 173.
[3] Macpherson, p. 61.

spirits, and in this hallowed place offerings to the gods are made.[1]

Here, then, among the lower races, is surely evidence enough to put on their true historic footing the rites of tree and grove which we find flourishing or surviving within the range of Semitic or Aryan culture. Mentions in the Old Testament record the Canaanitish Ashera-worship, the sacrifice under every green tree, the incense rising beneath oak and willow and shady terebinth, rites whose obstinate revival proves how deeply they were rooted in the old religion of the land.[2] The evidence of these Biblical passages is corroborated by other evidence from Semitic regions, as in the lines by Silius Italicus which mention the prayer and sacrifice in the Numidian holy groves, and the records of the council of Carthage which show that in the 5th century, an age after Augustine's time, it was still needful to urge that the relics of idolatry in trees and groves should be done away.[3] From the more precise descriptions which lie within the range of Aryan descent and influence, examples may be drawn to illustrate every class of belief and rite of the forest. Modern Hinduism is so largely derived from the religions of the non-Aryan indigenes, that we may fairly explain thus a considerable part of the tree-worship of modern India, as where in the Birbhûm district of Bengal a great annual pilgrimage is made to a shrine in the jungle, to make offerings of rice and money and sacrifice animals to a certain ghost who dwells in a bela-tree.[4] In thoroughly Hindu districts we may see the pippala (Ficus religiosa) planted as the village tree, the "chaityataru" of Sanskrit

---

[1] Dalton, 'Kols,' in 'Tr. Eth. Soc.' vol. vi. p. 34. Bastian, 'Oestl. Asien.' vol. i. p. 134, vol. iii. p. 252.
[2] Deut. xii. 3; xvi. 21. Judges vi. 25. 1 Kings xiv. 23; xv. 13; xviii. 19. 2 Kings xvii. 10; xxiii. 4. Is. lvii. 5. Jerem. xvii. 2. Ezek. vi. 13; xx. 28. Hos. iv. 13, etc. etc.
[3] Sil. Ital. Punica, iii. 675, 690. Harduin, Acta Conciliorum, vol. i. For further evidence as to Semitic tree-and-grove worship, see Movers, 'Phönizier,' vol. i. p. 560, etc.
[4] Hunter, 'Rural Bengal,' pp. 131, 194.

literature, while the Hindu in private life plants the banyan and other trees and worships them with divine honours.[1] Greek and Roman mythology give perfect types not only of the beings attached to individual trees, but of the dryads, fauns, and satyrs living and roaming in the forest—creatures whose analogues are our own elves and fairies of the woods. Above these graceful fantastic beings are the higher deities who have trees for shrines and groves for temples. Witness the description in Ovid's story of Erisichthon:—

"And Ceres' grove he ravaged with the axe,
They say, and shamed with iron the ancient glades.
There stood a mighty oak of age-long strength,
Festooned with garlands, bearing on its trunk
Memorial tablets, proofs of helpful vows.
Beneath, the dryads led their festive dance,
And circled hand-in-hand the giant bole."[2]

In more prosaic fashion, Cato instructs the woodman how to gain indemnity for thinning a holy grove; he must offer a hog in sacrifice with this prayer, "Be thou god or goddess to whom this grove is sacred, permit me, by the expiation of this pig, and in order to restrain the overgrowth of this wood, etc., etc."[3] Slavonic lands had their groves where burned the everlasting fire of Piorun the Heaven-god; the old Prussians venerated the holy oak of Romowe, with its drapery and images of the gods, standing in the midst of the sacred inviolate forest where no twig might be broken nor beast slain; and so on down to the elder-tree beneath which Pushkait was worshipped with offerings of bread and beer.[4] The Keltic Heaven-god, whose image was a mighty oak, the white-robed Druids

[1] Boehtlingk & Roth, s. v. 'chaityataru.' Ward, 'Hindoos,' vol. ii. p. 204.
[2] Ovid. Metamm. viii. 741.
[3] Cato de Re Rustica, 139; Plin. xvii. 47.
[4] Hanusch, 'Slaw. Myth.' pp. 98, 229. Hartknoch, part i. ch. v. vii.; Grimm, 'D. M.' p. 67.

climbing the sacred tree to cut the mistletoe, and sacrificing the two white bulls beneath, are types from another national group.[1] Teutonic descriptions begin with Tacitus, "Lucos ac nemora consecrant, deorumque nominibus adpellant secretum illud, quod sola reverentia vident," and the curious passage which describes the Semnones entering the sacred grove in bonds, a homage to the deity that dwelt there; many a century after, the Swedes were still holding solemn sacrifice and hanging the carcases of the slaughtered beasts in the grove hard by the temple of Upsal.[2] With Christianity comes a crusade against the holy trees and groves. Boniface hews down in the presence of the priests the huge oak of the Hessian Heaven-god, and builds of the timber a chapel to St. Peter. Amator expostulated with the hunters who hung the heads of wild beasts to the boughs of the sacred pear-tree of Auxerre, "Hoc opus idololatriæ culturæ est, non christianæ elegantissimæ disciplinæ;" but this mild persuasion not availing, he chopped it down and burned it. In spite of all such efforts, the old religion of the tree and grove survived in Europe often in most pristine form. Within the last two hundred years, there were old men in Gothland who would "go to pray under a great tree, as their forefathers had done in their time"; and to this day the sacrificial rite of pouring milk and beer over the roots of trees is said to be kept up on out-of-the-way Swedish farms.[3] In Russia, the Lyeshy or wood-demon still protects the birds and beasts in his domain, and drives his flocks of field-mice and squirrels from forest to forest, when we should say they are migrating. The hunter's luck depends on his treatment of the forest-spirit, wherefore he will leave him as a sacrifice the first game he kills, or some smaller offering of bread or salted pancake on a stump. Or if one falls ill on returning from the forest, it is known that this is the Lyeshy's doing, so

---

[1] Maxim. Tyr. viii. ; Plin. xvi. 95.
[2] Tacit. Germania, 9, 39, etc. ; Grimm, 'D. M.' I. 66.
[3] Hyltén-Cavallius, 'Wärend och Wirdarne,' part i. p. 142.

the patient carries to the wood some bread and salt in a clean rag, and leaving it with a prayer, comes home cured.[1] Names like *Holyoake* and *Holywood* record our own old memories of the holy trees and groves, memories long lingering in the tenacious peasant mind; and it was a great and sacred *linden*-tree with three stems, standing in the parish of Hvitaryd in South Sweden, which with curious fitness gave a name to the family of *Linnæus*. Lastly, Jacob Grimm even ventures to connect historically the ancient sacred inviolate wood with the later royal forest, an ethnological argument which would begin with the savage adoring the Spirit of the Forest, and end with the modern landowner preserving his pheasants.[2]

To the modern educated world, few phenomena of the lower civilization seem more pitiable than the spectacle of a man worshipping a beast. We have learnt the lessons of Natural History at last thoroughly enough to recognize our superiority to our "younger brothers," as the Red Indians call them, the creatures whom it is our place not to adore but to understand and use. By men at lower levels of culture, however, the inferior animals are viewed with a very different eye. For various motives, they have become objects of veneration ranking among the most important in the lower ranges of religion. Yet I must here speak shortly and slightly of Animal-worship, not as wanting in interest, but as over-abounding in difficulty. Wishing rather to bring general principles into view than to mass uninterpreted facts, all I can satisfactorily do is to give some select examples from the various groups of evidence, so as at once to display the more striking features of the subject, and to trace the ancient ideas upward from the savage level far into the higher civilization.

First and foremost, uncultured man seems capable of simply worshipping a beast as beast, looking on it as possessed of power, courage, cunning, beyond his own, and

[1] Ralston, 'Songs of Russian People,' p. 153, see 238.
[2] Grimm, 'D. M.' p. 62, etc.

animated like a man by a soul which continues to exist after bodily death, powerful as ever for good and harm. Then this idea blends with the thought of the creature as being an incarnate deity, seeing, hearing, and acting even at a distance, and continuing its power after the death of the animal body to which the divine spirit was attached. Thus the Kamchadals, in their simple veneration of all things that could do them harm or good, worshipped the whales that could overturn their boats, and the bears and wolves of whom they stood in fear. The beasts, they thought, could understand their language, and therefore they abstained from calling them by their names when they met them, but propitiated them with certain appointed formulas.[1] Tribes of Peru, says Garcilaso de la Vega, worshipped the fish and vicuñas that provided them food, the monkeys for their cunning, the sparrowhawks for their keen sight. The tiger and the bear were to them ferocious deities, and mankind, mere strangers and intruders in the land, might well adore these beings, its old inhabitants and lords.[2] How, indeed, can we wonder that in direct and simple awe, the Philippine islanders, when they saw an alligator, should have prayed him with great tenderness to do them no harm, and to this end offered him of whatever they had in their boats, casting it into the water.[3] Such rites display at least a partial truth in the famous apophthegm which attributes to fear the origin of religion: "Primos in orbe deos fecit timor."[4] In discussing the question of the souls of animals in a previous chapter, instances were adduced of men seeking to appease by apologetic phrase and rite the animals they killed.[5] It is instructive to observe how naturally such personal intercourse between man and animal may pass into full worship, when the creature is powerful

[1] Steller, 'Kamtschatka,' p. 276.
[2] Garcilaso de la Vega, 'Comentarios Reales,' i. ch. ix. etc.
[3] Marsden, 'Sumatra,' p. 303.
[4] Petron. Arb. Fragm.; Statius, iii. Theb. 661.
[5] See ante, ch. xi.

or dangerous enough to claim it. When the Stiêns of Kambodia asked pardon of the beast they killed, and offered sacrifice in expiation, they expressly did so through fear lest the creature's disembodied soul should come and torment them.[1] Yet, strange to say, even the worship of the animal as divine does not prevent the propitiatory ceremony from passing into utter mockery. Thus Charlevoix describes North American Indians who, when they had killed a bear, would set up its head painted with many colours, and offer it homage and praise while they performed the painful duty of feasting on its body.[2] So among the Ainos, the indigenes of Yesso, the bear is a great divinity. It is true they slay him when they can, but while they are cutting him up they salute him with obeisances and fair speeches, and set up his head outside the house to preserve them from misfortune.[3] In Siberia, the Yakuts worship the bear in common with the spirits of the forest, bowing toward his favourite haunts with appropriate phrases of prose and verse, in praise of the bravery and generosity of their "beloved uncle." Their kindred the Ostyaks swear in the Russian courts of law on a bear's head, for the bear, they say, is all-knowing, and will slay them if they lie. This idea actually serves the people as a philosophical, though one would say rather superfluous, explanation of a whole class of accidents: when a hunter is killed by a bear, it is considered that he must at some time have forsworn himself, and now has met his doom. Yet these Ostyaks, when they have overcome and slain their deity, will stuff its skin with hay, kick it, spit on it, insult and mock it till they have satiated their hatred and revenge, and are ready to set it up in a yurt as an object of worship.[4]

Whether an animal be worshipped as the receptacle or

[1] Mouhot, 'Indo-China,' vol. i. p. 252.
[2] Charlevoix, 'Nouvelle France,' vol. v. p. 443.
[3] W. M. Wood in 'Tr. Eth. Soc.' vol. iv. p. 36.
[4] Simpson, 'Journey,' vol. ii. p. 269; Erman, 'Siberia,' vol. i. p. 492; Latham, 'Descr. Eth.' vol. i. p. 456; 'Journ. Ind. Archip.' vol. iv. p. 590.

incarnation of an indwelling divine soul or other deity, or as one of the myriad representatives of the presiding god of its class, the case is included under and explained by the general theory of fetish-worship already discussed. Evidence which displays these two conceptions and their blending is singularly perfect in the islands of the Pacific. In the Georgian group, certain herons, kingfishers, and woodpeckers were held sacred and fed on the sacrifices, with the distinct view that the deities were embodied in the birds, and in this form came to eat the offered food and give the oracular responses by their cries.[1] The Tongans never killed certain birds, or the shark, whale, etc., as being sacred shrines in which gods were in the habit of visiting earth; and if they chanced in sailing to pass near a whale, they would offer scented oil or kava to him.[2] In the Fiji Islands, certain birds, fish, plants, and some men, were supposed to have deities closely connected with or residing in them. Thus the hawk, fowl, eel, shark, and nearly every other animal became the shrine of some deity, which the worshipper of that deity might not eat, so that some were even tabued from eating human flesh, the shrine of their god being a man. Ndengei, the dull and otiose supreme deity, had his shrine or incarnation in the serpent.[3] Every Samoan islander had his tutelary deity or "aitu," appearing in some animal, an eel, shark, dog, turtle, etc., which species became his fetish, not to be slighted or injured or eaten, an offence which the deity would avenge by entering the sinner's body and generating his proper incarnation within him till he died.[4] The "atua" of the New Zealander, corresponding with this in name, is a divine ancestral soul, and is also apt to appear in the body of an animal.[5] If we pass to Sumatra, we shall find that the veneration paid by the

---

[1] Ellis, 'Polyn. Res.' vol. i. p. 336.
[2] Farmer, 'Tonga,' p. 126; Mariner, vol. ii. p. 106.
[3] Williams, 'Fiji,' vol. i. p. 217, etc.
[4] Turner, 'Polynesia,' p. 238.
[5] Shortland, 'Trads. of N. Z.' ch. iv.

Malays to the tiger, and their habit of apologizing to it when a trap is laid, is connected with the idea of tigers being animated by the souls of departed men.[1] In other districts of the world, one of the most important cases connected with these is the worship paid by the North American Indian to his medicine-animal, of which he kills one specimen to preserve its skin, which thenceforth receives adoration and grants protection as a fetish.[2] In South Africa, as has been already mentioned, the Zulus hold that divine ancestral shades are embodied in certain tame and harmless snakes, whom their human kinsfolk receive with kindly respect and propitiate with food.[3] In West Africa, monkeys near a grave-yard are supposed to be animated by the spirits of the dead, and the general theory of sacred and worshipped crocodiles, snakes, birds, bats, elephants, hyænas, leopards, etc., is divided between the two great departments of the fetish-theory, in some cases the creature being the actual embodiment or personation of the spirit, and in other cases sacred to it or under its protection.[4] Hardly any region of the world displays so perfectly as this the worship of serpents as fetish-animals endowed with high spiritual qualities, to kill one of whom would be an offence unpardonable. For a single description of negro ophiolatry, may be cited Bosman's description from Whydah in the Bight of Benin; here the highest order of deities were a kind of snakes which swarm in the villages, reigned over by that huge chief monster, uppermost and greatest and as it were the grandfather of all, who dwelt in his snake-house beneath a lofty tree, and there received the royal offerings of meat and drink, cattle and money and stuffs. So heartfelt was

[1] Marsden, 'Sumatra,' p. 292.
[2] Loskiel, 'Ind. of N. A.' part i. p. 40; Catlin, 'N. A. Ind.' vol. i. p. 36; Schoolcraft, 'Tribes,' part i. p. 34, part v. p. 652; Waitz, vol. iii. p. 190.
[3] See antè, p. 8; Callaway, 'Rel. of Amazulu,' p. 196.
[4] Steinhauser, 'Religion des Negers,' l. c. p. 133. J. L. Wilson, 'W. Afr.' pp. 210, 218. Schlegel, 'Ewe-Sprache,' p. xv.

the veneration of the snakes, that the Dutchmen made it a means of clearing their warehouses of tiresome visitors ; as Bosman says, "If we are ever tired with the natives of this country, and would fain be rid of them, we need only speak ill of the snake, at which they immediately stop their ears and run out of doors."[1] Lastly, among the Tatar tribes of Siberia, Castrén finds the explanation of the veneration which the nomade pays to certain animals, in a distinct fetish-theory which he thus sums up: "Can he also contrive to propitiate the snake, bear, wolf, swan, and various other birds of the air and beasts of the field, he has in them good protectors, for in them are hidden mighty spirits."[2]

The cases of a divine ancestral soul worshipped as incarnate in an animal body, form a link between manesworship and beast-worship, and this connexion is made otherwise in another department of the religion of the lower races, the veneration of a particular species of animal by a particular family, clan, or tribe. It is well known that numerous tribes of mankind connect themselves with, call themselves by the name of, and even derive their mythic pedigree from, some animal, plant, or thing, but most often an animal. Among the Algonquin Indians of North America, the name of such a tribe-animal, as Bear, Wolf, Tortoise, Deer, Rabbit, etc., serves to designate each of a number of clans into which the race is divided, a man belonging to each such clan being himself actually spoken of as a bear, wolf, etc., and the figures of these creatures indicating his clan in the native picture-writing. Such creatures must be, so far as possible, distinguished from the mere patron-animal of an individual, the "medicine" just mentioned among the American Indians. The name

[1] Bosman, 'Guinea,' letter 19 ; in Pinkerton, vol. xvi. p. 499. See Burton, 'Dahome,' ch. iv., xvii. An account of the Vaudoux serpent-worship still carried on among the negroes of Hayti, in 'Lippincott's Magazine,' Philadelphia, March 1870.
[2] Castrén, 'Finn. Myth.' p. 196, see 228.

or symbol of an Algonquin clan-animal is called "dodaim," and this word, in its usual form of "totem," has become an accepted term among ethnologists to describe similar customary surnames over the world, the system of dividing tribes in this way being called Totemism. The origin of totemism of course comes within the domain of mythology, while the social divisions, marriage arrangements, and so forth, connected with it, form a highly important part of the law and custom of mankind at certain stages of culture. It only comes within the province of religion so far as the clan-animals, etc., are the subjects of religious observance, or are actually treated as patron-deities. To some extent this seems to happen among the Algonquins themselves, some accounts describing the totem-animal as being actually regarded as the sacred object or "medicine" or protector of the family bearing its name and symbol.[1] This is the case among certain Australian tribes; a family has some animal (or vegetable) as its "kobong," its friend or protector, and a mysterious connexion subsists between a man and his tribe-animal, of which species he is reluctant to kill one, for it might be his own protector, while if his kobong be a vegetable, there are restrictions on his gathering it.[2] So in South Africa the Bechuana people are divided into clans : Bakuena, men of the crocodile ; Batlapi, of the fish ; Bataung, of the lion ; Bamorara, of the wild vine. A man does not eat his tribe animal or clothe himself in its skin, and if he must kill it as hurtful, the lion for instance, he asks pardon of it and purifies himself from the sacrilege.[3] So in Asia, among the Kols of Chota-Nagpur, we find many of the Oraon and Munda clans named after animals, as Eel, Hawk, Crow, Heron, and they must not kill or eat what they are named after ; it is to be noticed, however, that this only forms one part of a system of tribal food-prohi-

[1] James, 'Long's Exp.' vol. i. ch. xv. ; John Long, 'Voyages and Travels,' p. 86. Waitz, vol. iii. p. 190. See 'Early History of Mankind,' p. 286.
[2] Grey, 'Australia,' vol. ii. p. 228.
[3] Casalis, 'Basutos,' p. 211 ; Livingstone, p. 13.

bitions.[1] Among the Yakuts of Siberia, again, each tribe looks on some particular animal as sacred, and abstains from eating it.[2] These facts seem to indicate not mere accidental peculiarities, but a wide-spread common principle acting among mankind in the lower culture. Mr. M'Lennan, in a remarkable investigation, has endeavoured to account for much of the wide-spread animal-worship of the world by considering it as inherited from an early "totem-stage of society."[3] If this view be more or less admitted as just, the question then arises, what is the origin of totemism? Sir John Lubbock, in his work on the Origin of Civilization,[4] and Mr. Herbert Spencer,[5] have favoured the idea of its springing from the really very general practice of naming individual men after animals, Bear, Deer, Eagle, etc., these becoming in certain cases hereditary tribe-names. It must be admitted as possible that such personal epithets might become family surnames, and eventually give rise to myths of the families being actually descended from the animals in question as ancestors, whence might arise many other legends of strange adventures and heroic deeds of ancestors, to be attributed to the quasi-human animals whose names they bore; at the same time, popular mystification between the great ancestor and the creature whose name he held and handed down to his race, might lead to veneration for the creature itself, and thence to full animal-worship. All this might indeed possibly happen, and when it did happen it might set an example which other families could imitate, and thus bring on the systematic division of a whole people into a number of totem-clans, each referred to a mythic animal ancestor. Yet, while granting that such a theory affords a rational interpretation of the obscure facts of totemism, we must treat it as a theory not vouched

---

[1] Dalton in 'Tr. Eth. Soc.' vol. vi. p. 36.
[2] Latham, 'Descr. Eth.' vol. i. p. 364.
[3] J. F. M'Lennan in 'Fortnightly Review,' 1869-70.
[4] Lubbock, 'Origin of Civilization,' p. 183.
[5] Spencer in 'Fortnightly Review,' 1870.

for by sufficient evidence, and within our knowledge liable to mislead, if pushed to extremes. It offers plausible yet quite unsound explanations of points of mythology and theology which seem to have direct and reasonable explanations of their own. We may well shrink from using too confidently a method of myth-interpretation which can account for solar and lunar nature-myths, by referring them to traditions of human heroes and heroines who chanced to bear the names of Sun and Moon. As to animal-worship, when we find men paying distinct and direct reverence to the lion, the bear, or the crocodile as mighty superhuman beings, or adoring other beasts, birds, or reptiles as incarnations of spiritual deities, we can hardly supersede such well-defined developments of animistic religion, by seeking their origin in personal names of deceased ancestors, who chanced to be called Lion, Bear, or Crocodile.

The three motives of animal-worship which have been described, viz., direct worship of the animal for itself, indirect worship of it as a fetish acted through by a deity, and veneration for it as a totem or representative of a tribe-ancestor, no doubt account in no small measure for the phenomena of Zoolatry among the lower races, due allowance being also made for the effects of myth and symbolism, of which we may gain frequent glimpses. Notwithstanding the obscurity and complexity of the subject, a survey of Animal-worship as a whole may yet justify an ethnographic view of its place in the history of civilization. If we turn from its appearances among the less cultured races to notice the shapes in which it has held its place among peoples advanced to the stage of national organization and stereotyped religion, we shall find a reasonable cause for its new position in the theory of development and survival, whereby ideas at first belonging to savage theology have in part continued to spread and solidify in their original manner, while in part they have been changed to accommodate them to more advanced ideas, or have been defended from the attacks of reason by being set up as sacred mysteries. Ancient Egypt was a land of sacred cats and jackals and hawks, whose mummies are among us to this day, but the reason of whose worship was a subject too sacred for the Father of History to discuss. Egyptian animal-worship seems to show, in a double line, traces of a savage ancestry extending into ages lying far behind even the remote antiquity of the Pyramids. Deities patronising special sacred animals, incarnate in their bodies, or represented in their figures, have nowhere better examples than the bull-dynasty of Apis, Horus wearing the head of his sacred hawk, Bubastis and her cat, Thoth and his cynocephalus and ibis, the cow-headed Hathor and the hippopotamus Typhon. Moreover, the local character of many of the sacred creatures, worshipped in certain nomes yet killed and eaten with impunity elsewhere, fits remarkably with that character of tribe-fetishes and deified totems with which Mr. M'Lennan's argument is concerned. See the men of Oxyrynchos reverencing and sparing the fish oxyrynchos, and those of Latopolis likewise worshipping the latos. At Apollinopolis men hated crocodiles and never lost a chance of killing them, while the people of the Arsinoite nome dressed geese and fish for these sacred creatures, adorned them with necklaces and bracelets, and mummified them sumptuously when they died.[1] In the modern world the most civilized people among whom animal-worship vigorously survives, lie within the range of Brahmanism, where the sacred animal, the deity incarnate in an animal or invested with or symbolized by its shape, may to this day be studied in clear example. The sacred cow is not merely to be spared, she is as a deity worshipped in annual ceremony, daily perambulated and bowed to by the pious Hindu, who offers her fresh grass and flowers; Hanuman the monkey-god has his temples and his idols, and in him Siva is incarnate, as Durga is in the jackal; the wise Ganesa wears the elephant's head;

[1] Herod. ii.; Plutarch, De Iside & Osiride; Strabo, xvii. 1; Wilkinson, 'Ancient Eg.' vol. i. ch. iv. etc. Bunsen, 'Egypt's Place in Univ. Hist.' 2nd Edition, with notes by Birch, vol. i.

the divine king of birds, Garuda, is Vishnu's vehicle; the forms of fish, and boar, and tortoise, were assumed in those avatar-legends of Vishnu which are at the intellectual level of the Red Indian myths they so curiously resemble.[1] The conceptions which underlie the Hindu creed of divine animals were not ill displayed by that Hindu who, being shown the pictures of Matthew, Mark, Luke, and John with their respective man, lion, ox, and eagle, explained these quite naturally and satisfactorily as the avatars or vehicles of the four evangelists.

In Animal-worship, some of the most remarkable cases of development and survival belong to a class from which striking instances have already been taken. Serpent-worship unfortunately fell years ago into the hands of speculative writers, who mixed it up with occult philosophies, Druidical mysteries, and that portentous nonsense called the "Arkite Symbolism," till now sober students hear the very name of Ophiolatry with a shiver. Yet it is in itself a rational and instructive subject of inquiry, especially notable for its width of range in mythology and religion. We may set out among the lower races, with such accounts as those of the Red Indian's reverence to the rattlesnake, as grandfather and king of snakes, as a divine protector able to give fair winds or cause tempests;[2] or of the worship of great snakes among the tribes of Peru before they received the religion of the Incas, as to whom an old author says, "They adore the demon when he presents himself to them in the figure of some beast or serpent, and talks with them."[3] Thenceforth such examples of direct Ophiolatry may be traced on into classic and barbaric Europe; the great serpent which defended the citadel of Athens and enjoyed its monthly honey-cakes;[4] the Roman genius loci appearing in the form of the snake (Nullus enim locus sine

[1] Ward, 'Hindoos,' vol. ii. p. 195, etc.
[2] Schoolcraft, part iii. p. 231; Brinton, p. 108, etc.
[3] Garcilaso de la Vega, 'Comentarios Reales,' i. 9.
[4] Herodot. viii. 41.

genio est, qui per anguem plerumque ostenditur);[1] the old Prussian serpent-worship and offering of food to the household snakes;[2] the golden viper adored by the Lombards, till Barbatus got it in his hands and the goldsmiths made it into paten and chalice.[3] To this day, Europe has not forgotten in nursery tales or more serious belief the snake that comes with its golden crown and drinks milk out of the child's porringer; the house-snake, tame and kindly but seldom seen, that cares for the cows and the children and gives omens of a death in the family; the pair of household snakes which have a mystic connexion of life and death with the husband and housewife themselves.[4] Serpent-worship, apparently of the directest sort, was prominent in the indigenous religions of Southern Asia. It now even appears to have maintained no mean place in early Indian Buddhism, for the sculptures of the Sanchi tope show scenes of adoration of the five-headed snake-deity in his temple, performed by a race of serpent-worshippers, figuratively represented with snakes growing from their shoulders, and whose raja himself has a five-headed snake arching hood-wise over his head. Here, moreover, the totem theory comes into contact with ophiolatry. The Sanskrit name of the snake, "nâga," becomes also the accepted designation of its adorers, and thus mythological interpretation has to reduce to reasonable sense legends of serpent-races who turn out to be simply serpent-worshippers, tribes who have from the divine reptiles at once their generic name of Nâgas, and with it their imagined ancestral descent from serpents.[5] In different ways, these Nâga tribes of South Asia are on the one hand analogues of the

[1] Servius ad Æn. v. 95.
[2] Hartknoch, 'Preussen,' part i. pp. 143, 162.
[3] Grimm, 'D. M.' p. 648.
[4] Grimm, 'D. M.' p. 650. Rochholz, 'Deutscher Glaube,' etc. vol. i. p. 146. Monnier, 'Traditions Populaires,' p. 644. Grohmann, 'Aberglauben aus Böhmen,' etc., p. 78. Ralston, 'Songs of Russian People,' p. 175.
[5] Fergusson, 'Tree and Serpent Worship,' p. 55, etc. pl. xxiv. M'Lennan, l. c. p. 563, etc.

Snake Indians of America, and on the other of the Ophio-genes or Serpent-race of the Troad, kindred of the vipers whose bite they could cure by touch, and descendants of an ancient hero transformed into a snake.[1] Serpents hold a prominent place in the religions of the world, as the incarnations, shrines, or symbols of high deities. Such were the rattlesnake worshipped in the Natchez temple of the Sun, and the snake belonging in name and figure to the Aztec deity Quetzalcoatl;[2] the snake as worshipped still by the Slave Coast negro, not for itself but for its indwelling deity;[3] the snake kept and fed with milk in the temple of the old Slavonic god Potrimpos;[4] the serpent-symbol of the healing deity Asklepios, who abode in or manifested himself through the huge tame snakes kept in his temples[5] (it is doubtful whether this had any original connexion with the adoption of the snake, from its renewal by casting its old slough, as the accepted emblem of new life or immortality in later symbolism); and lastly, the Phœnician serpent with its tail in its mouth, symbol of the world and of the Heaven-god Taaut, in its original meaning probably a mythic world-snake like the Scandinavian Midgard-worm, but in the changed fancy of later ages adapted into an emblem of eternity.[6] It scarcely seems proved that savage races, in all their mystic contemplations of the serpent, ever developed out of their own minds the idea, to us so familiar, of adopting it as a personification of evil.[7] In ancient times, we may ascribe this character perhaps to the monster whose well-known form is to be seen on the mummy-cases, the Apophis-serpent of the Egyptian

[1] Strabo, xiii. 1, 14.
[2] J. G. Müller, 'Amer. Urrel.' pp. 62, 585.
[3] J. B. Schlegel, 'Ewe-Sprache,' p. xiv.
[4] Hanusch, 'Slaw. Myth.' p. 217.
[5] Pausan. ii. 28 ; Ælian. xvi. 39. See Welcker, 'Griech. GötterL' vol. ii. p. 734.
[6] Macrob. Saturnal. i. 9. Movers, 'Phönizier,' vol. i. p. 500.
[7] Details such as in Schoolcraft, 'Ind. Tribes,' part i. pp. 38, 414, may be ascribed to Christian intercourse. See Brinton, p. 121.

Hades;- and it unequivocally belongs to the Wicked Serpent of the Zarathustrians, Aji Dahaka,[2] a figure which bears so remarkable a relation to that of the Semitic serpent of Eden, which may possibly stand in historical connexion with it. A wondrous blending of the ancient rites of Ophiolatry with mystic conceptions of Gnosticism appears in the cultus which tradition (in truth or slander) declares the semi-Christian sect of Ophites to have rendered to their tame snake, enticing it out of its chest to coil round the sacramental bread, and worshipping it as representing the great king from heaven who in the beginning gave to the man and woman the knowledge of the mysteries.[3] Thus the extreme types of religious veneration, from the soberest matter-of-fact to the dreamiest mysticism, find their places in the worship of animals.[4]

Hitherto in the study of animistic doctrine, our attention has been turned especially to those minor spirits whose functions concern the closer and narrower detail of man's life and its surroundings. In passing thence to the consideration of divine beings whose functions have a wider scope, the transition may be well made through a special group. An acute remark of Auguste Comte's calls attention to an important process of theological thought, which we may here endeavour to bring as clearly as possible before our minds. In his "Philosophie Positive," he defines deities proper as differing by their general and abstract character from pure fetishes (*i. e.*, animated objects), the humble fetish governing but a single object from which it is inseparable, while the gods administer a special order of phenomena at once in different bodies. When, he con-

---

[1] Lepsius, 'Todtenbuch' and Birch's transl. in Bunsen's 'Egypt,' vol. v.
[2] Spiegel, 'Avesta,' tr. by Bleek, vol. ii. p. 51, vol. iii. p. 35.
[3] Epiphan. Adv. Hæres. xxxvii. Tertullian. De Præscript. contra Hæreticos, 47.
[4] Further collections of evidence relating to Zoolatry in general may be found in Bastian, 'Das Thier in seiner mythologischen Bedeutung,' in Bastian and Hartmann's 'Zeitschrift für Ethnologie,' vol. i. ; Meiners, 'Geschichte der Religionen,' vol. i.

tinues, the similar vegetation of the different oaks of a forest led to a theological generalization from their common phenomena, the abstract being thus produced was no longer the fetish of a single tree, but became the god of the forest; here, then, is the intellectual passage from fetishism to polytheism, reduced to the inevitable preponderance of specific over individual ideas.[1] Now this observation of Comte's may be more immediately applied to a class of divine beings which may be accurately called species-deities. It is highly suggestive to study the crude attempts of barbaric theology to account for the uniformity observed in large classes of objects, by making this generalization from individual to specific ideas. To explain the existence of what we call a species, they would refer it to a common ancestral stock, or to an original archetype, or to a species-deity, or they combined these conceptions. For such speculations, classes of plants and animals offered perhaps an early and certainly an easy subject. The uniformity of each kind not only suggested a common parentage, but also the notion that creatures so wanting in individuality, with qualities so measured out as it were by line and rule, might not be independent arbitrary agents, but mere copies from a common model, or mere instruments used by controlling deities. Thus in Polynesia, as has been just mentioned, certain species of animals were considered as incarnations of certain deities, and among the Samoans we learn that the question as to the individuality of such creatures was actually asked and answered. If, for instance, a village god were accustomed to appear as an owl, and one of his votaries found a dead owl by the roadside, he would mourn over the sacred bird and bury it with much ceremony, but the god himself would not be thought to be dead, for he remains incarnate in all existing owls.[2] According to Father Geronimo Boscana, the Acagchemen tribe of Upper California furnish a curious parallel to this notion. They

[1] Comte, 'Philosophie Positive,' vol. v. p. 101.
[2] Turner 'Polynesia,' p. 242.

worshipped the "panes" bird, which seems to have been an eagle or vulture, and each year, in the temple of each village, one of them was solemnly killed without shedding blood, and the body burned. Yet the natives maintained and believed that it was the same individual bird they sacrificed each year, and more than this, that the same bird was slain by each of the villages.[1] Among the comparatively cultured Peruvians, Acosta describes another theory of celestial archetypes. Speaking of star-deities, he says that shepherds venerated a certain star called Sheep, another star called Tiger protected men from tigers, etc.: "And generally, of all the animals and birds there are on the earth, they believed that a like one lived in heaven, in whose charge were their procreation and increase, and thus they accounted of divers stars, such as that they call Chacana, and Topatorca, and Mamana, and Mizco, and Miquiquiray, and other such, so that in a manner it appears that they were drawing towards the dogma of the Platonic ideas."[2] The North American Indians also have speculated as to the common ancestors or deities of species. One missionary notes down their idea as he found it in 1634. "They say, moreover, that all the animals of each species have an elder brother, who is as it were the principle and origin of all the individuals, and this elder brother is marvellously great and powerful. The elder brother of the beavers, they told me, is perhaps as large as our cabin." Another early account is that each species of animals has its archetype in the land of souls; there exists, for example, a manitu or archetype of all oxen, which animates all oxen.[3] Here, again, occurs a noteworthy correspondence with the ideas of a distant race. In Buyán, the island paradise of Russian myth, there

[1] Brinton, 'Myths of New World,' p. 105.
[2] Acosta, 'Historia de las Indias,' book v. c. iv.; Rivero & Tschudi, pp. 161, 179; J. G. Müller, p. 365.
[3] Le Jeune in 'Rel. des Jes. dans la Nouvelle France,' 1634, p. 13. Lafitau, 'Mœurs des Sauvages,' vol. i. p. 370. See also Waitz, vol. iii. p. 194; Schoolcraft, part iii. p. 327.

are to be found the Snake older than all snakes, and the prophetic Raven, elder brother of all ravens, and the Bird, the largest and oldest of all birds, with iron beak and copper claws, and the Mother of Bees, eldest among bees.[1] Morgan's comparatively modern account of the Iroquois mentions their belief in a spirit of each species of trees and plants, as of oak, hemlock, maple, whortleberry, raspberry, spearmint, tobacco; most objects of nature being thus under the care of protecting spirits.[2] The doctrine of such species-deities is perhaps nowhere more definitely stated than by Castrén in his "Finnish Mythology." In his description of the Siberian nature-worship, the lowest level is exemplified by the Samoyeds, whose direct worship of natural objects for themselves may perhaps indicate the original religious condition of the whole Turanian race. But the doctrine of the comparatively cultured heathen Finns was at a different stage. Here every object in nature has a "haltia," a guardian deity or genius, a being which was its creator and thenceforth became attached to it. These deities or genii are, however, not bound to each single transitory object, but are free personal beings which have movement, form, body, and soul. Their existence in no wise depends on the existence of the individual objects, for although no object in nature is without its guardian deity, this deity extends to the whole race or species. This ash-tree, this stone, this house, has indeed its particular "haltia," yet these same "haltiat" concern themselves with other ash-trees, stones, and houses, of which the individuals may perish, but their presiding genii live on in the species.[3] It seems as though some similar view ran through the doctrine of more civilized races, as in the well-known

[1] Ralston, 'Songs of the Russian People,' p. 375. The Slavonic myth of Buyán, with its dripping oak and the snake Garafena lying beneath, is obviously connected with the Scandinavian myth of the dripping ash, Yggdrasill, the snake Nidhögg below, and the two Swans of the Urdhar-fount, parents of all swans.
[2] Morgan, 'Iroquois,' p. 162.
[3] Castrén, 'Finn. Myth.' pp. 106, 160, 189, etc.

Egyptian and Greek examples where whole species of animals, plants, or things, stand as symbolic of, and as protected by, particular deities. The thought appears with most perfect clearness in the Rabbinical philosophy which apportions to each of the 2100 species, of plants for instance, a presiding angel in heaven, and assigns this as the motive of the Levitical prohibition of mixtures among animals and plants.[1] The interesting likeness pointed out by Father Acosta between these crude theological conceptions and the civilized philosophical conceptions which have replaced them, was again brought into view in the last century by the President De Brosses, in comparing the Red Indians' archetypes of species with the Platonic archetypal ideas.[2] As for animals and plants, the desire of naturalists to ascend to primal unity to some extent finds satisfaction in a theory tracing each species to an origin in a single pair. And though this is out of the question with inanimate objects, our language seems in suggestive metaphor to lay hold on the same thought, when we say of a dozen similar swords, or garments, or chairs, that they have the same *pattern* (patronus, as it were father), whereby they were shaped from their *matter* (materia, or mother substance).

[1] Eisenmenger, 'Judenthum,' part ii. p. 376; Bastian, 'Mensch,' vol. iii. p. 194.
[2] De Brosses, 'Dieux Fétiches,' p. 58.

CHAPTER XVI.

ANIMISM—*continued.*

Higher Deities of Polytheism—Human characteristics applied to Deity—Lords of Spiritual Hierarchy—Polytheism : its course of development in lower and higher Culture—Principles of its investigation ; classification of Deities according to central conceptions of their significance and function —Heaven-god—Rain-god—Thunder-god—Wind-god—Earth-god—Water-god—Sea-god—Fire-god—Sun-god—Moon-god.

SURVEYING the religions of the world and studying the descriptions of deity among race after race, we may recur to old polemical terms in order to define a dominant idea of theology at large. Man so habitually ascribes to his deities human shape, human passions, human nature, that we may declare him an Anthropomorphite, an Anthropopathite, and (to complete the series) an Anthropophysite. In this state of religious thought, prevailing as it does through so immense a range among mankind, one of the strongest confirmations may be found of the theory here advanced concerning the development of Animism. This theory that the conception of the human soul is the very "fons et origo" of the conceptions of spirit and deity in general, has been already vouched for by the fact of human souls being held to pass into the characters of good and evil demons, and to ascend to the rank of deities. But beyond this, as we consider the nature of the great gods of the nations, in whom the vastest functions of the universe are vested, it will still be apparent that these mighty deities are modelled on human souls, that in great measure their feeling and sympathy, their character and habit, their will and action, even their material and form, display throughout their adaptations, exaggerations and distortions, characteristics shaped upon those of the human spirit. The key to investigation of the Dii Majorum Gentium of the world is the reflex of humanity, and as we behold their figures in their proper districts of theology, memory ever brings back the Psalmist's words, "Thou thoughtest I was altogether as thyself."

The higher deities of Polytheism have their places in the general animistic system of mankind. Among nation after nation it is still clear how, man being the type of deity, human society and goverment became the model on which divine society and government were shaped. As chiefs and kings are among men, so are the great gods among the lesser spirits. They differ from the souls and minor spiritual beings which we have as yet chiefly considered, but the difference is rather of rank than of nature. They are personal spirits, reigning over personal spirits. Above the disembodied souls and manes, the local genii of rocks and fountains and trees, the host of good and evil demons, and the rest of the spiritual commonalty, stand these mightier deities, whose influence is less confined to local or individual interests, and who, as it pleases them, can act directly within their vast domain, or control and operate through the lower beings of their kind, their servants, agents, or mediators. The great gods of Polytheism, whose dominion thus stretches far and wide over the world, are not, any more than the lower spirits, creations of a civilized theology. In the rudest religions of the lower races, their principal types were already cast, and thenceforward, for many an age of progressing or relapsing culture, it became the work of poet and priest, legend-monger and historian, theologian and philosopher, to develop and renew, to degrade and abolish, the mighty lords of the Pantheon.

With little exception, wherever a savage or barbaric system of religion is thoroughly described, reigning deities make their appearance in the spiritual world as distinctly as chiefs in the human tribe. These beings need by no means correspond in nature and function between tribe and tribe, yet for the most part each is a definite theological figure with a definite meaning and origin, and as such recurs in many districts, while its definition finds its proper pigeon-hole in the ethnographer's generalization. This state of things comes into view at a glance. Even among the Australians, above the swarming souls, nature-spirits, demons, there stand out mythic figures of higher divinity; Nguk-wonga, the Spirit of the Waters; Biam, who gives ceremonial songs and causes disease, and is perhaps the same as Baiame the creator; Nambajandi and Warrugura, lords of heaven and the nether world.[1] In South America, if we look into the theology of the Manaos (whose name is well known in the famous legend of El Dorado and the golden city of Manoa), we see Mauari and Saraua, who may be called the Good and Evil Spirit, and beside the latter the two Gamainhas, Spirits of the Waters and the Forest.[2] In North America the description of a solemn Algonquin sacrifice introduces us to twelve dominant manitus or gods; first the Great Manitu in heaven, then the Sun, Moon, Earth, Fire, Water, the House-god, the Indian corn, and the four Winds or Cardinal Points.[3] The Polynesian's crowd of manes, and the lower ranks of deities of earth, sea, and air, stand below the great gods of Peace and War, Oro and Tane the national deities of Tahiti and Huahine, Raitubu the Sky-producer, Hina who aided in the work of forming the world, her father Taaroa, the uncreate Creator who dwells in Heaven.[4] Among the Land Dayaks of Borneo, the commonalty of spirits consists of the souls of the departed, and of such beings as dwell in the noble old forests on the tops of lofty hills, or such as hover about villages and devour the stores of rice; above these are Tapa, creator and preserver of man, and Iang, who taught

[1] Eyre, 'Australia,' vol. ii. p. 362; Oldfield in 'Tr. Eth. Soc.' vol. iii. p. 228; Lang, 'Queensland,' p. 444.
[2] Martius, 'Ethnog. Amer.' vol. i. p. 583.
[3] Loskiel, 'Ind. of N. America,' part i. p. 43.
[4] Ellis, 'Polyn. Res.' vol. i. p. 322.

the Dayaks their religion, Jirong, whose function is the birth and death of men, and Tenabi, who made, and still causes to flourish, the earth and all things therein save the human race.[1] In West Africa, let us take an example from the theology of the Slave Coast, a systematic scheme of all nature as moved and quickened by spirits, kindly or hostile to mankind. These spirits dwell in field and wood, mountain and valley; they live in air and water; multitudes of them have been human souls, such ghosts hover about the graves and near the living, and have influence with the under-gods, whom they worship; among these "edrŏ" are the patron-deities of men and families and tribes; through these subordinate beings works the highest god, Mawu. The missionary who describes this negro hierarchy quite simply sees in it Satan and his Angels.[2] In Asia, the Samoyed's little spirits that are bound to his little fetishes, and the little elves of wood and stream, have greater beings above them, the Forest-Spirit, the River-Spirit, the Sun and Moon, the Evil Spirit and the Good Spirit above all.[3] The countless host of the local gods of the Khonds pervade the world, rule the functions of nature, and control the life of men, and these have their chiefs; above them rank the deified souls of men who have become tutelary gods of tribes; above these are the six great gods, the Rain-god, the goddess of Firstfruits, the god of Increase, the god of Hunting, the iron god of War, the god of Boundaries, with which group stands also the Judge of the Dead, and above all other gods, the Sun-god and Creator Boora Pennu, and his wife the mighty Earth-goddess, Tari Pennu.[4] The Spanish conquerors found in Mexico a complex and systematic hierarchy of spiritual beings; numberless were the little deities who had their worship in house and lane,

[1] St. John, 'Far East,' vol. i. p. 180.
[2] J. B Schlegel, 'Schüssel zur Ewe Sprache,' p. xii.; compaer Bowen, 'Yoruba Lang.' in 'Smithsonian Contrib.' vol. i. p. xvi.
[3] Samoiedia, in Pinkerton, vol. i. p. 531.
[4] Macpherson, p. 84, etc.

ANIMISM. 251

grove and temple, and from these the worshipper could pass to gods of flowers or of pulque, of hunters and goldsmiths, and then to the great deities of the nation and the world, the figures which the mythologist knows so well, Centeotl the Earth-goddess, Tlaloc the Water-God, Huitzilopochtli the War-god, Mictlanteuctli the Lord of Hades, Tonatiuh and Metztli the Sun and Moon.[1] Thus, starting from the theology of savage tribes, the student arrives at the polytheistic hierarchies of the Aryan nations. In ancient Greece, the cloud-compelling Heaven-god reigns over such deities as the god of War and the goddess of Love, the Sun-god and the Moon-goddess, the Fire-god and the ruler of the Under-world, the Winds and Rivers, the nymphs of wood and well and forest.[2] In modern India, Brahma-Vishnu-Siva reign prominent over a series of divinities, heterogeneous and often obscure in nature, but among whom stand out in clear meaning and purpose such figures as Indra of Heaven and Sûrya of the Sun, Agni of the Fire, Pavana of the Winds and Varuna of the Waters, Yama lord of the Under-world, Kâma god of Love and Kârttikeya of War, Panchânana who gives epilepsy and Manasâ who preserves from snake-bites, the divine Rivers, and below these the ranks of nymphs, elves, demons, ministering spirits of heaven and earth—Gandharvas, Apsaras, Siddhas, Asuras, Bhûtas, Râkshasas.[3]

The systematic comparison of polytheistic religions has been of late years worked with admirable results. These have been due to the adoption of comparatively exact methods, as where the ancient Aryan deities of the Veda have been brought into connexion with those of the Homeric poems, in some cases as clearly as where we Englishmen can study in the Scandinavian Edda the old gods of our own race, whose names stand in local names on the map of England, and serve as counters to reckon our days of the

[1] Clavigero, 'Messico,' vol. ii. ch. i.
[2] Gladstone, 'Juventus Mundi,' ch. vii. etc.
[3] Ward, 'Hindoos,' vol. ii.

week. Yet it need scarcely be said that to compare in full detail the deities even of closely connected nations, and à fortiori those of tribes not united in language and history, is still a difficult and unsatisfactory task. The old-fashioned identifications of the gods and heroes of different nations admitted most illusory evidence. Some had little more ground than similar-sounding names, as when the Hindu Brahma and Prajâpati were discovered to be the Hebrew Abraham and Japhet, and when even Sir William Jones identified Woden with Buddha. With not much more stringency, it is still often taken as matter of course that the Keltic Beal, whose bealtines correspond with a whole class of bonfire-customs among several branches of the Aryan race, is the Bel or the Baal of the Semitic cultus. Unfortunately, classical scholarship at the Renaissance started the subject on an unsound footing, by accepting the Greek deities with the mystified shapes and perverted names they had assumed in Latin literature. That there was a partial soundness in such comparisons, as in identifying Zeus and Jupiter, Hestia and Vesta, made the plan all the more misleading when Kronos came to figure as Saturn, Poseidon as Neptune, Athene as Minerva. To judge by example of the possible results of comparative theology worked on such principles, Thoth being identified with Hermes, Hermes with Mercury, and Mercury with Woden, there comes to pass the absurd transition from the Egyptian ibis-headed divine scribe of the gods, to the Teutonic heaven-dwelling driver of the raging tempest. It is not in this loose fashion that the mental processes are to be sought out, which led nations to arrange so similarly and yet so diversely their array of deities.

A twofold perplexity besets the soberest investigator on this ground, caused by the modification of deities by development at home and adoption from abroad. Even among the lower races, gods of long traditional legend and worship acquire a mixed and complex personality. The mythologist who seeks to ascertain the precise definition of the Red

ANIMISM. 253

Indian Michabu in his various characters of Heaven-god and Water-god, Creator of the Earth and first ancestor of Man, or who examines the personality of the Polynesian Maui in his relation to Sun, lord of Heaven or Hades, first Man, and South Sea Island hero, will sympathize with the Semitic or Aryan student bewildered among the heterogeneous attributes of Baal and Astarte, Herakles and Athene. Sir William Jones scarcely overstated the perplexity of the problem in the following remarkable forecast delivered more than eighty years ago, in the first anniversary discourse before the Asiatic Society of Bengal, at a time when glimpses of the relation of the Hindu to the Greek Pantheon were opening into a new broad view of comparative theology in his mind. "We must not be surprised," he says, "at finding, on a close examination, that the characters of all the Pagan deities, male and female, melt into each other and at last into one or two; for it seems a well-founded opinion, that the whole crowd of gods and goddesses in ancient Rome, and modern Váránes [Benares] mean only the powers of nature, and principally those of the Sun, expressed in a variety of ways and by a multitude of fanciful names." As to the travelling of gods from country to country, and the changes they are apt to suffer on the road, we may judge by examples of what has happened within our knowledge. It is not merely that one nation borrows a god from another with its proper figure and attributes and rites, as where in Rome the worshipper of the Sun might take his choice whether he would adore in the temple of the Greek Apollo, the Egyptian Osiris, the Persian Mithra, or the Syrian Elagabalus. The intercourse of races can produce quainter results than this. Any Orientalist will appreciate the wonderful hotchpot of Hindu and Arabic language and religion in the following details, noted down among rude tribes of the Malay Peninsula. We hear of Jin Bumi the Earth-god (Arabic jin = demon, Sanskrit bhûmi = earth); incense is burnt to Jewajewa (Sanskrit dewa = god) who intercedes with Pirman the supreme invisible deity above the sky (Brahma?); the Moslem Allah Táala, with his wife Nabi Mahamad (Prophet Mohammed), appear in the Hinduized characters of creator and destroyer of all things; and while the spirits worshipped in stones are called by the Hindu term of "dewa" or deity, Moslem conversion has so far influenced the mind of the stone-worshipper, that he will give to his sacred boulder the title of a Prophet Mohammed.[1] If we would have examples nearer home, we may trace the evil demon Aeshma Daeva of the ancient Persian religion becoming the Asmodeus of the book of Tobit, afterwards to find a place in the devilry of the middle ages, and to end his career as the Diable Boiteux of Le Sage. Even the Aztec war-god Huitzilopochtli may be found figuring as the demon Vizlipuzli in the popular drama of Doctor Faustus.

In ethnographic comparisons of the religions of mankind, unless there is evidence of direct relation between gods belonging to two peoples, the safe and reasonable principle is to limit the identification of deities to the attributes they have in common. Thus it is proper to compare the Dendid of the White Nile with the Aryan Indra, in so far as both are Heaven-gods and Rain-gods; the Aztec Tonatiuh with the Greek Apollo, in so far as both are Sun-gods; the Australian Baiame with the Scandinavian Thor, in so far as both are Thunder-gods. The present purpose of displaying Polytheism as a department of Animism does not require that elaborate comparison of systems which would be in place in a manual of the religions of the world. The great gods may be scientifically ranged and treated according to their fundamental ideas, the strongly-marked and intelligible conceptions which, under names often obscure and personalities often mixed and mystified, they stand to represent. It is enough to show the similarity of principle on which the theologic mind of the lower races shaped those old familiar types of deity, with which our first acquaintance was gained in the pantheon of classic mytho-

[1] 'Journ. Ind. Archip.' vol. i. pp. 33, 255, 275, 338, vol. ii. p. 692.

logy. It will be observed that not all, but the principal figures, belong to strict Nature-worship. These may be here first surveyed. They are Heaven and Earth, Rain and Thunder, Water and Sea, Fire and Sun and Moon, worshipped either directly for themselves, or as animated by their special deities, or these deities are more fully set apart and adored in anthropomorphic shape—a group of conceptions distinctly and throughout based on the principles of savage fetishism. True, the great Nature-gods are huge in strength and far-reaching in influence, but this is because the natural objects they belong to are immense in size or range of action, pre-eminent and predominant among lesser fetishes, though still fetishes themselves.

In the religion of the North American Indians, the Heaven-god displays perfectly the gradual blending of the material sky itself with its personal deity. In the early times of French colonization, Father Brebeuf describes the Hurons addressing themselves to the earth, rivers, lakes, and dangerous rocks, but above all to heaven, believing that it is all animated, and some powerful demon dwells therein. He describes them as speaking directly to heaven by its personal name "Aronhiaté!" Thus when they throw tobacco into the fire as sacrifice, if it is Heaven they address, they say "Aronhiaté! (Heaven!) behold my sacrifice, have pity on me, aid me!" They have recourse to Heaven in almost all their necessities, and respect this great body above all creatures, remarking in it particularly something divine. They imagine in the sky an "oki," *i. e.* demon or power, which rules the seasons of the year and controls the winds and waves. They dread its anger, calling it to witness when they make some important promise or treaty, saying, Heaven hears what we do this day, and fearing chastisement should their word be broken. One of their renowned sorcerers said, Heaven will be angry if men mock him; when they cry every day to Heaven, Aronhiaté! yet give him nothing, he will avenge himself. Etymology again suggests the divine sky as the inner meaning of the Iroquois supreme deity, Taronhiawagon the "sky-comer" or "sky-holder," who had his festival about the winter solstice, who brought the ancestral race out of the mountain, taught them hunting, marriage, and religion, gave them corn and beans, squashes and potatoes and tobacco, and guided them on their migrations as they spread over the land. Among the North American tribes, not only does the conception of the personal divine Heaven thus seem the fundamental idea of the "Master of Heaven," the Heaven-god, but it may expand into a yet more general thought of divinity in the Great Spirit in Heaven.[1] In South Africa, the Zulus speak of the Heaven as a person, ascribing to it the power of exercising a will, and they also speak of a Lord of Heaven, whose wrath they deprecate during a thunderstorm. In the native legends of the Zulu princess in the country of the Half-Men, the captive maiden expostulates personally with the Sky, for only acting in an ordinary way, and not in the way she wishes, to destroy her enemies:—

"Listen, yon heaven. Attend; mayoya, listen.
Listen, heaven. It does not thunder with loud thunder.
It thunders in an undertone. What is it doing?
It thunders to produce rain and change of season."

Thereupon the clouds gather tumultuously; the princess sings again and it thunders terribly, and the Heaven kills the Half-Men round about her, but she is left unharmed.[2] West Africa is another district where the Heaven-god reigns, in whose attributes may be traced the transition from the direct conception of the personal sky to that of the supreme creative deity. Thus in Bonny, one word serves for god, heaven, cloud; and in Aquapim, Yankupong is at once the highest god and the weather. Of this latter deity, the

[1] Brebeuf in 'Rel. des. Jes.', 1636. p. 107; Lafitau, 'Mœurs des Sauvages Amériquains,' vol. i. p. 132. Schoolcraft, 'Iroquois,' p. 36, etc. 237. Brinton, 'Myths of New World,' pp. 48, 172. J. G. Müller, 'Amer. Urrelig.' p. 119.
[2] Callaway, 'Zulu Tales,' vol. i. p. 203.

Nyankupon of the Oji nation, it is remarked by Riis : " The idea of him as a supreme spirit is obscure and uncertain, and often confounded with the visible heavens or sky, the upper world (sorro) which lies beyond human reach ; and hence the same word is used also for heavens, sky, and even for rain and thunder."[1] The same transition from the divine sky to its anthropomorphic deity shows out in the theology of the Tatar tribes. The rude Samoyed's mind scarcely if at all separates the visible personal Heaven from the divinity united with it under one and the same name, Num. Among the more cultured Finns, the cosmic attributes of the Heaven-god, Ukko the Old One, display the same original nature; he is the ancient of Heaven, the father of Heaven, the bearer of the Firmament, the god of the Air, the dweller on the Clouds, the Cloud-driver, the shepherd of the Cloud-lambs.[2] So far as the evidence of language, and document, and ceremony, can preserve the record of remotely ancient thought, China shows in the highest deity of the state religion a like theologic development. Tien, Heaven, is in personal shape the Shang-ti or Upper Emperor, the Lord of the Universe. The Chinese books may idealize this supreme divinity; they may say that his command is fate, that he rewards the good and punishes the wicked, that he loves and protects the people beneath him, that he manifests himself through events, that he is a spirit full of insight, penetrating, fearful, majestic. Yet they cannot refine him so utterly away into an abstract celestial deity, but that language and history still recognize him as what he was in the beginning, Tien, Heaven.[3]

With such evidence perfectly accords the history of the

[1] Waitz, 'Anthropologie,' vol. ii. p. 168, etc. ; Burton, ' W. & W. fr. W. Afr.' p. 76.
[2] Castrén, 'Finn. Myth.' p. 7, etc.
[3] Plath, 'Religion und Cultus der alten Chinesen,' part i. p. 18, etc. ; part ii. p. 32; Doolittle, 'Chinese,' vol. ii. p. 396. See Max Müller, 'Lectures,' 2d. S. p. 437 ; Legge, 'Confucius,' p. 100. For further evidence as to savage and barbaric worship of the Heaven as Supreme Deity, see chap. xvii.

Heaven-god among our Indo-European race. The being adored by the primitive Aryan was—

" . . . . the whole circle of the heavens, for him
A sensitive existence, and a God,
With lifted hands invoked, and songs of praise."

The evidence of Aryan language to this effect has been set forth with extreme clearness by Professor Max Müller. In the first stage, the Sanskrit Dyu (Dyaus), the bright sky, is taken in a sense so direct that it expresses the idea of day, and the storms are spoken of as going about in it; while Greek and Latin rival this distinctness in such terms as ἔνδιος, "in the open air," ἔνδιος, "well-skyed, calm," sub divo, "in the open air," sub Jove frigido, "under the cold sky," and that graphic description by Ennius of the bright firmament, Jove whom all invoke :—

"Aspice hoc sublime candens, quem invocant omnes Jovem."

In the second stage, Dyaus pitar, Heaven-father, stands in the Veda as consort of Prithivî mâtar, Earth-mother, ranked high or highest among the bright gods. To the Greek he is Ζεὺς πατήρ, the Heaven-father, Zeus the All-seer, the Cloud-compeller, King of Gods and Men. As Max Müller writes : "There was nothing that could be told of the sky that was not in some form or other ascribed to Zeus. It was Zeus who rained, who thundered, who snowed, who hailed, who sent the lightning, who gathered the clouds, who let loose the winds, who held the rainbow. It is Zeus who orders the days and nights, the months, seasons, and years. It is he who watches over the fields, who sends rich harvests, and who tends the flocks. Like the sky, Zeus dwells on the highest mountains; like the sky, Zeus embraces the earth; like the sky, Zeus is eternal, unchanging, the highest god. For good and for evil, Zeus the sky and Zeus the god are wedded together in the Greek mind, language triumphing over thought, tradition over religion." The same Aryan Heaven-father is Jupiter, in that original name and nature which he bore in Rome long before they arrayed him in the borrowed garments of Greek myth, and adapted him to the ideas of classic philosophy.[1] Thus, in nation after nation, took place the great religious development by which the Father-Heaven became the Father in Heaven.

The Rain-god is most often the Heaven-god exercising a special function, though sometimes taking a more distinctly individual form, or blending in characteristics with a general Water-god. The Dinkas of the White Nile, with a thought which travellers in their land can well understand, seem to identify their heaven-dwelling Creator with the all-producing Great Rain, under the name of Dendid; among the Damaras the highest deity is Omakuru the Rain-giver, who dwells in the far North; while to the negro of West Africa the Heaven-god is the rain-giver, and may pass in name into the rain itself.[2] Pachacamac, the Peruvian world-creator, has set the Rain-goddess to pour waters over the land, and send down hail and snow.[3] The Aztec Tlaloc was no doubt originally a Heaven-god, for he holds the thunder and lightning, but he has taken especially the attributes of Water-god and Rain-god; and so in Nicaragua the Rain-god Quiateot (Aztec quiahuitl=rain, teotl=god) to whom children were sacrificed to bring rain, shows his larger celestial nature by being also sender of thunder and lightning.[4] The Rain-god of the Khonds is Pidzu Pennu, whom the priests and elders propitiate with eggs and arrack and rice and a sheep, and invoke with quaintly pathetic prayers. They tell him how, if he will not give water, the

---

[1] Max Müller, 'Lectures,' 2nd Series, p. 425; Grimm, 'D. M.' ch. ix.; Cicero De Natura Deorum. iii. 4. Connexion of the Sanskrit Dyu with the Scandinavian Tyr and the Anglo Saxon Tiw is perhaps rather of etymology than definition.
[2] Lejean, 'Le Haut-Nil,' etc., in Rev. D. M. Apr. 1, 1862. Waitz, 'Anthropologie,' vol. ii. p. 169 (W. Afr.) p. 416 (Damaras).
[3] Markham, 'Quichua Gr. and Dic.' p. 9; J. G. Müller, 'Amer. Urrel.' pp. 318, 388.
[4] Ibid. pp. 496-9; Oviedo, 'Nicaragua,' pp. 40, 72.

land must remain unploughed, the seed will rot in the ground, they and their children and cattle will die of want, the deer and the wild hog will seek other haunts, and then of what avail will it be for the Rain-god to relent, how little any gift of water will avail, when there shall be left neither man, nor cattle, nor seed; so let him, resting on the sky, pour waters down upon them through his sieve, till the deer are drowned out of the forests and take refuge in the houses, till the soil of the mountains is washed into the valleys, till the cooking-pots burst with the force of the swelling rice, till the beasts gather so plentifully in the green and favoured land, that men's axes shall be blunted with cutting up the game.[1] With perfect meteorological fitness, the Kol tribes of Bengal consider their great deity Marang Buru, Great Mountain, to be the Rain-god. Marang Buru, one of the most conspicuous hills of the plateau near Lodmah in Chota-Nagpur, is the deity himself or his dwelling. Before the rains come on, the women climb the hill, led by the wives of the pahans, with girls drumming, to carry offerings of milk and bel-leaves, which are put on the flat rock at the top. Then the wives of the pahans kneel with loosened hair and invoke the deity, beseeching him to give the crops seasonable rain. They shake their heads violently as they reiterate this prayer, till they work themselves into a frenzy, and the movement becomes involuntary. They go on thus wildly gesticulating, till a cloud is seen; then they rise, take the drums, and dance the kurrun on the rock, till Marang Buru's response to their prayer is heard in the distant rumbling of thunder, and they go home rejoicing. They must go fasting to the mount, and stay there till there is "a sound of abundance of rain," when they get them down to eat and drink. It is said that the rain always comes before evening, but the old women appear to choose their own moment for beginning the fast.[2] It was to Ukko the

[1] Macpherson, 'India,' pp. 89, 355.
[2] Dalton, Kols, in 'Tr. Eth. Soc.' vol. vi. p. 34. Compare 1 Kings xviii.

Heaven-god, that in old days the Finn turned with such prayers :—

"Ukko, thou, O God above us,
Thou, O Father in the heavens,
Thou who rulest in the cloud-land,
And the little cloud-lambs leadest,
Send us down the rain from heaven,
Make the clouds to drop with honey,
Let the drooping corn look upward,
Let the grain with plenty rustle."[1]

Quite like this were the classic conceptions of Ζεὺς ὑέτιος, Jupiter Pluvius. They are typified in the famous Athenian prayer recorded by Marcus Aurelius, "Rain, rain, O dear Zeus, on the plough-lands of the Athenians, and the plains!"[2] and in Petronius Arbiter's complaint of the irreligion of his times, that now no one thinks heaven is heaven, no one keeps a fast, no one cares a hair for Jove, but all men with closed eyes reckon up their goods. Aforetime the ladies walked up the hill in their stoles with bare feet and loosened hair and pure minds, and entreated Jove for water; then all at once it rained bucketsfull, then or never, and they all went home wet as drowned rats.[3] In later ages, when drought parched the fields of the mediæval husbandman, he transferred to other patrons the functions of the Rain-god, and with procession and litany sought help from St. Peter or St. James, or, with more of mythological consistency, from the Queen of Heaven. As for ourselves, we have lived to see the time when men shrink from addressing even to Supreme Deity the old customary rain-prayers, for the rainfall is passing from the region of the supernatural, to join the tides and seasons in the realm of physical science.

[1] Castrén, 'Finn. Myth.' p. 36; Kalewala, Rune ii. 317.
[2] Marc. Antonin. v. 7. " Εὐχὴ 'Αθηναίων, ὗσον, ὗσον, ὦ φίλε Ζεῦ, κατὰ τῆς ἀρούρας τῶν 'Αθηναίων καὶ τῶν πεδίων."
[3] Petron. Arbiter. Sat xliv. "Antea stolatæ ibant nudis pedibus in clivum, passis capillis, mentibus puris, et Jovem aquam exorabant. Itaque statim urceatim plovebat: aut tunc aut nunquam; et omnes redibant udi tanquam mures." See Grimm, 'D. M.' p. 160.

The place of the Thunder-god in polytheistic religion is similar to that of the Rain-god, in many cases even to entire coincidence. But his character is rather of wrath than of beneficence, a character which we have half lost the power to realize, since the agonizing terror of the thunderstorm which appals savage minds has dwindled away in ours, now that we behold in it not the manifestation of divine wrath, but the restoration of electric equilibrium. North American tribes, as the Mandans, heard in the thunder and saw in the lightning the clapping wings and flashing eyes of that awful heaven-bird which belongs to, or even is, the Great Manitu himself.[1] The Dacotas could show at a place called Thunder-tracks, near the source of the St. Peter's River, the footprints of the thunder-bird five and twenty miles apart. It is to be noticed that these Sioux, among their varied fancies about thunder-birds and the like, give unusually well a key to the great thunderbolt-myth which recurs in so many lands. They consider the lightning entering the ground to scatter there in all directions thunderbolt-stones, which are flints, &c., their reason for this notion being the very rational one, that these siliceous stones actually produce a flash when struck.[2] In an account of certain Carib deities, who were men and are now stars, occurs the name of Savacou, who was changed into a great bird; he is captain of the hurricane and thunder, he blows fire through a tube and that is lightning, he gives the great rain. Rochefort describes the effect of a thunderstorm on the partly Europeanized Caribs of the West Indies two centuries ago. When they perceive its approach, he says, they quickly betake themselves to their cabins, and range themselves in the kitchen on their little seats near the fire; hiding their faces and leaning their heads in their hands and on their knees, they fall to weeping and lamenting in their jargon "Maboya mouche fache contre Caraïbe," *i.e.*,

[1] Pr. Max v. Wied, 'N. Amer.' vol. ii. pp. 152, 223; J. G. Müller, p. 120; Waitz, vol. iii. p. 179.

[2] Keating, 'Narr.' vol. i. p. 407; Eastman, 'Dahcotah,' p. 71; Brinton, p. 150, etc.; see M'Coy, 'Baptist Indian Missions,' p. 363.

Maboya (the evil demon) is very angry with the Caribs. This they say also when there comes a hurricane, not leaving off this dismal exercise till it is over, and there is no end to their astonishment that the Christians on these occasions manifest no such affliction and fear.[1] The Tupi tribes of Brazil are an example of a race among whom the Thunder or the Thunderer, Tupan, flapping his celestial wings and flashing with celestial light, was developed into the very representative of highest deity, whose name still stands among their Christian descendants as the equivalent of God.[2] In Peru, a mighty and far-worshipped deity was Catequil the Thunder-god, child of the Heaven-god, he who set free the Indian race from out of the ground by turning it up with his golden spade, he who in thunder-flash and clap hurls from his sling the small round smooth thunderstones, treasured in the villages as fire-fetishes and charms to kindle the flames of love. How distinct in personality and high in rank was the Thunder and Lightning (Chuqui yllayllapa) in the religion of the Incas, may be judged from his huaca or fetish-idol standing on the bench beside the idols of the Creator and the Sun at the great Solar festival in Cuzco, when the beasts to be sacrificed were led round them, and the priests prayed thus: "O Creator, and Sun, and Thunder, be for ever young! do not grow old. Let all things be at peace! let the people multiply, and their food, and let all other things continue to increase."[3]

In Africa, we may contrast the Zulu, who perceives in thunder and lightning the direct action of Heaven or Heaven's lord, with the Yoruba, who assigns them not to Olorun the Lord of Heaven, but to a lower deity, Shango the Thunder-god, whom they call also Dzakuta the Stone-caster, for it is he who (as among so many other peoples

[1] De la Borde, 'Caraïbes,' p. 530; Rochefort, 'Iles Antilles,' p. 481.
[2] De Laet, 'Novus Orbis,' xv. 2. Waitz, vol. iii. p. 417; J. G. Müller, p. 270; also 421 (thunderstorms by anger of Sun, in Cumana, etc.).
[3] Brinton, p. 153; Herrera, 'Indias Occidentales,' Dec. v. 4. J. G. Müller, p. 327. 'Rites and Laws of the Yncas,' tr. & ed. by C. R. Markham, p. 16, see 81; Prescott, 'Peru,' vol. i. p. 86.

who have forgotten their Stone Age) flings down from heaven the stone hatchets which are found in the ground, and preserved as sacred objects.[1] In the religion of the Kamchadals, Billukai, the hem of whose garment is the rainbow, dwells in the clouds with many spirits, and sends thunder and lightning and rain.[2] Among the Ossetes of the Caucasus the Thunderer is Ilya, in whose name mythologists trace a Christian tradition of Elijah, whose fiery chariot seems indeed to have been elsewhere identified with that of the Thunder-god, while the highest peak of Ægina, once the seat of Pan-hellenic Zeus, is now called Mount St. Elias. Among certain Moslem schismatics, it is even the historical Ali, cousin of Mohammed, who is enthroned in the clouds, where the thunder is his voice, and the lightning the lash wherewith he smites the wicked.[3] Among the Turanian or Tatar race, the European branch shows most distinctly the figure of the Thunder-god. To the Lapps, Tiermes appears to have been the Heaven-god, especially conceived as Aija the Thunder-god; of old they thought the Thunder (Aija) to be a living being, hovering in the air and hearkening to the talk of men, smiting such as spoke of him in an unseemly way; or, as some said, the Thunder-god is the foe of sorcerers, whom he drives from heaven and smites, and then it is that men hear in thunder-peals the hurtling of his arrows, as he speeds them from his bow, the Rainbow. In Finnish poetry, likewise, Ukko the Heaven-god is portrayed with such attributes. The Runes call him Thunderer, he speaks through the clouds, his fiery shirt is the lurid storm-cloud, we hear of his stones and his hammer, he flashes his fiery sword and it lightens, or he draws his mighty rainbow, Ukko's bow, to shoot his fiery copper arrows, wherewith men would invoke him to

---

[1] Bowen, 'Yoruba Lang.' p. xvi. in 'Smithsonian Contr.' vol. i. See Burton, 'Dahome,' vol. ii. p. 142. Details as to thunder-axes, etc. in 'Early Hist. of Mankind,' ch. viii.

[2] Steller, ' Kamtschatka,' p. 266.

[3] Klemm, ' C. G.' vol. iv. p. 85. (Ossetes, etc.) See Welcker, vol. i. p. 170 ; Grimm, ' D. M.' p. 158. Bastian, 'Mensch,' vol. ii. p. 423 (Ali-sect.).

smite their enemies. Or when it is dark in his heavenly house he strikes fire, and that is lightning. To this day the Finlanders call a thunderstorm an "ukko," or an "ukkonen," that is, "a little ukko," and when it lightens they say, "There is Ukko striking fire!"[1]

What is the Aryan conception of the Thunder-God, but a poetic elaboration of thoughts inherited from the savage state through which the primitive Aryans had passed? The Hindu Thunder-god is the Heaven-god Indra, Indra's bow is the rainbow, Indra hurls the thunderbolts, he smites his enemies, he smites the dragon-clouds, and the rain pours down on earth, and the sun shines forth again. The Veda is full of Indra's glories: "Now will I sing the feats of Indra, which he of the thunderbolt did of old. He smote Ahi, then he poured forth the waters; he divided the rivers of the mountains. He smote Ahi by the mountain; Tvashtar forged for him the glorious bolt."—" Whet, O strong Indra, the heavy strong red weapon against the enemies!" —" May the axe (the thunderbolt) appear with the light; may the red one blaze forth bright with splendour!"— "When Indra hurls again and again his thunderbolt, then they believe in the brilliant god." Nor is Indra merely a great god in the ancient Aryan pantheon, he is the very patron-deity of the invading Aryan race in India, to whose help they look in their conflicts with the dark-skinned tribes of the land. "Destroying the Dasyus, Indra protected the Aryan colour"—"Indra protected in battle the Aryan worshipper, he subdued the lawless for Manu, he conquered the black skin."[2] This Hindu Indra is the offspring of Dyaus the Heaven. But in the Greek religion, Zeus is himself Zeus Kerauneios, the wielder of the thunderbolt, and thunders from the cloud-capped tops of Ida or Olympos. In like manner the Jupiter Capitolinus of Rome is himself Jupiter Tonans:

[1] Castren, 'Finn. Myth.' p. 39, etc.
[2] 'Rig-Veda,' i. 32. 1, 55. 5, 130. 8, 165; iii. 34. 9; vi. 20; x. 43. 9, 89, 9. Max Müller, 'Lectures,' 2nd S. p. 427; 'Chips,' vol. i. p. 42, vol. ii. p. 323. See Muir, 'Sanskrit Texts.'

"Ad penetrale Numæ, Capitolinumque Tonantem."[1]

Thus, also, it was in accurate language that the old Slavonic nations were described as adoring Jupiter Tonans as their highest god. He was the cloud-dwelling Heaven-god, his weapon the thunder-bolt, the lightning-flash, his name Perun the Smiter (Perkun, Perkunas). In the Lithuanian district, the thunder itself is Perkun; in past times the peasant would cry when he heard the thunder peal "Dewe Perkune apsaugog mus!—God Perkun spare us!" and to this day he says, "Perkunas gravja!—Perkun is thundering!" or "Wezzajs barrahs!—the Old One growls!"[2] The old German and Scandinavian theology made Thunder, Donar, Thor, a special deity to rule the clouds and rain, and hurl his crushing hammer through the air. He reigned high in the Saxon heaven, till the days came when the Christian convert had to renounce him in solemn form, "ec forsacho Thunare!—I forsake Thunder!" Now, his survival is for the most part in mere verbal form, in the etymology of such names as Donnersberg, Thorwaldsen, Thursday.[3]

In the polytheism of the lower as of the higher races, the Wind-gods are no unknown figures. The Winds themselves, and especially the Four Winds in their four regions, take name and shape as personal divinities, while some deity of wider range, a Wind-god, Storm-god, Air-god, or the mighty Heaven-god himself, may stand as compeller or controller of breeze and gale and tempest. We have already taken as examples from the Algonquin mythology of North America the four winds whose native legends have been versified in "Hiawatha;" Mudjekeewis the West Wind, Father of the Winds of Heaven, and his children, Wabun the East Wind, the morning-bringer, the lazy Shawondasee the South Wind, the wild and cruel North

[1] Homer, Il. viii. 170, xvii. 595. Ovid. Fast. ii. 69. See Max Müller, 'Lectures,' l. c.; Welcker, 'Griech. Götterl.' vol. ii. p. 194.
[2] Hanusch, 'Slaw. Myth.' p. 257.
[3] Grimm, 'Deutsche Myth.' ch. viii. Edda; Gylfaginning, 21, 44.

Wind, the fierce Kabibonokka. Viewed in their religious aspect, these mighty beings correspond with four of the great manitus sacrificed to among the Delawares, the West, South, East, and North; while the Iroquois acknowledged a deity of larger grasp, Gäoh, the Spirit of the Winds, who holds them prisoned in the mountains in the Home of the Winds.[1] The Polynesian Wind-gods are thus described by Ellis: "The chief of these were Veromatautoru and Tairibu, brother and sister to the children of Taaroa, their dwelling was near the great rock, which was the foundation of the world. Hurricanes, tempests, and all destructive winds, were supposed to be confined within them, and were employed by them to punish such as neglected the worship of the gods. In stormy weather their compassion was sought by the tempest-driven mariner at sea, or the friends of such on shore. Liberal presents, it was supposed, would at any time purchase a calm. If the first failed, subsequent ones were certain of success. The same means were resorted to for procuring a storm, but with less certainty. Whenever the inhabitants of one island heard of invasion from those of another, they immediately carried large offerings to these deities, and besought them to destroy by tempest the hostile fleet whenever it might put to sea. Some of the most intelligent people still think evil spirits had formerly great power over the winds, as they say there have been no such fearful storms since they abolished idolatry, as there were before." Or, again, the great deity Maui adds a new complication to his enigmatic solar-celestial character by appearing as a Wind-god. In Tahiti he was identified with the East Wind; in New Zealand he holds all the winds but the west in his hands, or he imprisons them with great stones rolled to the mouths of their caves, save the West Wind

[1] Schoolcraft, 'Algic. Res.' vol. i. p. 139, vol. ii. p. 214; Loskiel, part i. p. 43; Waitz, vol. iii. p. 190. Morgan, 'Iroquois,' p. 157; J. G. Müller, p. 56; Further American evidence in Brinton, 'Myths of New World,' pp. 50, 74; Cranz, 'Grönland,' 267 (Sillagiksartok, Weather-spirit); De la Borde, 'Caraïbes,' p. 530 (Carib-Star Curumon, makes the billows and upsets canoes).

which he cannot catch or prison, so that it almost always blows.[1] To the Kamchadal, it is Billukai the Heaven-god who comes down and drives his sledge on earth, and men see his traces in the wind-drifted snow.[2] To the Finn, while there are traces of subordinate Wind-gods in his mythology, the great ruler of wind and storm is Ukko the Heaven-god;[3] while the Esth looked rather to Tuule-ema, Wind's Mother, and when the gale shrieks he will still say "Wind's mother wails, who knows what mothers shall wail next."[4] Such instances from Allophylian mythology[5] show types which are found developed in full vigour by the Aryan races. In the Vedic hymns, the Storm-gods, the Maruts, toss the clouds across the surging sea; Indra the Heaven-god, with the swift Maruts who break through the stronghold, finds in their hiding places the bright cows, the days.[6] No effort of the Red Indian's personifying fancy in the tales of the dancing Pauppuk-keewis the Whirlwind, or that fierce and shifty hero, Manabozho the North-West Wind, can more than match the description in the Iliad, of Achilles calling on Boreas and Zephyros with libations and vows of sacrifice, to blow into a blaze the funeral pyre of Patroklos—

" . . . . his prayer
Swift Iris heard, and bore it to the Winds.
They in the hall of gusty Zephyrus
Were gathered round the feast; in haste appearing,
Swift Iris on the stony threshold stood.
They saw, and rising all, besought her each
To sit beside him; she with their requests
Refused compliance, and addressed them thus," &c.

[1] Ellis, 'Polyn. Res.' vol. i. p. 329, (compare with the Maori Tempest-god Tawhirimatea, Grey, 'Polyn. Myth.' p. 5); Schirren, 'Wandersage der Neuseeländer,' etc. p. 85; Yate, 'New Zealand,' p. 144. See also Mariner, 'Tonga. Is.' vol. ii. p. 115.
[2] Steller, 'Kamschatka,' p. 266.
[3] Castrén, 'Finn. Myth.' pp. 37, 68.
[4] Boecler, pp. 106, 147.
[5] See also Klemm, 'Cultur-Gesch.' vol. iv. p. 85 (Circassian Water-god and Wind-god).
[6] 'Rig-Veda,' tr. by Max Müller, i. 6. 5, 19. 7.

Æolus with the winds imprisoned in his cave has the office of the Red Indian Spirit of the Winds, and of the Polynesian Maui. With quaint adaptation to nature-myth and even to moral parable, the Harpies, the Storm-gusts that whirl and snatch and dash and smirch with eddying dust-clouds, become the loathsome bird-monsters sent to hover over the table of Phineus to claw and defile his dainty viands.[1] If we are to choose an Aryan Storm-god for ideal grandeur, we must seek him in

". . . . the hall where Runic Odin
Howls his war-song to the gale."

Jacob Grimm has defined Odin or Woden as "the all-penetrating creative and formative power." But we can hardly ascribe such abstract conceptions to his barbaric worshippers. As little may we seek his real nature among the legends which degrade him to a historical king of Northern men, an " Othinus rex." See the All-father sitting cloud-mantled on his heaven-seat, overlooking the deeds of men, and we must discern in him the attributes of the Heaven-god. Hear the peasant say of the raging tempest, that it is " Odin faring by ; " trace the mythological transition from Woden's tempest to the " Wütende Heer," the " Wild Huntsman " of our own grand storm-myth, and we shall recognize the old Teutonic deity in his function of cloud-compeller, of Tempest-god.[2] The " rude Carinthian boor " can show a relic from a yet more primitive stage of mental history, when he sets up a wooden bowl of various meats on a tree before his house, to fodder the wind that it may do no harm. In Swabia, Tyrol, and the Upper Palatinate, when the storm rages, they will fling a spoonful or a handful of meal in the face of the gale, with this formula in the last-named district, " Da Wind, hast du Mehl für dein Kind, aber aufhören musst du ! "[3]

[1] Homer, Il. xxiii. 192 (Lord Derby's trans.) Odys. xx. 66, 77; Apollon. Rhod. Argonautica ; Apollodor. i. 9. 21 ; Virg. Æn. i. 56 ; Welcker, 'Griech. Götterl.' vol. i. p. 707, vol. iii. p. 67.
[2] Grimm, 'Deutsche Myth.' pp. 121, 871.
[3] Wuttke, 'Deutsche Volksabergl.' p. 86.

The Earth-deity takes an important place in polytheistic religion. The Algonquins would sing medicine-songs to Mesukkummik Okwi, the Earth, the Great-Grandmother of all. In her charge (and she must be ever at home in her lodge) are left the animals whose flesh and skins are man's food and clothing, and the roots and medicines of sovereign power to heal sickness and kill game in time of hunger; therefore good Indians never dig up the roots of which their medicines are made, without depositing an offering in the earth for Mesukkummik Okwi.[1] In the list of fetish-deities of Peruvian tribes, the Earth, adored as Mamapacha, Mother Earth, took high subordinate rank below Sun and Moon in the pantheon of the Incas, and at harvest-time ground corn and libations of chicha were offered to her that she might grant a good harvest.[2] Her rank is similar in the Aquapim theology of West Africa; first the Highest God in the firmament, then the Earth as universal mother, then the fetish. The negro, offering his libation before some great undertaking, thus calls upon the triad: "Creator! come drink! Earth, come drink! Bosumbra, come drink!"[3]

Among the indigenes of India, the Bygah tribes of Seonee show a well-marked worship of the Earth. They call her "Mother Earth" or Dhurteemah, and before praying or eating their food, which is looked on always as a daily sacrifice, they invariably offer some of it to the earth, before using the name of any other god.[4] Of all religions of the world, perhaps that of the Khonds of Orissa gives the Earth-goddess her most remarkable place and function. Boora Pennu or Bella Pennu, the Light-god or Sun-god, created Tari Pennu the Earth-goddess for his

[1] Tanner's 'Narrative,' p. 193; Loskiel, l. c. See also Rochefort, 'Iles Antilles,' p. 414; J. G. Müller, p. 178 (Antilles).
[2] Garcilaso de la Vega, 'Commentarios Reales,' i. 10; Rivero & Tschudi, p. 161; J. G. Müller, p. 369.
[3] Waitz, 'Anthropologie,' vol. ii. p. 170.
[4] 'Report of Ethnological Committee, Jubbulpore Exhibition,' 1866-7. Nagpore, 1868, part ii. p. 54.

consort, and from them were born the other great gods. But strife arose between the mighty parents, and it became the wife's work to thwart the good creation of her husband, and to cause all physical and moral ill. Thus to the Sun-worshipping sect she stands abhorred on the bad eminence of the Evil Deity. But her own sect, the Earth-worshipping sect, seem to hold ideas of her nature which are more primitive and genuine. The functions which they ascribe to her, and the rites with which they propitiate her, display her as the Earth-mother, raised by an intensely agricultural race to an extreme height of divinity. It was she who with drops of her blood made the soft muddy ground harden into firm earth; thus men learnt to offer human victims, and the whole earth became firm; the pastures and ploughed fields came into use, and there were cattle and sheep and poultry for man's service; hunting began, and there were iron and ploughshares and harrows and axes, and the juice of the palm-tree; and love arose between the sons and daughters of the people, making new households, and society with its relations of father and mother, and wife and child, and the bonds between ruler and subject. It was the Khond Earth-goddess who was propitiated with those hideous sacrifices, the suppression of which is matter of recent Indian history. With dances and drunken orgies, and a mystery play to explain in dramatic dialogue the purpose of the rite, the priest offered Tari Pennu her sacrifice, and prayed for children and cattle and poultry and brazen pots and all wealth; every man and woman wished a wish, and they tore the slave-victim piecemeal, and spread the morsels over the fields they were to fertilize.[1] In Northern Asia, also, among the Tatar races, the office of the Earth-deity is strongly and widely marked. Thus in the nature-worship of the Tunguz and Buraets, Earth stands among the greater divinities. It is especially interesting to notice among the Finns a transition like that just observed from the god

[1] Macpherson, 'India,' chap. vi.

Heaven to the Heaven-god. In the designation of Maaemä, Earth-mother, given to the earth itself, we seem to trace survival from the stage of direct nature-worship, while the passage to the conception of a divine being inhabiting and ruling the material substance, is marked by the use of the name Maan emo, Earth's mother, for the ancient subterranean goddess whem men would ask to make the grass shoot thick and the thousandfold ears mount high, or might even entreat to rise in person out of the earth to give them strength. The analogy of other mythologies agrees with the definition of the divine pair who reign in Finn theology: as Ukko the Grandfather is the Heaven-god, so his spouse Akka the Grandmother is the Earth-goddess.[1] Thus in the ancient nature-worship of China, the personal Earth holds a place below the Heaven. Tien and Tu are closely associated in the national rites, and the idea of the pair as universal parents, if not an original conception in Chinese theology, is at any rate developed in Chinese classic symbolism. Heaven and Earth receive their solemn sacrifices not at the hands of common mortals but of the Son of Heaven, the Emperor, and his great vassals and mandarins. Yet their adoration is national; they are worshipped by the people who offer incense to them on the hill-tops at their autumn festival, they are adored by successful candidates in competitive examination, and, especially and appropriately, the prostration of bride and bridegroom before the father and mother of all things, the "worshipping of Heaven and Earth," is the all-important ceremony of a Chinese marriage.[2]

The Vedic hymns commemorate the goddess Prithivî, the broad Earth, and in their ancient strophes the modern Brahmans still pray for benefits to mother Earth and father Heaven, side by side:—

[1] Georgi, 'Reise im Russ. Reich.' vol. i. pp. 275, 317. Castrén, 'Finn. Myth.' p. 86, etc.
[2] Plath, 'Religion der Alten Chinesen,' part i. pp. 36, 73, part ii. p. 32. Doolittle, 'Chinese,' vol. i. pp. 86, 354, 413, vol. ii. pp. 67, 380, 455.

"Tanno Vâto mayobhu vâtu bheshajam tanmâtâ Prithivî tatpitâ Dyauh."[1]
Greek religion shows a transition to have taken place like that among the Turanian tribes, for the older simpler nature-deity Gaia, Γῆ πάντων μήτηρ, Earth the All-Mother, seems to have faded into the more anthropomorphic Dēmētēr, Earth-Mother, whose eternal fire burned in Mantinēa, and whose temples stood far and wide over the land which she made kindly to the Greek husbandman.[2] The Romans acknowledged her plain identity as Terra Mater, Ops Mater.[3] Tacitus could rightly recognize this deity of his own land among German tribes, worshippers of "Nerthum (or, Hertham), id est Terram matrem," Mother Earth, whose holy grove stood in an ocean isle, whose chariot drawn by cows passed through the land making a season of peace and joy, till the goddess, satiated with mortal conversation, was taken back by her priest to her temple, and the chariot and garments and even the goddess herself were washed in a secret lake, which forthwith swallowed up the ministering slaves—"hence a mysterious terror and sacred ignorance, what that should be which only the doomed to perish might behold."[4] If in these modern days we seek in Europe traces of Earth-worship, we may find them in curiously distinct survival in Germany, if no longer in the Christmas food-offerings buried in and for the earth up to early in this century,[5] at any rate among Gypsy hordes. Dewel, the great god in heaven (dewa, deus), is rather feared than loved by these weatherbeaten outcasts, for he harms them on their wanderings with his thunder and lightning, his snow and rain, and his stars interfere with their dark doings. Therefore they curse him foully when misfortune falls on them, and when a child dies, they say that Dewel has eaten it. But Earth, Mother of all good,

[1] 'Rig-Veda,' i. 89. 4, etc. etc.
[2] Welcker, 'Griech. Götterl.' vol. i. p. 385, etc.
[3] Varro de Ling. Lat. iv.
[4] Tacit. Germania, 40. Grimm, 'Deutsche Myth.' p. 229, etc.
[5] Wuttke, 'Deutsche Volksabergl.' p. 87.

self-existing from the beginning, is to them holy, so holy that they take heed never to let the drinking-cup touch the ground, for it would become too sacred to be used by men.[1]

Water-worship, as we have seen, may be classified as a special department of religion. It by no means follows, however, that savage water-worshippers should necessarily have generalized their ideas, and passed beyond their particular water-deities to arrive at the conception of a general deity presiding over water as an element. Divine springs, streams, and lakes, water-spirits, deities concerned with the clouds and rain, are frequent, and many details of them are cited here, but I have not succeeded in finding among the lower races any divinity whose attributes, fairly criticized, will show him or her to be an original and absolute elemental Water-god. Among the deities of the Dakotas, Unktahe the fish-god of the waters is a master-spirit of sorcery and religion, the rival even of the mighty Thunder-bird.[2] In the Mexican pantheon, Tlaloc god of rain and waters, fertilizer of earth and lord of paradise, whose wife is Chalchihuitlicue, Emerald-Skirt, dwells among the mountain-tops where the clouds gather and pour down the streams.[3] Yet neither of these mythic beings approaches the generality of conception that belongs to full elemental deity, and even the Greek Nēreus, though by his name he should be the very personification of water ($\nu\eta\rho\acute{o}s$), seems too exclusively marine in his home and family to be cited as the Water-god. Nor is the reason of this hard to find. It is an extreme stretch of the power of theological generalization to bring water in its myriad forms under one divinity, though each individual body of water, even the smallest stream or lake, can have its personal individuality or indwelling spirit.

[1] Liebich, 'Die Zigeuner,' pp. 30, 84.
[2] Schoolcraft, Indian Tribes,' part iii. p. 485 ; Eastman, 'Dahcotah,' p. i. 118, 161.
[3] Clavigero, vol. ii. p. 14.

Islanders and coast-dwellers indeed live face to face with mighty water-deities, the divine Sea and the great Sea-gods. What the sea may seem to an uncultured man who first beholds it, we may learn among the Lampongs of Sumatra: "The inland people of that country are said to pay a kind of adoration to the sea, and to make to it an offering of cakes and sweetmeats on their beholding it for the first time, deprecating its power of doing them mischief."[1] The higher stage of such doctrine is where the sea, no longer itself personal, is considered as ruled by indwelling spirits. Thus Tuaraatai and Ruahatu, principal among marine deities of Polynesia, send the sharks to execute their vengeance. Hiro descends to the depths of the ocean and dwells among the monsters, they lull him to sleep in a cavern, the Wind-god profits by his absence to raise a violent storm to destroy the boats in which Hiro's friends are sailing, but, roused by a friendly spirit-messenger, the Sea-god rises to the surface and quells the tempest.[2] This South Sea Island myth might well have been in the Odyssey. We may point to the Guinea Coast as a barbaric region where Sea-worship survives in its extremest form. It appears from Bosman's account, about 1700, that in the religion of Whydah, the Sea ranked only as younger brother in the three divine orders, below the Serpents and Trees. But at present, as appears from Captain Burton's evidence, the religion of Whydah extends through Dahome, and the divine Sea has risen in rank. "The youngest brother of the triad is Hu, the ocean or sea. Formerly it was subject to chastisement, like the Hellespont, if idle or useless. The Huno, or ocean priest, is now considered the highest of all, a fetish king, at Whydah, where he has 500 wives. At stated times he repairs to the beach, begs 'Agbwe,' the . . . ocean god, not to be boisterous, and throws in rice and corn, oil and beans, cloth, cowries, and other valuables. . . . At times the King sends as an ocean sa-

[1] Marsden, 'Sumatra,' p. 301 ; see also 303 (Tagals).
[2] Ellis, 'Polyn. Res.' vol. i. p. 328.

crifice from Agbome a man carried in a hammock, with the dress, the stool, and the umbrella of a caboceer; a canoe takes him out to sea, where he is thrown to the sharks."[1] While in these descriptions the individual divine personality of the sea is so well marked, an account of the closely related Slave Coast religion states that a great god dwells in the sea, and it is to him, not to the sea itself, that offerings are cast in.[2] In South America the idea of the divine Sea is clearly marked in the Peruvian worship of Mamacocha, Mother Sea, giver of food to men.[3] Eastern Asia, both in its stages of lower and higher civilization, contributes members to the divine group. In Kamchatka, Mitgk the Great Spirit of the Sea, fish-like himself, sends the fish up the rivers.[4] Japan deifies separately on land and at sea the lords of the waters; Midsuno Kami, the Water-god, is worshipped during the rainy season; Jebisu, the Sea-god, is younger brother of the Sun.[5]

Among barbaric races we thus find two conceptions current, the personal divine Sea and the anthropomorphic Sea-god. These represent two stages of development of one idea—the view of the natural object as itself an animated being, and the separation of its animating fetish-soul as a distinct spiritual deity. To follow the enquiry into classic times shows the same distinction as strongly marked. When Kleomenes marched down to Thyrea, having slaughtered a bull to the sea (σφαγιασάμενος δὲ τῇ θαλάσσῃ ταῦρον) he embarked his army in ships for the Tirynthian land and Nauplia.[6] Cicero makes Cotta remark to Balbus that "our generals, embarking on the sea, have been accustomed to immolate a victim to the waves," and he goes on to argue,

---

[1] Bosman, 'Guinea,' letter xix.; in Pinkerton, vol. xvi. p. 494. Burton, 'Dahome,' vol. ii. p. 141. See also below, chap. xviii. (sacrifice).
[2] Schlegel, 'Ewe Sprache,' p. xiv.
[3] Garcilaso de la Vega, 'Commentarios Reales,' i. 10, vi. 17; Rivero & Tschudi, 'Peru,' p. 161.
[4] Steller, 'Kamtschatka,' p. 265.
[5] Siebold, 'Nippon,' part v. p. 9.
[6] Herod. vi. 76.

not unfairly, that if the Earth herself is a goddess, what is she other than Tellus, and "if the Earth, the Sea too, whom thou saidst to be Neptune."[1] Here is direct nature-worship in its extremest sense of fetish-worship. But in the anthropomorphic stage appear that dim præ-Olympian figure of Nēreus the Old Man of the Sea, father of the Nereids in their ocean caves, and the Homeric Poseidōn the Earth-shaker, who stables his coursers in his cave in the Ægean deeps, who harnesses the gold-maned steeds to his chariot and drives through the dividing waves, while the subject sea-beasts come up at the passing of their lord, a king so little bound to the element he governs, that he can come from the brine to sit in the midst of the gods in the assembly on Olympos, and ask the will of Zeus.[2]

Fire-worship brings into view again, though under different aspects and with different results, the problems presented by water-worship. The real and absolute worship of fire falls into two great divisions, the first belonging rather to fetishism, the second to polytheism proper, and the two apparently representing an earlier and later stage of theological ideas. The first is the rude barbarian's adoration of the actual flame which he watches writhing, roaring, devouring like a live animal; the second belongs to an advanced generalization, that any individual fire is a manifestation of one general elemental being, the Fire-god. Unfortunately, evidence of the exact meaning of fire-worship among the lower races is scanty, while the transition from fetishism to polytheism seems a gradual process of which the stages elude close definition. Moreover, it must be borne in mind that rites performed with fire are, though often, yet by no means necessarily, due to worship of the fire itself. Authors who have indiscriminately mixed up such rites as the new fire, the perpetual fire, the passing

[1] Cicero, De Natura Deorum, iii. 20.
[2] Homer, Il. i. 538, xiii. 18, xx. 13. Gladstone, 'Juventus Mundi.' Welcker, 'Griech. Götterl.' vol. i. p. 616, etc. Cox, 'Mythology of Aryan Nations,' vol. ii. ch. vi.

through the fire, classing them as acts of fire-worship, without proper evidence as to their meaning in any particular case, have added to the perplexity of a subject not too easy to deal with, even under strict precautions. Two sources of error are especially to be noted. On the one hand, fire happens to be a usual means whereby sacrifices are transmitted to departed souls and deities in general; and on the other hand, the ceremonies of earthly fire-worship are habitually and naturally transferred to celestial fire-worship in the religion of the Sun.

It may best serve our present purpose to carry a line of some of the best-defined facts which seem to bear on fire-worship proper, from savagery on into the higher culture. In the last century, Loskiel, a missionary among the North American Indians, remarks that "In great danger, an Indian has been observed to lie prostrate on his face, and throwing a handful of tobacco into the fire, to call aloud, as in an agony of distress, 'There, take and smoke, be pacified, and don't hurt me.'" Of course this may have been a mere sacrifice transmitted to some other spiritual being through fire, but we have in this region explicit statements as to a distinct fire-deity. The Delawares, it appears from the same author, acknowledged the Fire-manitu, first parent of all Indian nations, and celebrated a yearly festival in his honour, when twelve manitus, animal and vegetable, attended him as subordinate deities.[1] In North-West America, in Washington Irving's account of the Chinooks and other Columbia River tribes, mention is made of the spirit which inhabits fire. Powerful both for evil and good, and seemingly rather evil than good in nature, this being must be kept in good humour by frequent offerings. The Fire-spirit has great influence with the winged aërial supreme deity, wherefore the Indians implore him to be their interpreter, to procure them success in hunting and fishing, fleet horses, obedient wives, and male children.[2] In the elaborately

[1] Loskiel, 'Ind. of N. A.' part i. pp. 41, 45. See also J. G. Müller, p. 55.
[2] Irving, 'Astoria,' vol. ii. ch. xxii.

systematic religion of Mexico, there appears in his proper place a Fire-god, closely related to the Sun-god in character, but keeping well marked his proper identity. His name was Xiuhteuctli, Fire-lord, and they called him likewise Huehueteotl, the old god. Great honour was paid to this god Fire, who gives them heat, and bakes their cakes, and roasts their meat. Therefore at every meal the first morsel and libation were cast into the fire, and every day the deity had incense burnt to him. Twice in the year were held his solemn festivals. At the first, a felled tree was set up in his honour, and the sacrificers danced round his fire with the human victims, whom afterwards they cast into a great fire, only to drag them out half roasted for the priests to complete the sacrifice. The second was distinguished by the rite of the new fire, so well known in connexion with solar worship; the friction-fire was solemnly made before the image of Xiuhteuctli in his sanctuary in the court of the great teocalli, and the game brought in at the great hunt which began the festival was cooked at the sacred fire for the banquets that ended it.[1] Polynesia well knows from the mythological point of view Mahuika the Fire-god, who keeps the volcano-fire on his subterranean hearth, whither Maui goes down (as the Sun into the Underworld) to bring up fire for man; but in the South Sea islands there is scarcely a trace of actual rites of fire-worship.[2] In West Africa, among the gods of Dahome is Zo the fire-fetish; a pot of fire is placed in a room, and sacrifice is offered to it, that fire may "live" there, and not go forth to destroy the house.[3]

Asia is a region where distinct fire-worship may be peculiarly well traced through the range of lower and higher civilization. The rude Kamchadals, worshipping all things

[1] Torquemada, 'Monarquia Indiana,' vi. c. 28, x. c. 22, 30; Brasseur 'Mexique,' vol. iii. pp. 492, 522, 536.
[2] Schirren, 'Wandersage der Neuseeländer,' etc. p. 32; Turner, 'Polynesia,' pp. 252, 527.
[3] Burton, 'Dahome,' vol. ii. p. 148; Schlegel, 'Ewe Sprache,' p. x v.

that did them harm or good, worshipped the fire, offering to it noses of foxes and other game, so that one might tell by looking at furs whether they had been taken by baptized or heathen hunters.[1] The Ainos of Yesso have many gods, but Fire is the principal deity, to whom, not to Sun, Moon, and Stars, they pray for all they need.[2] Turanian tribes likewise hold fire a sacred element, many Tunguz, Mongol, and Turk tribes sacrifice to Fire, and some clans will not eat meat without first throwing a morsel upon the hearth. The following passage from a Mongol wedding-song to the personified Fire, seems curiously to acknowledge the precedence of the ancient friction-fire made by the wooden drill, over that made by the more modern flint and steel. "Mother Ut, Queen of Fire, thou who art made from the elm that grows on the mountain-tops of Changgai-Chan and Burchatu-Chan, thou who didst come forth when heaven and earth divided, didst come forth from the footsteps of Mother Earth, and wast formed by the King of Gods. Mother Ut, whose father is the hard steel, whose mother is the flint, whose ancestors are the elm-trees, whose shining reaches to the sky and pervades the earth. Goddess Ut, we bring thee yellow oil for offering, and a white wether with yellow head, thou who hast a manly son, a beauteous daughter-in-law, bright daughters. To thee, Mother Ut, who ever lookest upward, we bring brandy in bowls, and fat in both hands. Give prosperity to the King's son (the bridegroom), to the King's daughter (the bride), and to all the people!"[3] As an analogue to Hephaistos the Greek divine smith, may stand the Circassian Fire-god, Tleps, patron of metal-workers and the peasants whom he has provided with plough and hoe.[4]

[1] Steller, 'Kamtschatka,' p. 276.
[2] Bickmore, Ainos, in 'Tr. Eth. Soc.' vol. vii. p. 20.
[3] Castrén, 'Finn. Myth.' p. 57 ; Billings, 'N. Russia,' p. 123 (Yakuts); Bastian, 'Vorstellungen von Wasser und Feuer,' in 'Zeitschr. für Ethnologie,' vol. i. p. 383 (Mongols).
[4] Klemm, 'Cultur-Gesch.' vol. vi. p. 85 (Circassia). Welcker, vol. i. p. 663.

The fire-worship of Assyria, Chaldæa, Phœnicia, is famous in history, the fire-pillars, the temple of the Tyrian Baal where stood no image but the eternal fire burning on the hearth, the Canaanitish Moloch to whom (whether in actual or symbolic sacrifice) children passed through the fire. "And they built the high places of Baal, in the valley of the son of Hinnom, to cause their sons and their daughters to pass through to Moloch."[1] But the records which have reached us of these ancient deities are obscure and complex in their definition, and their study is perhaps more valuable in compiling the history than in elucidating the principles of religion. For this scientific purpose, the more full and minute documents of Aryan religion can give a better answer. In various forms and under several names, the Fire-god is known. Nowhere does he carry his personality more distinctly than under his Sanskrit name of Agni, a word which keeps its quality, though not its divinity, in the Latin "ignis." The name of Agni is the first word of the first hymn of the Rig-Veda: "Agnim île puro-hitam yajnasya devam ritvijam!—Agni I entreat, divine appointed priest of sacrifice!" The sacrifices which Agni receives go to the gods, he is the mouth of the gods, but he is no lowly minister, as it is said in another hymn:

"No god indeed, no mortal is beyond the might of thee, the mighty one, with the Maruts come hither, O Agni!"

Such the mighty Agni is among the gods, yet he comes within the peasant's cottage to be protector of the domestic hearth. His worship has survived the transformation of the ancient patriarchal Vedic religion of nature into the priest-ridden ritualistic Hinduism of our own day, where Agni still, as among the ruder Mongol hordes north of the Himalaya, is new-born of the twirling fire-sticks, and receives the melted butter of the sacrifice.[2] Among the

[1] 2 Kings, xxiii. 10; Jerem. xxxii. 35; etc. Movers, 'Phönizier,' vol. i. p. 327 etc., 337 etc., 401.
[2] 'Rig-Veda,' i. 1. 1, 19 2, iii. 1. 18, etc.; Max Müller, vol. i. p. 39 Ward, 'Hindoos,' vol. ii. p. 53.

records of fire-worship in Asia, is the account in Jonas Hanway's 'Travels,' dating from about 1740, of the everlasting fire at the burning wells near Baku, on the Caspian. At the sacred spot stood several ancient stone temples, mostly arched vaults 10 to 15 feet high. One little temple was still used for worship, near the altar of which, about three feet high, a large hollow cane conveyed the gas up from the ground, burning at the mouth with a blue flame. Here were generally forty or fifty poor devotees, come on pilgrimage from their country to make expiation for themselves and others, and subsisting on wild celery, etc. These pilgrims are described as marking their foreheads with saffron, and having great veneration for a red cow; they wore little clothing, and the holiest of them kept one arm on their heads, or continued unmoved in some other posture; they are described as Ghebers, or Gours, the usual Moslem term for Fire-worshippers.[1]

In general, this name of Ghebers is applied to the Zoroastrians or Parsis, whom a modern European would all but surely point to if asked to instance a modern race of Fire-worshippers. Classical accounts of the Persian religion set down fire-worship as part and parcel of it; the Magi, it is recorded, hold the gods to be Fire and Earth and Water; and again, the Persians reckon the Fire to be a god ($\theta\epsilon o\phi o\rho o\hat{v}\sigma\iota\nu$).[2] On the testimony of the old religious books of the Parsis themselves, Fire, as the greatest Ized, as giver of increase and health, as craving for wood and scents and fat, seems to take the distinctest divine personality. Their doctrine that Ardebehist, the presiding angel or spirit of fire, is adored, but not the material object he belongs to, is a perfect instance of the development of the idea of an elemental divinity from that of an animated fetish. When, driven by Moslem persecution from Persia,

[1] Hanway, 'Journal of Travels,' London, 1753. vol. i. ch. lvii.
[2] Diog. Laert. Proœm. ii. 6. Sextus Empiricus adv. Physicos, ix.; Strabo, xv. 3, 13.

Parsi exiles landed in Gujarat, they described their religion in an official document as being the worship of Agni or Fire, thus claiming for themselves a place among recognized Hindu sects.[1] In modern times, though for the most part the Parsis have found toleration and prosperity in India, yet an oppressed remnant of the race still keeps up the everlasting fires at Yezd and Kirman, in their old Persian land. The modern Parsis, as in Strabo's time, scruple to defile the fire or blow it with their breath, they abstain from smoking out of regard not to themselves but to the sacred element, and they keep up consecrated ever-burning fires before which they do worship. Nevertheless, Prof. Max Müller is able to say of the Parsis of our own day: "The so-called Fire-worshippers certainly do not worship the fire, and they naturally object to a name which seems to place them on a level with mere idolators. All they admit is, that in their youth they are taught to face some luminous object while worshipping God, and that they regard the fire, like other great natural phenomena, as an emblem of the Divine power. But they assure us that they never ask assistance or blessings from an unintelligent material object, nor is it even considered necessary to turn the face to any emblem whatever in praying to Ormuzd."[2] Now, admitting this view of fire-worship as true of the more intelligent Parsis, and leaving aside the question how far among the more ignorant this symbolism may blend (as in such cases is usual) into actual adoration, we may ask what is the history of ceremonies which thus imitate, yet are not, fire-worship. The ethnographic answer is clear and instructive. The Parsi is the descendant of a race in this respect represented by the modern Hindu, a race who did simply and actually worship Fire. But the development of the more philosophic Zarathustrian doctrines has led to a result common in the history of religion, that the ancient distinctly

---

[1] John Wilson, 'The Parsi Religion,' ch. iv; 'Avesta,' tr. by Spiegel & Bleek, Yaçna, i. lxi.
[2] Max Müller, 'Chips,' vol. i. p. 169.

meant rite has dwindled to a symbol, to be preserved with changed sense in a new theology.

Somewhat of the same kind may have taken place among the European race who seem in some respects the closest relatives of the old Persians. Slavonic history possibly keeps up some trace of direct and absolute fire-worship, as where in Bohemia the Pagans are described as worshipping fires, groves, trees, stones. But though the Lithuanians and Old Prussians and Russians are among the nations whose especial rite it was to keep up sacred everlasting fires, yet it seems that their fire-rites were in the symbolic stage, ceremonies of their great celestial-solar religion, rather than acts of direct worship to a Fire-god.[1] Classical religion, on the other hand, brings prominently into view the special deities of fire. Hēphaistos, Vulcan, the divine metallurgist who had his temples on Ætna and Lipari, stands in especial connexion with the subterranean volcanic fire, and combines the nature of the Polynesian Mahuika and the Circassian Tleps. The Greek Hestia, the divine hearth, the ever-virgin venerable goddess, to whom Zeus gave fair office instead of wedlock, sits in the midst of the house, receiving fat:—

"Τῇ δὲ πατὴρ Ζεὺς δῶκε καλὸν γέρας ἀντὶ γάμοιο,
Καί τε μέσῳ οἴκῳ κατ' ἄρ' ἕζετο πῖαρ ἑλοῦσα."

In the high halls of gods and men she has her everlasting seat, and without her are no banquets among mortals, for to Hestia first and last is poured the honey-sweet wine:—

"'Εστίη, ἡ πάντων ἐν δώμασιν ὑψηλοῖσιν
Ἀθανάτων τε θεῶν χαμαὶ ἐρχομένων τ' ἀνθρώπων
Ἕδρην ἀΐδιον ἔλαχε, πρεσβηΐδα τιμήν,
Καλὸν ἔχουσα γέρας καὶ τίμιον· οὐ γὰρ ἄτερ σοῦ
Εἰλαπίναι θνητοῖσιν, ἵν' οὐ πρώτῃ πυμάτῃ τε
Ἑστίῃ ἀρχόμενος σπένδει μελιηδέα οἶνον."[2]

In Greek civil life, Hestia sat in house and assembly as

[1] Hanusch, 'Slaw. Myth.' pp. 88, 98.
[2] Homer. Hymn. Aphrod. 29, Hestia 1. Welcker, 'Griech. Götterl.' vol. ii. pp. 686, 691.

representative of domestic and social order. Like her in name and origin, but not altogether in development, is Vesta with her ancient Roman cultus, and her retinue of virgins to keep up her pure eternal fire in her temple, needing no image, for she herself dwelt within :—

" Esse diu stultus Vestæ simulacra putavi:
Mox didici curvo nulla subesse tholo.
Ignis inextinctus templo celatur in illo.
Effigiem nullam Vesta nec ignis habet." [1]

The last lingering relics of fire-worship in Europe reach us, as usual, both through Turanian and Aryan channels of folklore. The Esthonian bride consecrates her new hearth and home by an offering of money cast into the fire, or laid on the oven for Tule-ema, Fire-mother.[2] The Carinthian peasant will " fodder " the fire to make it kindly, and throw lard or dripping to it, that it may not burn his house. To the Bohemian it is a godless thing to spit into the fire, " God's fire " as he calls it. It is not right to throw away the crumbs after a meal, for they belong to the fire. Of every kind of dish some should be given to the fire, and if some runs over it is wrong to scold, for it belongs to the fire. It is because these rites are now so neglected that harmful fires so often break out.[3]

What the Sea is to Water-worship, in some measure the Sun is to Fire-worship. From the doctrines and rites of earthly fire, various and ambiguous in character, generalized from many phenomena, applied to many purposes, we pass to the religion of heavenly fire, whose great deity has a perfect definiteness from his embodiment in one great individual fetish, the Sun.

Rivalling in power and glory the all-encompassing Heaven, the Sun moves eminent among the deities of nature, no mere cosmic globe affecting distant material worlds by force

[1] Ovid. Fast. vi. 295.
[2] Boecler, 'Ehsten Abergl. Gebr.' p. 29, etc.
[3] Wuttke, 'Volksabergl.' p. 86. Gröhmann, 'Aberglauben aus Böhmen,' p. 41.

in the guise of light and heat and gravity, but a living reigning Lord:—

"O thou, that with surpassing glory crown'd,
Look'st from thy sole dominion like the God
Of this new world."

It is no exaggeration to say, with Sir William Jones, that one great fountain of all idolatry in the four quarters of the globe was the veneration paid by men to the sun: it is no more than an exaggeration to say with Mr. Helps of the sun-worship in Peru, that it was inevitable. Sun-worship is by no means universal among the lower races of mankind, but manifests itself in the upper levels of savage religion in districts far and wide over the earth, often assuming the prominence which it keeps and develops in the faiths of the barbaric world. Why some races are sun-worshippers and others not, is indeed too hard a question to answer in general terms. Yet one important reason is obvious, that the Sun is not so evidently the god of wild hunters and fishers, as of the tillers of the soil, who watch him day by day giving or taking away their wealth and their very life. On the geographical significance of sun-worship, D'Orbigny has made a remark, suggestive if not altogether sound, connecting the worship of the sun not so much with the torrid regions where his glaring heat oppresses man all day long, and drives him to the shade for refuge, as with climates where his presence is welcomed for his life-giving heat, and nature chills at his departure. Thus while the low sultry forests of South America show little prominence of Sun-worship, this is the dominant organized cultus of the high table-lands of Peru and Cundinamarca.[1] The theory is ingenious, and if not carried too far may often be supported. We may well compare the feelings with which the sun-worshipping Massagetæ of Tartary must have sacrificed their horses to the deity who freed them from the miseries of winter, with the thoughts of men in those burn-

[1] D'Orbigny, 'L'Homme Américain,' vol. i. p. 242.

ing lands of Central Africa where, as Sir Samuel Baker says, "the rising of the sun is always dreaded . . . the sun is regarded as the common enemy," words which recall Herodotus' old description of the Atlantes or Atarantes who dwelt in the interior of Africa, who cursed the sun at his rising, and abused him with shameful epithets for afflicting them with his burning heat, them and their land.[1]

The details of Sun-worship among the native races of America give an epitome of its development among mankind at large. Among many of the ruder tribes of the northern continent, the Sun is looked upon as one of the great deities, as representative of the greatest deity, or as that greatest deity himself. Indian chiefs of Hudson's Bay smoked thrice to the rising sun. In Vancouver Island men pray in time of need to the sun as he mounts toward the zenith. Among the Delawares the sun received sacrifice as second among the twelve great manitus; the Virginians bowed before him with uplifted hands and eyes as he rose and set; the Pottawatomis would climb sometimes at sunrise on their huts, to kneel and offer to the luminary a mess of Indian corn; his likeness represented the Great Manitu in Algonquin picture-writings. Father Hennepin, whose name is well known to geologists as the earliest visitor to the Falls of Niagara, about 1678, gives an account of the native tribes, Sioux and others, of this far-west region. He describes them as venerating the Sun, " which they recognize, though only in appearance, as the Maker and Preserver of all things;" to him first they offer the calumet when they light it, and to him they often present the best and most delicate of their game in the lodge of the chief, " who profits more by it than the Sun." The Creeks regarded the Sun as symbol or minister of the Great Spirit, sending toward him the first puff of the calumet at treaties, and bowing reverently toward him in confirming their council talk or haranguing their warriors to battle.[2] Among the

[1] Herod. i. 216, iv. 184. Baker, 'Albert Nyanza,' vol. i. p. 144.
[2] Waitz, 'Anthropologie,' vol. iii. p. 181 (Hudson's B., Pottawatomies), 205

rude Botocudos of Brazil, the idea of the Sun as the great good deity seems not unknown; the Araucanians are described as bringing offerings to him as highest deity; the Puelches as ascribing to the sun, and praying to him for, all good things they possess or desire; the Diaguitas of Tucuman as having temples dedicated to the Sun, whom they adored, and to whom they consecrated birds' feathers, which they then brought back to their cabins and sprinkled from time to time with the blood of animals.[1]

Such accounts of Sun-worship appearing in the lower native culture of America, may be taken to represent its first stage. It is on the whole within distinctly higher culture that its second stage appears, where it has attained to full development of ritual and appurtenance, and become in some cases even the central doctrine of national religion and statecraft. Sun-worship had reached this level among the Natchez of Louisiana, with whom various other tribes of this district stood in close relation. Every morning at sunrise the great Sun-chief stood at the house-door facing the east, shouted and prostrated himself thrice, and smoked first toward the sun, and then toward the other three quarters. The Sun-temple was a circular hut some thirty feet across and dome-roofed : here in the midst was kept up the everlasting fire, here prayer was offered thrice daily, and here were kept images and fetishes and the bones of dead chiefs. The Natchez government was a solar hierarchy. At its head stood the great chief, called the Sun or the

(Virginians). J. G. Müller, 'Amer. Urrel.' p. 117 (Delawares, Sioux, Mingos, etc.). Sproat, 'Ind. of Vancouver's I.' in 'Tr. Eth. Soc.' vol. v. p. 253. Loskiel, 'Ind. of N. A.' part i. p. 43 (Delawares). Hennepin, 'Voyage dans l'Amérique,' p. 302 (Sioux), etc. Bartram, 'Creek and Cherokee Ind.' in 'Tr. Amer. Eth. Soc.' vol. iii. part i. pp. 20, 26 ; see also Schoolcraft, 'Ind. Tribes,' part ii. p. 127 (Comanches, etc.); Morgan, 'Iroquois,' p. 164; Gregg, vol. ii. p. 238 (Shawnees) ; but compare the remarks of Brinton, 'Myths of New World,' p. 141.
[1] Martius, 'Ethnog. Amer.' vol. i. p. 327 (Botocudos). Waitz, vol iii. p. 518 (Araucanians). Dobrizhoffer, vol. ii. p. 89 (Puelches). Charlevoix, 'Hist. du Paraguay,' vol. i. p. 331 (Diaguitas). J. G. Müller, p. 255 (Botocudos, Aucas, Diaguitas).

Sun's brother, high priest and despot over his people. By his side stood his sister or nearest female relative, the female chief who of all women was alone permitted to enter the Sun-temple. Her son, after the custom of female succession common among the lower races, would succeed to the primacy and chiefship; and the solar family took to themselves wives and husbands from the plebeian order, who were their inferiors in life, and were slain to follow them as attendants in death.[1] Another nation of sun-worshippers were the Apalaches of Florida, whose daily service was to salute the Sun at their doors as he rose and set. The Sun, they said, had built his own conical mountain of Olaimi, with its spiral path leading to the cave-temple, in the east side. Here, at the four solar festivals, the worshippers saluted the rising sun with chants and incense as his rays entered the sanctuary, and again when at midday the sunlight poured down upon the altar through the hole or shaft pierced for this purpose in the rocky vault of the cave; through this passage the sun-birds, the tonatzuli, were let fly up sunward as messengers, and the ceremony was over.[2] Day by day, in the temples of Mexico, the rising sun was welcomed with blast of horns, and incense, and offering of a little of the officiators' own blood drawn from their ears, and a sacrifice of quails. Saying, the Sun has risen, we know not how he will fulfil his course nor whether misfortune will happen, they prayed to him—"Our Lord, do your office prosperously." In distinct and absolute personality, the divine Sun in Aztec theology was Tonatiuh, whose huge pyramid-mound stands on the plain of Teotihuacan, a witness of his worship for future ages. Beyond this, the religion of Mexico, in its complex system or congeries of great gods, such as results from the mixture and alliance of the deities of several nations, shows the solar element rooted deeply and widely in other personages

---

[1] Charlevoix, 'Nouvelle France,' vol. vi. p. 172; Waitz, vol. iii. p. 217.
[2] Rochefort, 'Iles Antilles,' book ii. ch. viii.

of its divine mythology, and attributes especially to the Sun the title of Teotl, God.[1] Again, the high plateau of Bogota in New Granada was the seat of the semi-civilized Chibchas or Muyscas, of whose mythology and religion the leading ideas were given by the Sun. The Sun was the great deity to whom the human sacrifices were offered, and especially that holiest sacrifice, the blood of a pure captive youth daubed on a rock on a mountain-top for the rising sun to shine on. In native Muysca legend, the mythic civilizer of the land, the teacher of agriculture, the founder of the theocracy and institutor of sun-worship, is a figure in whom we cannot fail to discern the personal Sun himself.[2] It is thus, lastly, in the far more celebrated native theocracy to the south. In the great religion of Peru, the Sun was at once ancestor and founder of the dynasty of Incas, who reigned as his representatives and almost in his person, who took wives from the convent of virgins of the Sun, and whose descendants were the solar race, the ruling aristocracy. The Sun's innumerable flocks of llamas grazed on the mountains, and his fields were tilled in the valleys, his temples stood throughout the land, and first among em the "Place of Gold" in Cuzco, where his new fire was kindled at the annual solar festival of Raymi, and where his splendid golden disc with human countenance looked forth to receive the first rays of its divine original. Sun-worship was ancient in Peru, but it was the Incas who made it the great state religion, imposing it wherever their wide conquests reached, till it became the central idea of Peruvian life.[3]

[1] Torquemada, 'Monarquia Indiana,' ix. c. 34; Sahagun, 'Hist. de Nueva España,' ii. App. in Kingsborough, 'Antiquities of Mexico;' Waitz. vol. iv. p. 138; J. G. Müller, p. 474, etc.; Brasseur, 'Mexique,' vol. iii. p. 487; Tylor, 'Mexico,' p. 141.

[2] Piedrahita, 'Hist. Gen. de las Conquistas del Nuevo Reyno de Granada,' Antwerp, 1688 : part i. book i. c. iii. iv.; Humboldt, 'Vues des Cordillères;' Waitz, vol. iv. p. 352, etc.; J. G. Müller, p. 432, etc.

[3] Garcilaso de la Vega, 'Commentarios Reales,' lib. i. c. 4, iii. c. 20; v. c. 2, 6; 'Rites and Laws of the Yncas,' tr. & ed. by C. R. Markham, (Hakluyt Soc., 1873) p. 84; Prescott, 'Peru,' book i. ch. iii.; Waitz, vol. iv. p. 447, etc.; J. G. Müller, p. 362, etc.

The culture of the Old World never surpassed this highest range of Sun-worship in the New. In Australia and Polynesia the place of the solar god or hero is rather in myth than in religion. In Africa, though found in some districts,[1] Sun-worship is not very conspicuous out of Egypt. In tracing its Old World development, we begin among the ruder Allophylian tribes of Asia, and end among the great polytheistic nations. The northeast quarter of India shows the doctrine well defined among the indigenous stocks. The Bodo and Dhimal place the Sun in the pantheon as an elemental god, though in practical rank below the sacred rivers.[2] The Kol tribes of Bengal, Mundas, Oraons, Santals, know and worship as supreme, Sing-bonga, the Sun-god; to him some tribes offer white animals in token of his purity, and while not regarding him as author of sickness or calamity, they will resort to him when other divine aid breaks down in sorest need.[3] Among the Khonds, Bura Pennu the Light-god, or Bella Pennu the Sun-god, is creator of all things in heaven and earth, and great first cause of good. As such, he is worshipped by his own sect above the ranks of minor deities whom he brought into being to carry out the details of the universal work.[4] The Tatar tribes with much unanimity recognize as a great god the Sun, whose figure may be seen beside the Moon's on their magic drums, from Siberia to Lapland. Castrén, the ethnologist, speaking of the Samoyed expression for heaven or deity in general (jilibeambaertje) tells an anecdote from his travels, which gives a lively idea of the thorough simple nature-religion still possible to the wanderers of the steppes. "A Samoyed woman," he says, "told me it was her habit every morning and evening to step out of her tent and bow down before the sun; in the morning

[1] Meiners, 'Gesch. der Rel.' vol. i. p. 383. Burton, 'Central Afr.' vol. ii. p. 346; 'Dahome,' vol. ii. p. 147.
[2] Hodgson, 'Abor. of India,' pp. 167, 175 (Bodos, etc.).
[3] Dalton, 'Kols,' in 'Tr. Eth. Soc.' vol. vi. p. 33 (Oraons, etc.); Hunter, 'Annals of Rural Bengal,' p. 184 (Santals).
[4] Macpherson, 'India,' p. 84, etc. (Khonds).

saying, 'When thou Jilibeambaertje risest, I too rise from my bed!' in the evening, 'When thou Jilibeambaertje sinkest down, I too get me to rest!' The woman brought this as a proof of her assertion that even among the Samoyeds they said their morning and evening prayers, but she added with pity that there were also among them wild people who never sent up a prayer to God." Mongol hordes may still be met with whose shamans invoke the Sun, and throw milk up into the air as an offering to him, while the Karagas Tatars would bring to him as a sacrifice the head and heart of bear or stag. Tunguz, Ostyaks, Woguls, worship him in a character blending with that of their highest deity and Heaven-god; while among the Lapps, Baiwe the Sun, though a mighty deity, stood in rank below Tiermes the Thunder-god, and the great celestial ruler who had come to bear the Norwegian name of Storjunkare.[1]

In direct personal nature-worship like that of Siberian nomades of our day, the solar cultus of the ancient pastoral Aryans had its source. The Vedic bards sing of the great god Sûrya, knower of beings, the all-revealer before whom the stars depart with the nights like thieves. We approach Sûrya (they say) shining god among the gods, light most glorious. He shines on the eight regions, the three worlds, the seven rivers; the golden-handed Savitar, all-seeing, goes between heaven and earth. To him they pray, "On thy ancient paths, O Savitar, dustless, well made, in the air, on those good-going paths this day preserve us and bless us, O God!" Modern Hinduism is full of the ancient Sun-worship, in offerings and prostrations, in daily rite and appointed festival, and it is Savitar the Sun who is invoked in the "gâyatrî," the time-honoured formula repeated day by day since long-past ages by every Brahman: "Tat Savitur varenyam bhargo devasya dhîmahi, dhiyo yo nah prakodayât.—Let us meditate on the desirable

[1] Castrén, 'Finn. Myth.' pp. 16, 51, etc. Meiners, l. c. Georgi, 'Reise im Russ. Reich.' vol. i. pp. 275, 317. Klemm, 'Cultur-Geschichte,' vol. iii. p. 87. Sun-Worship in Japan, Siebold, 'Nippon,' part. v. p. 9. For further evidence as to savage and barbaric worship of the Sun as Supreme Deity, see chap. xvii.

light of the divine Sun; may he rouse our minds!" Every morning the Brahman worships the sun, standing on one foot and resting the other against his ankle or heel, looking towards the east, holding his hands open before him in a hollow form, and repeating to himself these prayers: "The rays of light announce the splendid fiery sun, beautifully rising to illumine the universe."—"He rises, wonderful, the eye of the sun, of water, and of fire, collective power of gods; he fills heaven, earth, and sky with his luminous net; he is the soul of all that is fixed or locomotive."—"That eye, supremely beneficial, rises pure from the east; may we see him a hundred years; may we live a hundred years; may we hear a hundred years."—"May we, preserved by the divine power, contemplating heaven above the region of darkness, approach the deity, most splendid of luminaries!"[1] A Vedic celestial deity, Mitra the Friend, came to be developed in the Persian religion into that great ruling divinity of light, the victorious Mithra, lord of life and head of all created beings. The ancient Persian Mihr-Yasht invokes him in the character of the dawning sun-light, Mithra with wide pastures, whom the lords of the regions praise at early dawn, who as the first heavenly Yazata rises over Hara before the sun, the immortal with swift steeds, who first with golden form seizes the fair summits, then surrounds the whole Aryan region. Mithra came to be regarded as the very Sun, as where Bacchus addresses the Tyrian Bel, "εἴτε σὺ Μίθρης, Ἥλιος Βαβυλῶνος." His worship spread from the East across the Roman empire, and in Europe he takes rank among the great solar gods absolutely identified with the personal Sun, as in this inscription on a Roman altar dating from Trajan's time—"Deo Soli Mithræ."[2]

[1] 'Rig-Veda,' i. 35, 50; iii. 62, 10. Max Müller, 'Lectures,' 2nd Ser. pp. 378, 411; 'Chips,' vol. i. p. 19. Colebrooke, 'Essays,' vol. i. pp. 30, 133 Ward, 'Hindoos,' vol. ii. p. 42.
[2] 'Khordah-Avesta,' xxvi. in Avesta tr. by Spiegel, vol. iii. See Cox. 'Mythology of Aryan Nations,' vol. i. p. 334, vol. ii. p. 354. Strabo, xv. 3, 13. Nonnus, xl. 400. Movers, 'Phönizier,' vol. i. p. 180 : "'Ηλίῳ Μίθρᾳ ἀνικήτῳ;" "Διὸς ἀνικήτου Ἡλίου."

The earlier Sun-worship of Europe, upon which this new Oriental variety was intruded, in certain of its developments shows the same clear personality. The Greek Helios, to whom horses were sacrificed on the mountain-top of Taugetos, was that same personal Sun to whom Sokrates, when he had staid rapt in thought till daybreak, offered a prayer before he departed (ἔπειτ' ᾤχετ' ἀπιὼν προσευξάμενος τῷ ἡλίῳ).[1] Cæsar devotes to the German theology of his time three lines of his Commentaries. They reckon in the number of the gods, he says, those only whom they perceive and whose benefits they openly enjoy, Sun and Vulcan and Moon, the rest they know not even by report.[2] It is true that Cæsar's short summary does no justice to the real number and quality of the deities of the German pantheon, yet his forcible description of nature-worship in its most primitive stage may probably be true of the direct adoration of the sun and moon, and possibly of fire. On the other hand, European sun-worship leads into the most perplexing problems of mythology. Well might Cicero exclaim, "How many suns are set forth by the theologians!"[3] The modern student who shall undertake to discriminate among the Sun-gods of European lands, to separate the solar and non-solar elements of the Greek Apollo and Herakles, or the Slavonic Perun and Swatowit, has a task before him complicate with that all but hopeless difficulty which besets the study of myth, the moment that the clue of direct comparison with nature falls away.

The religion of ancient Egypt is one of which we know much, yet little—much of its temples, rites, names of deities, liturgical formulas, but little of the esoteric religious ideas which lay hidden within these outer manifestations. Yet it is clear that central solar conceptions as it

[1] Plat. Sympos. xxxvi. See Welcker, 'Griech. Gotterlehre,' vol. i. pp. 400, 412.
[2] Cæsar de Bello Gallico, vi. 21 : "Deorum numero eos solos ducunt, quos cernunt et quorum aperte opibus juvantur, Solem et Vulcanum et Lunam, reliquos ne fama quidem acceperunt."
[3] Cicero de Natura Deorum, iii. 21.

were radiate through the Egyptian theology. Ra, who traverses in his boat the upper and lower regions of the universe, is the Sun himself in plain cosmic personality. And to take two obvious instances of solar characters in other deities, Osiris the manifester of good and truth, who dies by the powers of darkness and becomes judge of the dead in the west-land of Amenti, is solar in his divine nature, as is also Har-p-chroti (Harpokrates) the new-born Sun of the winter solstice.[1] In the religions of the Semitic race, the place of the Sun is marked through a long range of centuries. The warning to the Israelites lest they should worship and serve sun, moon, and stars, and the mention of Josiah taking away the horses that the Kings of Judah had given to the sun, and burning the chariots of the sun with fire,[2] agree perfectly with the recognition in Palmyra of the Lord Sun, Baal-Shemesh, and with the identification of the Assyrian Bel and the Tyrian Baal with the Sun. Syrian religion, like Persian, introduced a new phase of Sun-worship into Rome, the cultus of Elagabal, and the vile priest-emperor who bore this divine name made it more intelligible to classic ears as Heliogabalus.[3] Eusebius is a late writer as regards Semitic religion, but with such facts as these before us we need not withhold our confidence from him when he describes the Phœnicians and Egyptians as holding Sun, Moon, and Stars to be gods, sole causes of the generation and destruction of all things.[4]

The widely spread and deeply rooted religion of the Sun naturally offered strenuous resistance to the invasion of Christianity, and it was one of the great signs of the religious change of the civilized world when Constantine, that ardent votary of the Sun, abandoned the faith of Apollo for that of Christ. Amalgamation even proved possible be-

[1] See Bunsen, 'Egypt's Place in Univ. Hist. ;' Wilkinson, 'Ancient Egyptians,' etc.
[2] Deut. iv. 19, xvii. 3 ; II Kings xxiii. 11.
[3] Mövers, 'Phönizier,' vol. i. pp. 162, 180, etc. Lamprid. Heliogabal. L
[4] Euseb. Præparat. Evang. i. 6.

tween the doctrines of Sabæism and Christianity, and in and near Armenia a sect of Sun-worshippers have lasted on into modern times under the profession of Jacobite Christians;[1] a parallel case within the limits of Mohammedanism being that of Beduin Arabs who still continue the old adoration of the rising sun, in spite of the Prophet's expressed command not to bow before the sun or moon, and in spite of the good Moslem's dictum, that "the sun rises between the devil's horns."[2] Actual worship of the sun in Christendom soon shrank to the stage of survival. In Lucian's time the Greeks kissed their hands as an act of worship to the rising sun; and Tertullian had still to complain of many Christians that with an affectation of adoring the heavenly bodies they would move their lips toward the sunrise (Sed et plerique vestrum affectatione aliquando et cœlestia adorandi ad solis ortum labia vibratis).[3] In the 5th century, Leo the Great complains of certain Christians who, before entering the Basilica of St. Peter, or from the top of a hill, would turn and bow to the rising sun; this comes, he says, partly of ignorance and partly of the spirit of paganism.[4] To this day, in the Upper Palatinate, the peasant takes off his hat to the rising sun; and in Pomerania, the fever-stricken patient is to pray thrice turning toward the sun at sunrise, "Dear Sun, come soon down, and take the seventy-seven fevers from me. In the name of God the Father, etc."[5]

For the most part, the ancient rites of solar worship are represented in modern Christendom in two ways; by the ceremonies connected with turning to the east, of which an account is given in an ensuing chapter under the heading of Orientation; and in the continuance of the great sun-

[1] Neander, 'Church History,' vol. vi. p. 341. Carsten Niebuhr, 'Reisebeschr.' vol. ii. p. 396.
[2] Palgrave, 'Arabia,' vol. i. p. 9; vol. ii. p. 258. See Koran, xli. 37.
[3] Tertullian. Apolog. adv. Gentes, xvi. See Lucian de Saltat. xvii.; compare Job xxxi. 26.
[4] Leo. I. Serm. viii. in Natal. Dom.
[5] Wuttke, 'Volksaberglaube,' p. 150.

festivals, countenanced by or incorporated in Christianity. Spring-tide, reckoned by so many peoples as New-Year, has in great measure had its solar characteristics transferred to the Paschal festival. The Easter bonfires with which the North German hills used to be ablaze mile after mile, are not altogether given up by local custom. On Easter morning in Saxony and Brandenburg, the peasants still climb the hill-tops before dawn, to see the rising sun give his three joyful leaps, as our forefathers used to do in England in the days when Sir Thomas Browne so quaintly apologized for declaring that "the sun doth not dance on Easter Day." The solar rite of the New Fire, adopted by the Roman Church as a Paschal ceremony, may still be witnessed in Europe, with its solemn curfew on Easter Eve, and the ceremonial striking of the new holy fire. On Easter Eve, under the solemn auspices of the Greek Church, a mob of howling fanatics crush and trample to death the victims who faint and fall in their struggles to approach the most shameless imposture of modern Christendom, the miraculous fire from heaven which descends into the Holy Sepulchre.[1] Two other Christian festivals have not merely had solar rites transferred to them, but seem distinctly themselves of solar origin. The Roman winter-solstice festival, as celebrated on December 25 (VIII. Kal. Jan.) in connexion with the worship of the Sun-god Mithra, appears to have been instituted in this special form by Aurelian about A.D. 273, and to this festival the day owes its apposite name of Birthday of the Unconquered Sun, "Dies Natalis Solis invicti." With full symbolic appropriateness, though not with historical justification, the day was adopted in the Western Church, where it appears to have been generally introduced by the 4th century, and whence in time it passed to the Eastern Church, as the solemn anniversary of the birth of Christ, the Christian Dies Natalis, Christmas Day.

[1] Grimm, 'Deutsche Myth.' p. 581, etc. Wuttke, pp. 17, 93. Brand, 'Pop. Ant.' vol. i. p. 157, etc. 'Early Hist. of Mankind,' p. 260. Murray's 'Handbook for Syria and Palestine,' 1868, p. 162.

Attempts have been made to ratify this date as matter of history, but no valid nor even consistent early Christian tradition vouches for it. The real solar origin of the festival is clear from the writings of the Fathers after its institution. In religious symbolism of the material and spiritual sun, Augustine and Gregory of Nyssa discourse on the glowing light and dwindling darkness that follow the Nativity, while Leo the Great, among whose people the earlier solar meaning of the festival evidently remained in strong remembrance, rebukes in a sermon the pestiferous persuasion, as he calls it, that this solemn day is to be honoured not for the birth of Christ, but for the rising, as they say, of the new sun.[1] As for modern memory of the sun-rites of mid-winter, Europe recognizes Christmas as a primitive solar festival by bonfires which our "yule-log," the "souche de Noël," still keeps in mind; while the adaptation of ancient solar thought to Christian allegory is as plain as ever in the Christmas service chant, "Sol novus oritur."[2] The solar Christmas festival has its pendant at Midsummer. The summer solstice was the great season of fire-festivals throughout Europe, of bonfires on the heights, of dancing round and leaping through the fires, of sending blazing fire-wheels to roll down from the hills into the valleys in sign of the sun's descending course. These ancient rites attached themselves in Christendom to St. John's Eve.[3] It seems as though the same train of symbolism which had adapted the midwinter festival to the Nativity, may have suggested the dedication of the midsummer festival to John the Baptist, in clear allusion to his words, "He must increase, but I must decrease."

[1] See Pauly, 'Real-Encyclop.' s. v. 'Sol;' Bingham, 'Antiquities of Christian Church,' book xx. ch. iv.; Neander, 'Church Hist.' vol. iii. p. 437; Beausobre, 'Hist. de Manichée,' vol. ii. p. 691; Gibbon, ch. xxii.; Creuzer, 'Symbolik,' vol. i. p. 761, etc.
[2] Grimm, 'D. M.' pp. 593, 1223. Brand, 'Popular Antiquities,' vol. i. p. 467. Monnier, 'Traditions Populaires,' p. 188.
[3] Grimm, 'D. M.' p. 583; Brand, vol. i. p. 298; Wuttke, pp. 14, 140. Beausobre, l. c.

ANIMISM. 299

Moon-worship, naturally ranking below Sun-worship in importance, ranges through nearly the same district of culture. There are remarkable cases in which the Moon is recognized as a great deity by tribes who take less account, or none at all, of the Sun. The rude savages of Brazil seem especially to worship or respect the moon, by which they regulate their time and festivals, and draw their omens. They would hold up their hands to the moon with wonder-struck exclamations of teh! teh! they would have children smoked by the sorcerers to preserve them from moon-given sickness, or the women would hold up their babes to the luminary. The Botocudos are said to give the highest rank among the heavenly bodies to Taru the Moon, as causing thunder and lightning and the failure of vegetables and fruits, and as even sometimes falling to the earth, whereby many men die.[1] An old account of the Caribs describes them as esteeming the Moon more than the Sun, and at new moon coming out of their houses crying "Behold the Moon!"[2] The Ahts of Vancouver's Island, it is stated, worship the Sun and Moon, particularly the full moon and the sun ascending to the zenith. Regarding the Moon as husband and the Sun as wife, their prayers are more generally addressed to the Moon as the superior deity; he is the highest object of their worship, and they speak of him as "looking down upon the earth in answer to prayer, and seeing everybody."[3] With a somewhat different turn of mythic fancy, the Hurons seem to have considered Aataentsic the Moon as maker of the earth and man, and grandmother of Iouskeha the Sun, with whom she governs the world.[4] In Africa, Moon-worship is prominent in an immense district where Sun-worship is unknown or insignificant. Among south-central tribes, men will watch for the

[1] Spix and Martius, 'Reise in Brasilien,' vol. i. pp. 377, 381; Martius, 'Ethnog. Amer.' vol. i. p. 327; Pr. Max. v. Wied, vol. ii. p. 58; J. G. Müller, pp. 218, 254; also Musters, 'Patagonians,' pp. 58, 179.
[2] De la Borde, 'Caraibes,' p. 525.
[3] Sproat, 'Savage Life,' p. 206; 'Tr. Eth. Soc.' vol. v. p. 253.
[4] Brebeuf in 'Rel. des Jes.' 1635, p. 34.

first glimpse of the new Moon, which they hail with shouts of kua! and vociferate prayers to it; on such an occasion Dr. Livingstone's Makololo prayed, "Let our journey with the white man be prosperous!" etc.[1] These people keep holiday at new-moon, as indeed in many countries her worship is connected with the settlement of periodic festivals. Negro tribes seem almost universally to greet the new Moon, whether in delight or disgust. The Guinea people fling themselves about with droll gestures, and pretend to throw firebrands at it; the Ashango men behold it with superstitious fear; the Fetu negroes jumped thrice into the air with hands together and gave thanks.[2] The Congo people fell on their knees, or stood and clapped their hands, crying, "So may I renew my life as thou art renewed!"[3] The Hottentots are described early in the last century as dancing and singing all night at new and full moon, calling the Moon the Great Captain, and crying to him "Be greeted!" "Let us get much honey!" "May our cattle get much to eat and give much milk!" With the same thought as that just noticed in the district north-west of them, the Hottentots connect the Moon in legend with that fatal message sent to Man, which ought to have promised to the human race a moon-like renewal of life, but which was perverted into a doom of death like that of the beast who brought it.[4]

The more usual status of the Moon in the religions of the world is, as nature suggests, that of a subordinate companion deity to the Sun, such a position as is acknowledged in the precedence of Sunday to Monday. Their various mutual relations as brother and sister, husband and wife, have already been noticed here as matter of mythology. As wide-lying rude races who place them thus side by side in their theology, it is enough to mention the Delawares of

[1] Livingstone, 'S. Afr.' p. 235; Waitz, vol. ii. pp. 175, 342.
[2] Römer, 'Guinea,' p. 84; Du Chaillu, 'Ashango-land,' p. 428; see Purchas, vol. v. p. 766. Müller, 'Fetu,' p. 47.
[3] Merolla, 'Congo,' in Pinkerton, vol. xvi. p. 273.
[4] Kolbe, 'Beschryving van de Kaap de Goede Hoop,' part i. xxix. See ante, vol. i. p. 355.

North-America,[1] the Ainos of Yesso,[2] the Bodos of North-East India,[3] the Tunguz of Siberia.[4] This is the state of things which continues at higher levels of systematic civilization. Beside the Mexican Tonatiuh the Sun, Metztli the Moon had a smaller pyramid and temple;[5] in Bogota, the Moon, identified in local myth with the Evil Deity, had her place and figure in the temple beside the Sun her husband;[6] the Peruvian Mother-Moon, Mama-Quilla, had her silver disc-face to match the golden one of her brother and husband the Sun, whose companion she had been in the legendary civilizing of the land.[7] In the ancient Kami-religion of Japan, the supreme Sun-god ranks high above the Moon-god, who was worshipped under the form of a fox.[8] Among the historic nations of the Old World, documents of Semitic culture show Sun and Moon side by side. For one, we may take the Jewish law, to stone with stones till they died the man or woman who "hath gone and served other gods, and worshipped them, either the sun, or moon, or any of the host of heaven." For another, let us glance over the curious record of the treaty-oath between Philip of Macedon and the general of the Carthaginian and Libyan army, which so well shows how the original identity of nature-deities may be forgotten in their different local shapes, so that the same divinity may come twice or even three times over in as many national names and forms. Herakles and Apollo stand in company with the personal Sun, and as well as the personal Moon there is to be seen the Carthaginian goddess, whom there is good reason to look on as herself wholly or partly of lunar nature. This is the list of deities invoked: "Before Zeus and Hera and

[1] Loskiel, 'Ind. of N. A.' part i. p. 43.
[2] Bickmore, 'Ainos,' in 'Tr. Eth. Soc.' vol. vii. p. 20.
[3] Hodgson, 'Abor. of India,' p. 167.
[4] Georgi, 'Reise im Russ. R.' vol. i. p. 275.
[5] Clavigero, 'Messico,' vol. ii. pp. 9, 35; Tylor, 'Mexico,' l. c.
[6] Waitz, vol. iv. p. 362.
[7] Prescott, 'Peru,' vol. i. p. 90. But compare Garcilaso de la Vega, iii. 21.
[8] Siebold, 'Nippon,' part v. p. 9.

Apollo; before the goddess of the Carthaginians (δαίμονος Καρχηδονίων) and Herakles and Iolaos; before Ares, Triton, Poseidon; before the gods who fought with the armies, and Sun and Moon and Earth; before the rivers and meadows and waters; before all the gods who rule Macedonia and the rest of Greece; before all the gods who were at the war, they who have presided over this oath."[1] When Lucian visited the famous temple of Hierapolis in Syria, he saw the images of the other gods, "but only of the Sun and Moon they show no images." And when he asked why, they told him that the forms of other gods were not seen by all, but Sun and Moon are altogether clear, and all men see them.[2] In Egyptian theology, not to discuss other divine beings to whom a lunar nature has been ascribed, it is at least certain that Aah is the Moon in absolute personal divinity.[3] In Aryan theology, the personal Moon stands as Selēnē beside the more anthropomorphic forms of Hekatē and Artemis,[4] as Luna beside the less understood Lucina, and Diana with her borrowed attributes,[5] while our Teutonic forefathers were content with his plain name of Moon.[6] As for lunar survivals in the higher religions, they are much like the solar. Monotheist as he is, the Moslem still claps his hands at sight of the new moon, and says a prayer.[7] In Europe in the 15th century it was matter of complaint that some still adored the new moon with bended knee, or hood or hat removed, and to this day we may still see a hat raised to her, half in conservatism and half in jest. It is with reference to silver as the lunar metal, that money is turned when the act of adoration

---

[1] Deuteron. xvii. 3; Polyb. vii. 9; see Movers, 'Phönizier,' pp. 159, 536, 605.
[2] Lucian de Syria Dea, iv. 34.
[3] Wilkinson, 'Ancient Egyptians,' vol. iv. p. 239, vol. v. p. 5. Bunsen, 'Egypt,' vol. iv. See Plutarch. Is. et Osir.
[4] Welcker, 'Griech. Götterl.' vol. i. p. 550, etc.
[5] Cic. de Nat. Deor. ii. 27.
[6] Grimm, 'D. M.' ch. xxii.
[7] Akerblad, 'Lettre à Italinsky.' Burton, 'Central Afr.' vol. ii. p. 346. Mungo Park, 'Travels,' in 'Pinkerton,' vol. xvi. p. 875.

is performed, while practical peasant wit dwells on the ill-luck of having no piece of silver when the new moon is first seen.[1]

Thus, in tracing the development of Nature-Worship, it appears that though Fire, Air, Earth, and Water are not yet among the lower races systematized into a quaternion of elements, their adoration, with that of Sun and Moon, shows already arising in primitive culture the familiar types of those great divinities, who received their further development in the higher Polytheism.

[1] Grimm, 'D. M.' pp. 29, 667; Brand, vol. iii. p. 146; Forbes Leslie, Early Races of Scotland,' vol. i. p. 136.

CHAPTER XVII.

ANIMISM—(continued).

Polytheism comprises a class of Great Deities, ruling the course of Nature and the life of Man—Childbirth-god—Agriculture-god—War-god—God of the Dead—First Man as Divine Ancestor—Dualism ; its rudimentary and unethical nature among low races; its development through the course of culture—Good and Evil Deity—Doctrine of Divine Supremacy, distinct from, while tending towards, the doctrine of Monotheism—Idea of Supreme Deity evolved in various forms among the lower races ; its place as completion of the Polytheistic system and outcome of the Animistic philosophy ; its continuance and development among higher nations—General survey of Animism as a Philosophy of Religion—Recapitulation of the theory advanced as to its development through successive stages of culture ; its primary phases best represented among the lower races, while survivals of these among the higher races mark the transition from savage through barbaric to civilized faiths—Transition of Animism in the History of Religion ; its earlier and later stages as a Philosophy of the Universe ; its later stages as the principle of a Moral Institution.

POLYTHEISM acknowledges, beside great fetish-deities like Heaven and Earth, Sun and Moon, another class of great gods whose importance lies not in visible presence, but in the performance of certain great offices in the course of Nature and the life of Man. The lower races can furnish themselves with such deities, either by giving the recognised gods special duties to perform, or by attributing these functions to beings invented in divine personality for the purpose. The creation of such divinities is however carried to a much greater extent in the complex systems of the higher polytheism. For a compact group of examples showing to what different ideas men will resort for a deity to answer a special end, let us take the deity presiding over

Childbirth. In the West Indies, a special divinity occupied with this function took rank as one of the great indigenous fetish-gods;[1] in the Samoan group, the household god of the father's or mother's family was appealed to;[2] in Peru the Moon takes to this office,[3] and the same natural idea recurs in Mexico;[4] in Esthonian religion the productive Earth-mother appropriately becomes patroness of human birth;[5] classic theology carries on both these ideas, in so far as the Greek Hēra represents the Earth[6] and the Roman Lucina the Moon;[7] and to conclude the list, the Chinese work out the problem from the manes-worshipper's point of view, for the goddess whom they call "Mother" and propitiate with many a ceremony and sacrifice to save and prosper their children, is held to have been in human life a skilful midwife.[8]

The deity of Agriculture may be a cosmic being affecting the weather and the soil, or a mythic giver of plants and teacher of their cultivation and use. Thus among the Iroquois, Heno the Thunder, who rides through the heavens on the clouds, who splits the forest-trees with the thunder-bolt-stones he hurls at his enemies, who gathers the clouds and pours out the warm rains, was fitly chosen as patron of husbandry, invoked at seed-time and harvest, and called Grandfather by his children the Indians.[9] It is interesting to notice again on the southern continent the working out of this idea in the Tupan of Brazilian tribes; Thunder and Lightning, it is recorded, they call Tupan, considering themselves to owe to him their hoes and the profitable

---

[1] Herrera, 'Indias Occidentales,' Dec. i. 3, 3; J. G. Müller, 'Amer. Urrel.' pp. 175, 221.
[2] Turner, 'Polynesia,' p. 174.
[3] Rivero and Tschudi, 'Peru,' p. 160.
[4] Kingsborough, 'Mexico,' vol. v. p. 179.
[5] Castrén, 'Finn. Myth.' p. 89.
[6] Welcker, 'Griech. Götterl.' vol. i. p. 371.
[7] Ovid. Fast. ii. 449.
[8] Doolittle, 'Chinese,' vol. i. p. 264.
[9] Morgan, 'Iroquois,' p. 158.

art of tillage, and therefore acknowledging him as a deity.[1] Among the Guarani race, Tamoi the Ancient of Heaven had no less rightful claim, in his character of heaven-god, to be venerated as the divine teacher of agriculture to his people.[2] In Mexico, Centeotl the Grain-goddess received homage and offerings at her two great festivals, and took care of the growth and keeping of the corn.[3] In Polynesia, we hear in the Society Islands of Ofanu the god of husbandry, in the Tonga Islands of Alo Alo the fanner, god of wind and weather, bearing office as god of harvest, and receiving his offering of yams when he had ripened them.[4] A picturesque figure from barbaric Asia is Pheebee Yau, the Ceres of the Karens, who sits on a stump and watches the growing and ripening corn, to fill the granaries of the frugal and industrious.[5] The Khonds worship at the same shrine, a stone or tree near the village, both Būrbi Pennu the goddess of new vegetation, and Pidzu Pennu the rain-god.[6] Among Finns and Esths it is the Earth-mother who appropriately undertakes the task of bringing forth the fruits.[7] And so among the Greeks it is the same being, Dēmētēr the Earth-mother, who performs this function, while the Roman Ceres who is confused with her is rather, as in Mexico, a goddess of grain and fruit.[8]

The War-god is another being wanted among the lower races, and formed or adapted accordingly. Areskove the Iroquois War-god seems to be himself the great celestial deity; for his pleasant food they slaughtered human victims, that he might give them victory over their enemies; as a pleasant sight for him they tortured the war-captives; on

[1] De Laet, 'Novus Orbis,' xv. 2; Waitz, vol. iii. p. 417; Brinton, pp. 152, 185; J. G. Müller, p. 271, etc.
[2] D'Orbigny, 'L'Homme Américain,' vol. ii. p. 319.
[3] Clavigero, 'Messico,' vol. ii. pp. 16, 68, 75.
[4] Ellis, 'Polyn. Res.' vol. i. p. 333. Mariner, 'Tonga Is.' vol. ii. p. 115.
[5] Cross, in 'Journ. Amer. Oriental Soc.' vol. iv. p. 316; Mason, p. 215.
[6] Macpherson, 'India,' pp. 91, 355.
[7] Castrén, 'Finn. Myth.' p. 89.
[8] Welcker, 'Griech. Götterl.' vol. ii. p. 467. Cox, 'Mythology of Aryan Nations, vol. ii. p. 308.

him the war-chief called in solemn council, and the warriors, shouting his name, rushed into the battle he was surveying from on high. Canadian Indians before the fight would look toward the sun, and their leader prayed to the Great Spirit; Floridan Indians prayed to the Sun before their wars.[1] Araucanians of Chili entreated Pillan the Thundergod that he would scatter their enemies, and thanked him amidst their cups after a victory.[2] The very name of Mexico seems derived from Mexitli, the national War-god, identical or identified with the hideous gory Huitzilopochtli. Not to attempt a general solution of the enigmatic nature of this inextricable compound parthenogenetic deity, we may notice the association of his principal festival with the winter-solstice, when his paste idol was shot through with an arrow, and being thus killed, was divided into morsels and eaten, wherefore the ceremony was called the teoqualo or "god-eating." This and other details tend to show Huitzilopochtli as originally a nature-deity, whose life and death were connected with the year's, while his functions of War-god may be of later addition.[3] Polynesia is a region where quite an assortment of war-gods may be collected. Such, to take but one example, was Tairi, war-god of King Kamehameha of the Sandwich Islands, whose hideous image, covered with red feathers, shark-toothed, mother-of-pearl-eyed, with helmet-crest of human hair, was carried into battle by his special priest, distorting his own face into hideous grins, and uttering terrific yells which were considered to proceed from the god.[4] Two examples from Asia may show what different original conceptions may serve to shape such deities as these upon. The Khond War-god, who entered into all weapons, so that from instruments of peace they became weapons of war, who gave edge to the

---

[1] J. G. Müller, 'Amer. Urrel.' pp. 141, 271, 274, 591, etc.
[2] Dobrizhoffer, 'Abipones,' vol. ii. p. 90.
[3] Clavigero, 'Messico,' vol. ii. pp. 17, 81.
[4] Ellis, 'Polyn. Res.' vol. i. p. 326; vol. iv. p. 158. See also Mariner, 'Tonga Is.' vol. ii. p. 112; Williams, 'Fiji,' vol. i. p 218.

axe and point to the arrow, is the very personified spirit of tribal war, his token is the relic of iron and the iron weapons buried in his sacred grove which stands near each group of hamlets, and his name is Loha Pennu or Iron-god.[1] The Chinese War-god, Kuang Tä, on the other hand, is an ancient military ghost; he was a distinguished officer, as well as a "faithful and honest courtier," who flourished during the wars of the Han dynasty, and emperors since then have delighted to honour him by adding to his usual title more and more honorary distinctions.[2] Looking at these selections from the army of War-gods of the different regions of the world, we may well leave their classic analogues, Arēs and Mars, as beings whose warlike function we recognize, but not so easily their original nature.[3]

It would be easy, going through the religious systems of Polynesia and Mexico, Greece and Rome, India and China, to give the names and offices of a long list of divinities, patrons of hunting and fishing, carpentering and weaving, and so forth. But studying here rather the continuity of polytheistic ideas than the analysis of polytheistic divinities, it is needless to proceed farther in the comparison of these deities of special function, as recognized to some extent in the lower civilization, before their elaborate development became one of the great features of the higher.

The great polytheistic deities we have been examining, concerned as they are with the earthly course of nature and human life, are gods of the living. But even in savage levels man began to feel an intellectual need of a God of the Dead, to reign over the souls of men in the next life, and this necessity has been supplied in various ways. Of the deities set up as lords of Deadman's Land, some are beings whose original meaning is obscure. Some are distinctly nature-deities appointed to this office, often for local reasons,

[1] Macpherson, 'India,' pp. 90, 360.
[2] Doolittle, 'Chinese,' vol. i. p. 267.
[3] Welcker, 'Griech. Götterl.' vol. i. p. 413. Cox, 'Myth. of Aryan N.', vol. ii. pp. 254, 311.

as happening to belong to the regions where the dead take up their abode. Some, again, are as distinctly the deified souls of men. The two first classes may be briefly instanced together in America, where the light-side and shadow-side (as Dr. J. G. Müller well calls them) of the conception of a future life are broadly contrasted in the definitions of the Lord of the Dead. Among the Northern Indians this may be Tarenyawagon the Heaven-god, or the Great Spirit who receives good warriors in his happy hunting-grounds, or his grandmother, the bloodthirsty Death-goddess Atahentsic.[1] In Brazil, the Underworld-god, who places good warriors and sorcerers in Paradise, contrasts with Aygnan the evil deity who takes base and cowardly Tupi souls,[2] much as the Mexican Tlaloc, Water-god and lord of the earthly paradise, contrasts with Mictlanteuctli, ruler of the dismal dead-land in the shades below.[3] In Peru there seems to have existed a belief that the spirits of the departed went to be with the Creator and Teacher of the World—"Bring us too near to thee . . . that we may be fortunate, being near to thee, O Uira-cocha!" There are also statements as to an underworld of shades, the land of the demon Supay.[4] Accounts of this class must often be suspected of giving ideas mis-stated under European influence, or actually adopted from Europeans, but there is in some a look of untouched genuineness. Thus in Polynesia, the idea of a Devil borrowed from colonists or missionaries may be suspected in such a figure as the evil deity Wiro, chief of Reigna, the New Zealander's western world of departed souls. But few conceptions of deity are more quaintly original than that of the Samoan deity Saveasiuleo, at once

[1] J. G. Müller, 'Amer. Urrel.' pp. 137, etc. 272, 286, etc. 500, etc. See Sproat, p. 213 (Ahts), cited ante, p. 85. Chay-her signifies not only the world below, but Death personified as a boneless greybeard who wanders at night stealing men's souls away.
[2] Lery, 'Bresil,' p. 234.
[3] Clavigero, vol. ii. pp. 14, 17; Brasseur, 'Mexique,' vol. iii. p. 495.
[4] 'Rites and Laws of Yncas,' tr. and ed. by C. R. Markham, pp. 32, 48 (prayer from M.S. communication by C. R. M.); Garcilaso de la Vega, lib. ii. c. 2, 7; Brinton, 'Myths of New World,' p. 251.

ruler of destinies of war and other affairs of men, and chief of the subterranean Bulotū, with the human upper half of his body reclining in his great house in company with the spirits of departed chiefs, while his tail or extremity stretched far away into the sea, in the shape of an eel or serpent. Different in name and nature, yet not so different as to be in either beyond recognition, this being reappears in the kindred myths of the neighbouring group, the Tonga Islands. The Tongan Hikuleo has his home in the spirit-land of Bulotū, here conceived as out in the far western sea. Here we are told the use of his tail. His body goes away on journeys, but his tail remains watching in Bulotū, and thus he is aware of what goes on in more places than one. Hikuleo used to carry off the first-born sons of Tongan chiefs, to people his island of the blest, and he so thinned the ranks of the living that at last the other gods were moved to compassion. Tangaloa and Maui seized Hikuleo, passed a strong chain round him, and fastened one end to heaven and the other to earth. Another god of the dead, of well-marked native type, is the Rarotongan Tiki (no doubt a solar deity, a Maui), to whose long house, a place of unceasing joys, the dead are to find their way.[1] Among Turanian tribes, there are Samoyeds who believe in a deity called A', dwelling in impenetrable darkness, sending disease and death to men and reindeer, and ruling over a crowd of spirits which are manes of the dead. Tatars tell of the nine Irle-Chans, who in their gloomy subterranean kingdom not only rule over souls of the dead, but have at their command a multitude of ministering spirits, visible and invisible. In the gloomy under-world of the Finns reigns Mana or Tuoni, a being whose nature is worked out by personification from the dismal dead-land or death itself.[2] Much the

---

[1] Turner, 'Polynesia,' p. 237; Farmer, 'Tonga,' p. 126. Yate, 'New Zealand,' p. 140; J. Williams, 'Missionary Enterprise,' p. 145. See Schirren, 'Wandersagen der Neuseeländer,' p. 89; Williams, 'Fiji,' vol. i. p. 246.

[2] Castrén, 'Finn. Myth.' pp. 128, 147, 155; Waitz, vol. ii. p. 171 (Africa).

same may be said of the Greek Aidēs, Hades, and the Scandinavian Hel, whose names, perhaps not so much by confusion as with a sense of their latent significance, have become identified in language with the doleful abodes over which a personifying fancy set them to preside.[1] As appropriately, though working out a different idea, the ancient Egyptians conceived their great solar deity to rule in the regions of his western under-world—Osiris is Lord of the Dead in Amenti.[2]

In the world's assembly of great gods, an important place must be filled up by the manes-worshipper in logical development of his special system. The theory of family manes, carried back to tribal gods, leads to the recognition of superior deities of the nature of Divine Ancestor or First Man, and it is of course reasonable that such a being, if recognized, should sometimes fill the place of lord of the dead, whose ancestral chief he is. There is an anecdote among the Mandans told by Prince Maximilian von Wied, which brings into view conceptions lying in the deepest recesses of savage religion, the idea of the divine first ancestor, the mythic connexion of the sun's death and descent into the under-world with the like fate of man, and the nature of the spiritual intercourse between man's own soul and his deity. The First Man, it is said, promised the Mandans to be their helper in time of need, and then departed into the West. It came to pass that the Mandans were attacked by foes. One Mandan would send a bird to the great ancestor to ask for help, but no bird could fly so far. Another thought a look would reach him, but the hills walled him in. Then said a third, thought must be the safest way to reach the First Man. He wrapped himself in his buffalo-robe, fell down, and spoke, "I think—I have thought—I come back." Throwing off the fur, he was bathed in sweat. The divine helper he had called on in his

[1] Welcker, 'Griech. Götterl,' vol. i. p. 395. Grimm, 'Deutsch. Myth.' p. 288.
[2] 'Book of Dead,' tr. by Birch, in Bunsen, 'Egypt,' vol. v.

distress appeared.[1] There is instructive variety in the ways in which the lower American races work out the conception of the divine forefather. The Mingo tribes revere and make offerings to the First Man, he who was saved at the great deluge, as a powerful deity under the Master of Life, or even as identified with him; some Mississippi Indians said that the first Man ascended into heaven, and thunders there; among the Dog-ribs, he was creator of sun and moon;[2] Tamoi, the grandfather and ancient of heaven of the Guaranis, was their first ancestor, who dwelt among them and taught them to till the soil, and rose to heaven in the east, promising to succour them on earth, and at death to carry them from the sacred tree into a new life where they should all meet again, and have much hunting.[3]

Polynesia, again, has thoroughly worked the theory of divine ancestors into the native system of multiform and blending nature-deities. Men are sprung from the divine Maui, whom Europeans have therefore called the "Adam of New Zealand," or from the Rarotongan Tiki, who seems his equivalent (Mauitiki), and who again is the Tii of the Society Islands; it is, however, the son of Tii, who precisely represents a Polynesian Adam, for his name is Taata, *i.e.*, Man, and he is the ancestor of the human race. There is perhaps also reason to identity Maui and the First Man with Akea, first King of Hawaii, who at his earthly death descended to rule over his dark subterranean kingdom, where his subjects are the dead who recline under the spreading kou-trees, and drink of the infernal rivers, and feed on lizards and butterflies.[4] In the mythology of Kamchatka, the relation between the Creator and the First Man is one not of identity but of parentage. Among the sons of

[1] Pr. Max v. Wied, 'N. Amerika,' vol. ii. p. 157.
[2] J. G. Müller, 'Amer. Urrel.' pp. 133, etc. 228, 255. Catlin, 'N. A. Ind.' vol. i. pp. 159, 177; Pr. Max v. Wied, vol. ii. pp. 149, etc. Compare Sproat, 'Savage Life,' p. 179 (Quawteaht the Great Spirit is also First Man).
[3] D'Orbigny, 'L'Homme Américain,' vol. ii. p. 319.
[4] Schirren, 'Wandersagen der Neuseeländer,' p. 64, etc., 88, etc.; Ellis, 'Polyn. Res.' vol. i. p. 111, vol. iv. pp. 145, 366.

Kutka the Creator is Haetsh the First Man, who dwelt on earth, and died, and descended into Hades to be chief of the under-world; there he receives the dead and new-risen Kamchadals, to continue a life like that of earth in his pleasant subterranean land where mildness and plenty prevail, as they did in the regions above in the old days when the Creator was still on earth.[1] Among all the lower races who have reasoned out this divine ancestor, none excel those consistent manes-worshippers, the Zulus. Their worship of the manes of the dead has not only made the clan-ancestors of a few generations back into tribal deities (Unkulunkulu), but beyond these, too far off and too little known for actual worship, yet recognized as the original race-deity and identified with the Creator, stands the First Man, he who "broke off in the beginning," the Old-Old-One, the great Unkulunkulu. While the Zulu's most intense religious emotions are turned to the ghosts of the departed, while he sacrifices his beloved oxen and prays with agonising entreaty to his grandfather, and carries his tribal worship back to those ancestral deities whose praise-giving names are still remembered, the First Man is beyond the reach of such rites. "At first we saw that we were made by Unkulunkulu. But when we were ill we did not worship him, nor ask anything of him. We worshipped those whom we had seen with our eyes, their death and their life among us. . . . . Unkulunkulu had no longer a son who could worship him; there was no going back to the beginning, for people increased, and were scattered abroad, and each house had its own connections; there was no one who said, 'For my part I am of the house of Unkulunkulu.'" Nay more, the Zulus who would not dare to affront an "idhlozi," a common ghost, that might be angry and kill them, have come to make open mock of the name of the great first ancestor. When the grown-up people wish to talk privately or eat something by themselves, it is the regular thing to send the children out to

[1] Steller, 'Kamtschatka,' p. 271.

call at the top of their voices for Unkulunkulu. "The name of Unkulunkulu has no respect paid to it among black men; for his house no longer exists. It is now like the name of a very old crone, which has no power to do even a little thing for herself, but sits continually where she sat in the morning till the sun sets. And the children make sport of her, for she cannot catch them and flog them, but only talk with her mouth. Just so is the name of Unkulunkulu when all the children are told to go and call him. He is now a means of making sport of children." [1]

In Aryan religion, the savage divinities just described give us analogues for the Hindu Yama, throughout his threefold nature as Sun, as First Man, as Judge of the Dead. Professor Max Müller thus depicts his solar origin, which may indeed be inferred from his being called the child of Vivasvat, himself the very Sun: "The sun, conceived as setting or dying every day, was the first who had trodden the path of life from East to West—the first mortal—the first to show us the way when our course is run, and our sun sets in the far West. Thither the fathers followed Yama; there they sit with him rejoicing, and thither we too shall go when his messengers (day and night) have found us out. . . . . Yama is said to have crossed the rapid waters, to have shown the way to many, to have first known the path on which our fathers crossed over." It is a perfectly consistent myth-formation, that the solar Yama should become the first of mortals who died and discovered the way to the other world, who guides other men thither and assembles them in a home which is secured to them for ever. As representative of death, Yama had even in early Aryan times his aspects of terror, and in later Indian theology he becomes not only the Lord but the awful Judge of the Dead, whom some modern Hindus are said to worship alone of all the gods, alleging that their future state is to be determined only by Yama, and that they have nothing therefore to hope or fear from any beside him. In these

Callaway, 'Religion of Amazulu,' pp. 1-104.

days, Hindu and Parsi in Bombay are learning from scholars in Europe the ancient connexion of their long antagonistic faiths, and have to hear that Yama son of Vivasvat sitting on his awful judgment-seat of the dead, to reward the good and punish the wicked with hideous tortures, and Yima son of Vivanhâo who in primæval days reigned over his happy deathless kingdom of good Zarathustrian men, are but two figures developed in the course of ages out of one and the same Aryan sun-myth.[1] Within the limits of Jewish, Christian, and Moslem theology, the First Man scarcely occupies more than a place of precedence among the human race in Hades or in Heaven, not the high office of Lord of the Dead. Yet that tendency to deify an ideal ancestor, which we observe to act so strongly on lower races, has taken effect also here. The Rabbinical Adam is a gigantic being reaching from earth to heaven, for the definition of whose stature Rabbi Eliezer cites Deuteronomy iv. 32, "God made man (Adam) upon the earth, and from one end of heaven to the other."[2] It is one of the familiar episodes of the Koran, how the angels were bidden to bow down before Adam, the regent of Allah upon earth, and how Eblis (Diabolus) swelling with pride, refused the act of adoration.[3] Among the Gnostic sect of the Valentinians, Adam the primal man in whom the Deity had revealed himself, stood as earthly representative of the Demiurge, and was even counted among the Æons.[4]

The figures of the great deities of Polytheism, thus traced in outline according to the determining idea on which each is shaped, seem to show that conceptions originating under rude and primitive conditions of human thought and passing thence into the range of higher culture,

[1] 'Rig-Veda,' x. 'Atharva-Veda,' xviii. Max Müller, 'Lectures,' 2nd Ser. p. 514. Muir, 'Yama,' etc. in 'Journ. As. Soc. N. S.' vol. i. 1865. Roth in 'Ztschr. Deutsch. Morgenl. G.' vol. iv. p. 426. Ward, 'Hindoos,' vol. ii. p. 60. Avesta: 'Vendidad,' ii. Pictet, 'Origines Indo-Europ.' part ii. p. 621.
[2] Eisenmenger, part i. p. 365.
[3] Koran, ii. 28, vii. 10, etc.
[4] Neander, 'Hist. of Chr.' vol. ii. pp. 81, 109, 174.

may suffer in the course of ages the most various fates, to be expanded, elaborated, transformed, or abandoned. Yet the philosophy of modern ages still to a remarkable degree follows the primitive courses of savage thought, even as the highways of our land so often follow the unchanging tracks of barbaric roads. Let us endeavour timidly and circumspectly to trace onward from savage times the courses of vast and pregnant generalization which tend towards the two greatest of the world's schemes of religious doctrine, the systems of Dualism and Monotheism.

Rudimentary forms of Dualism, the antagonism of a Good and Evil Deity, are well known among the lower races of mankind. The investigation of these savage and barbaric doctrines, however, is a task demanding peculiar caution. The Europeans in contact with these rude tribes since their discovery, themselves for the most part holding strongly dualistic forms of Christianity, to the extent of practically subjecting the world to the contending influences of armies of good and evil spirits under the antagonistic control of God and Devil, were liable on the one hand to mistake and exaggerate savage ideas in this direction, so that their records of native religion can only be accepted with reserve, while on the other hand there is no doubt that dualistic ideas have been largely introduced and developed among the savages themselves, under this same European influence. For instance, among the natives of Australia, we hear of the great deity Nambajandi who dwells in his heavenly paradise, where the happy shades of black men feast and dance and sing for evermore; over against him stands the great evil being Warrūgūra, who dwells in the nethermost regions, who causes the great calamities which befal mankind, and whom the natives represent with horns and tail, although no horned beast is indigenous in the land.[1] There may be more or less native substratum in all this, but the hints borrowed from popular Christian ideas are unmistake-

[1] Oldfield in 'Tr. Eth. Soc.' vol. iii. p. 228. See also Eyr vcl. ii. p. 356; Lang, 'Queensland.' p. 444.

able. Thus also, among the North American Indians, the native religion was modified under the influence of ideas borrowed from the white men, and there arose a full dualistic scheme, of which Loskiel, a Moravian missionary conversant especially with Algonquin and Iroquois tribes, gives the following suggestive particulars, dating from 1794. "They (the Indians) seem to have had no idea of the *Devil*, as the Prince of Darkness, before the Europeans came into the country. They consider him now as a very powerful spirit, but unable to do good, and therefore call him *The Evil One*. Thus they now believe in two Beings, the one supremely good, and the other altogether evil. To the former they ascribe all good, and to the latter all evil. About thirty years ago, a great change took place in the religious opinions of the Indians. Some preachers of their own nation pretended to have received revelations from above, to have travelled into heaven, and conversed with God. They gave different accounts of their exploits on the journey, but all agreed in this, that no one could enter into heaven without great danger; for the road, say they, runs close by the gates of hell. There the Devil lies in ambush, and snatches at every one who is going to God. Now those who have passed by this dangerous place unhurt, come first to the Son of God, and through him to God himself, from whom they pretend to have received a commandment, to instruct the Indians in the way to heaven. By these preachers the Indians were informed that heaven was the dwelling of God, and hell that of the Devil. Some of their preachers confessed that they had not reached the dwelling of God, but had however approached near enough to hear the cocks crow, and to see the smoke of the chimneys in heaven, &c., &c."[1]

Such unequivocal proofs that savage tribes can adopt and work into the midst of their native beliefs the European doctrine of the Good and Evil Spirit, must induce us to criticize keenly all recorded accounts of the religion of un-

[1] Loskiel, 'Indians in North America,' part i. p. 34.

cultured tribes, lest we should mistake the confused reflexion of Christendom for the indigenous theology of Australia or Canada. It is the more needful to bring this state of things into the clearest light, in order that the religion of the lower tribes may be placed in its proper relation to the religion of the higher nations. Genuine savage faiths do in fact bring to our view what seem to be rudimentary forms of ideas which underlie dualistic theological schemes among higher nations. It is certain that even among rude savage hordes, native thought has already turned toward the deep problem of good and evil. Their crude though earnest speculation has already tried to solve the great mystery which still resists the efforts of moralists and theologians. But as in general the animistic doctrine of the lower races is not yet an ethical institution, but a philosophy of man and nature, so savage dualism is not yet a theory of abstract moral principles, but a theory of pleasure or pain, profit or loss, affecting the individual man, his family, or at the utmost stretch, his people. This narrow and rudimentary distinction between good and evil was not unfairly stated by the savage who explained that if anybody took away his wife, that would be bad, but if he himself took someone's else, that would be good. Now by the savage or barbarian mind, the spiritual beings which by their personal action account for the events of life and the operations of nature, are apt to be regarded as kindly or hostile, sometimes or always, like the human beings on whose type they are so obviously modelled. In such a case, we may well judge by the safe analogy of disembodied human souls, and it appears that these are habitually regarded as sometimes friends and sometimes foes of the living. Nothing could be more conclusive in this respect than an account of the three days' battle between two factions of Zulu ghosts for the life of a man and wife whom the one spiritual party desired to destroy and the other to save; the defending spirits prevailed, dug up the bewitched charm-bags which had been buried to cause sympathetic disease, and flung these objects into the midst of the assembly of the people watching in silence, just as the spirits now fling real flowers at a table-rapping séance.[1] For spirits less closely belonging to the definition of ghosts, we may take Rochefort's remark in the 17th century as to the two sorts of spirits, good and bad, recognized by the Caribs of the West Indies. This writer declares that their good spirits or divinities are in fact so many demons who seduce them and keep them enchained in their damnable servitude; but nevertheless, he says, the people themselves do distinguish them from their evil spirits.[2] Nor can we pronounce this distinction of theirs unreasonable, learning from other authorities that it was the office of some of these spirits to attend men as familiar genii, and of others to inflict diseases. After the numerous details which have incidentally been cited in the present volumes, it will be needless to offer farther proof that spiritual beings are really conceived by savages and barbarians as ranged in antagonistic ranks as good and evil, *i. e.*, friendly and hostile to themselves. The interesting inquiry on which it is here desirable to collect evidence, is this: how far are the docrines of the higher nations anticipated in principle among the lower tribes, in the assignment of the conduct of the universe to two mighty hostile beings, in whom the contending powers of good and evil are personified, the Good Deity and the Evil Deity, each the head and ruler of a spiritual host like-minded? The true answer seems to be that savage belief displays to us primitive conceptions which, developed in systematic form and attached to ethical meaning, have their place in religious systems of which the Zoroastrian is the type.

First, when in district after district two special deities with special native names are contrasted in native religion as the Good and Evil Deity, it is in many cases easier to explain these beings as native at least in origin, than to suppose that foreign intercourse should have exerted the

[1] Callaway, 'Rel. of Amazulu,' p. 348.
[2] Rochefort, 'Iles Antilles,' p. 416. See J. G. Müller, p. 207.

consistent and far-reaching influence needed to introduce them. Second, when the deities in question are actually polytheistic gods, such as Sun, Moon, Heaven, Earth, considered as of good or evil, *i. e.*, favourable or unfavourable aspect, this looks like native development, not innovation derived from a foreign religion ignoring such divinities. Third, when it is held that the Good Deity is remote and otiose, but the Evil Deity present and active, and worship is therefore directed especially to the propitiation of the hostile principle, we have here a conception which appears native in the lower culture, rather than derived from the higher culture to which it is unfamiliar and even hateful. Now Dualism, as prevailing among the lower races, will be seen in a considerable degree to assert its originality by satisfying one or more of these conditions.

There have been recorded among the Indians of North America a group of mythic beliefs, which display the fundamental idea of dualism in the very act of germinating in savage religion. Yet the examination of these myths leads us first to destructive criticism of a picturesque but not ancient member of the series. An ethnologist, asked to point out the most striking savage dualistic legend of the world, would be likely to name the celebrated Iroquois myth of the Twin Brethren. The current version of this legend is that set down in 1825 by the Christian chief of the Tuscaroras, David Cusick, as the belief of his people. Among the ancients, he relates, there were two worlds, the lower world in darkness and possessed by monsters, the upper world inhabited by mankind. A woman near her travail sank from this upper region to the dark world below. She alighted on a Tortoise, prepared to receive her with a little earth on his back, which Tortoise became an island. The celestial mother bore twin sons into the dark world, and died. The Tortoise increased to a great island, and the twins grew up. One was of gentle disposition, and was called Enigorio, the Good Mind; the other was of insolent character, and was named Enigonhahetgea, the Bad Mind (or Brinton's translations, Ugly Spirit and Beautiful Spirit, may be more accurate). The Good Mind, not contented to remain in darkness, wished to create a great light; the Bad Mind desired that the world should remain in its natural state. The Good Mind took his dead mother's head and made it the sun, and of a remnant of her body he made the moon. These were to give light to the day and to the night. Also he created many spots of light, now stars: these were to regulate the days, nights, seasons, years. Where the light came upon the dark world, the monsters were displeased, and hid themselves in the depths, lest man should find them. The Good Mind continued the creation, formed many creeks and rivers on the Great Island, created small and great beasts to inhabit the forests, and fishes to inhabit the waters. When he had made the universe, he doubted concerning beings to possess the Great Island. He formed two images of the dust of the ground in his own likeness, male and female, and by breathing into their nostrils gave them living souls, and named them Ea-gwe-howe, that is "real people;" and he gave the Great Island all the animals of game for their maintenance; he appointed thunder to water the earth by frequent rains; the island became fruitful, and vegetation afforded to the animals subsistence. The Bad Mind went throughout the island and made high mountains and waterfalls and great steeps, and created reptiles injurious to mankind; but the Good Mind restored the island to its former condition. The Bad Mind made two clay images in the form of man, but while he was giving them existence they became apes; and so on. The Good Mind accomplished the works of creation, notwithstanding the imaginations of the Bad Mind were continually evil; thus he attempted to enclose all the animals of game in the earth away from mankind, but his brother set them free, and traces of them were made on the rocks near the cave where they were shut in. At last the brethren came to single combat for the mastery of the universe. The Good Mind falsely persuaded the Bad Mind that whipping with flags would destroy his own life, but he himself used the deer-horns, the instrument of death. After a two days' fight, the Good Mind slew his brother and crushed him in the earth; and the last words of the Bad Mind were that he would have equal power over men's souls after death; then he sank down to eternal doom and became the Evil Spirit. The Good Mind visited the people, and then retired from the earth.[1]

This is a graphic tale. Its version of the cosmic myth of the World-Tortoise, and its apparent philosophical myth of fossil footprints, have much mythological interest. But its Biblical copying extends to the very phraseology, and only partial genuineness can be allowed to its main theme. Dr. Brinton has profitably criticized this, referring to early American writers to show how much dualistic fancy has sprung up since the times of first intercourse between natives and white men, and pointing out the habit of European narrators to make distinctions between good and evil spirits in ways foreign to Indian thought. When we compare this legend, he says, with the version of the same legend given by Father Brebeuf, missionary to the Hurons in 1636, we find its whole complexion altered; the moral dualism vanishes; the names of Good and Bad Mind do not appear; it is the story of Ioskeha the White One, with his brother Tawiscara the Dark One, and we at once perceive that Christian influence in the course of two centuries had given the tale a meaning foreign to its real intent.

Brinton's tracing of the myth to its earlier stage is quite just, and in great measure also his view as to the development of its dualism. Yet if we go back to the earliest sources and examine this myth of the White One and the Dark One, we shall find it to be itself one of the most perfect examples the world can show of the rise of primitive dualism in the savage mind. Father Brebeuf's story is as

[1] Schoolcraft, 'Indian Tribes,' part v. p. 632; see part i. p. 316, part vi. p. 166; 'Iroquois,' p. 36, see 237; Brinton, 'Myths of New World,' p. 63.

follows: Aataentsic the Moon fell from heaven on earth, and bore two sons, Taouiscaron and Iouskeha, who being grown up quarrelled; judge, he says, if there be not in this a touch of the death of Abel. They came to combat, but with very different weapons. Iouskeha had a stag-horn, Taouiscaron contented himself with some wild-rose berries, persuading himself that as soon as he should thus smite his brother, he would fall dead at his feet; but it fell out quite otherwise than he had promised himself, and Iouskeha struck him so heavy a blow in the side that the blood gushed forth in streams. The poor wretch fled, and from his blood which fell upon the land came the flints which the savages still call Taouiscara, from the victim's name. From this we see it to be true that the original myth of the two brothers, the White One and the Dark One, had no moral element. It seems mere nature-myth, the contest between Day and Night, for the Hurons knew that Iouskeha was the Sun, even as his mother or grandmother Aataentsic was the Moon. Yet in the contrast between these two, the Huron mind had already come to the rudimentary contrast of the Good and Evil Deity. Iouskeha the Sun, it is expressly said, seemed to the Indians their benefactor; their kettle would not boil were it not for him; it was he who learnt from the Tortoise the art of making fire; without him they would have no luck in hunting; it is he who makes the corn to grow. Iouskeha the Sun takes care for the living and all things concerning life, and therefore, says the missionary, they say he is good. But Aataentsic the Moon, the creatress of earth and man, makes men die and has charge of their departed souls, and they say she is evil. Iouskeha and Taouiskaron, the Sun and Moon, dwell together in their cabin at the end of the earth, and thither it was that the four Indians made the mythic journey of which various episodes have been more than once cited here; true to their respective characters, the Sun receives the travellers kindly and saves them from the harm the beauteous but hurtful Moon would have done them. Another missionary of still earlier time identifies Iouskeha with the supreme deity Atahocan: "Iouskeha," he says, "is good and gives growth and fair weather; his grandmother Eatahentsic is wicked and spoils."[1] Thus in early Iroquois legend, the Sun and Moon, as god and goddess of Day and Night, had already acquired the characters of the great friend and enemy of man, the Good and Evil Deity. And as to the related cosmic legend of Day and Night, contrasted in the persons of the two brothers, the White One and the Dark One, though this was originally pure unethic nature-myth, yet it naturally took the same direction among the half-Europeanized Indians of later times, becoming a moral myth of Good and Evil. We have thus before us the profoundly interesting fact, that the rude North American Indians have more than once begun the same mythologic transition which in ancient Asia shaped the contrast of light and darkness into the contrast of righteousness and wickedness, by following out the same thought which still in the European mind arrays in the hostile forms of Light and Darkness the contending powers of Good and Evil.

Judging by such evidence as this, at once of the rudimentary dualism springing up in savage animism, and of the tendency of this to amalgamate with similar thought brought in by foreign intercourse, we may fairly account for many systems of this class found in the native religions of America. While the character and age of the evidence may lead us to agree with Waitz that the North American Indian dualism, the most distinct and universal feature of their religion, is not to be referred to a modern and Christian origin, yet we shall be cautious in claiming anything that may be borrowed civilized theology, as being genuine evidence of primitive development. The Algonquin's belief recognizes the antagonistic Kitchi Manitu and Matchi Manitu, the Great Spirit and Evil Spirit, who preside over the

[1] Brebeuf in 'Rel. des Jesuites dans la Nouvelle France,' 1635, p. 34, 1636, p. 100. Sagard, 'Histoire du Canada,' Paris, 1636, p. 490. L. H. Morgan, 'Iroquois,' p. 156.

spiritual contending hosts which fill the world and struggle for the mastery over it. They are especially associated, the one with light and warmth, the other with damp and darkness, while some tribes identify them with Sun and Moon. Here the nature-religion of the savage may have been developed, but was not set on foot, by the foreigner. In the extreme north-west, we may doubt any native origin in the semi-Christianized Kodiak's definition of Shljem Shoá the creator of heaven and earth, to whom offerings were made before and after the hunt, as contrasted with Ijak the bad spirit dwelling in the earth. In the extreme south-east, we may find more originality among the Floridan Indians two or three centuries ago, for they are said to have paid solemn worship to the Bad Spirit Toia who plagued them with visions, but to have had small regard for the Good Spirit, who troubles himself little about mankind.[1] On the southern continent, Martius makes this characteristic remark as to the rude tribes of Brazil: "All Indians have a lively conviction of the power of an evil principle over them; in many there dawns also a glimpse of the good; but they revere the one less than they fear the other. It might be thought that they hold the Good Being weaker in relation to the fate of man than the Evil." This generalization is to some extent supported by statements as to particular tribes. The Macusis are said to recognize the good creator Macunaima, "he who works by night," and his evil adversary Epel or Horiuch: of these people it is observed that "All the powers of nature are products of the Good Spirit, when they do not disturb the Indian's rest and comfort, but the work of evil spirits when they do." Uauüloa and Locozy, the good and evil deity of the Yumanas, live above the earth and toward the sun : the Evil Deity is feared by these

[1] Waitz, 'Anthropologie,' vol. iii. pp. 182, 330, 335, 345; La Potherie, 'Hist. de l'Amér. Septentrionale,' Paris, 1722, vol. i. p. 121; J. G. Müller, p. 149, etc. Schoolcraft, 'Indian Tribes,' part i. p. 35, etc., 320, 412; Catlin, vol. i. p. 156; Gregg, 'Commerce of Prairies,' vol. ii. pp. 238, 305; Cranz, 'Grönland,' p. 263.

savages, but the Good Deity will come to eat fruit with the departed and take their souls to his dwelling, wherefore they bury the dead each doubled up in his great earthen pot, with fruit in his lap, and looking toward the sunrise. Even the rude Botocudos are thought to recognize antagonistic principles of good and evil in the persons of the Sun and Moon.[1] This idea has especial interest from its correspondence on the one hand with that of the Iroquois tribes, and on the other with that of the comparatively civilized Muyscas of Bogota, whose good deity is unequivocally a mythic Sun, thwarted in his kindly labours for man by his wicked wife Huythaca the Moon.[2] The native religion of Chili is said to have placed among the subaltern deities Meulen, the friend of man, and Huecuvu the bad spirit and author of evil. These people can hardly have learnt from Christianity to conceive their evil spirit as simply and fully the general cause of misfortune: if the earth quakes, Huecuvu has given it a shock; if a horse tires, Huecuvu has ridden him; if a man falls sick, Huecuvu has sent the disease into his body, and no man dies but that Huecuvu suffocates him.[3]

In Africa, again, rudimentary dualism is not ill represented in native religion. An old account from Loango describes the natives as theoretically recognizing Zambi the supreme deity, creator of good and lover of justice, and over against him Zambi-anbi the destroyer, the counsellor of crime, the author of loss and accident, of disease and death. But when it comes to actual worship, as the good God will always be favourable, it is the god of evil who must be appeased, and it is for his satisfaction that men

[1] Martius, 'Ethnog. Amer.' vol. i. pp. 327, 485, 583, 645, see 247, 393, 427, 696. See also J. G. Müller, 'Amer. Urrelig.' pp. 259, etc., 403, 423; D'Orbigny, 'L'Homme Américain,' vol. i. p. 405, vol. ii. p. 257; Falkner, 'Patagonia,' p. 114; Musters, 'Patagonians,' p. 179; Fitzroy, 'Voy. of Adventure and Beagle,' vol. i. pp. 180, 190.
[2] Piedrahita, 'Hist. de Nuev. Granada,' part i. book i. ch. 3.
[3] Molina, 'Hist. of Chili,' vol. ii. p. 84; Febres, 'Diccionario Chileno,' s. v.

abstain some from one kind of food and some from another.[1] Among accounts of the two rival deities in West Africa, one describes the Guinea negroes as recognizing below the Supreme Deity two spirits (or classes of spirits), Ombwiri and Onyambe, the one kind and gentle, doing good to men and rescuing them from harm, the other hateful and wicked, whose seldom mentioned name is heard with uneasiness and displeasure.[2] It would be scarcely profitable, in an inquiry where accurate knowledge of the doctrine of any insignificant tribe is more to the purpose than vague speculation on the theology of the mightiest nation, to dwell on the enigmatic traces of ancient Egyptian dualism. Suffice it to say that the two brother-deities Osiris and Seti, Osiris the beneficent solar divinity whose nature the blessed dead took on them, Seti perhaps a rival national god degraded to a Typhon, seem to have become the representative figures of a contrasted scheme of light and darkness, good and evil; the sculptured granite still commemorates the contests of their long-departed sects, where the hieroglyphic square-eared beast of Seti has been defaced to substitute for it the figure of Osiris.[3]

The conception of the light-god as the good deity in contrast to a rival god of evil, is one plainly suggested by nature, and naturally recurring in the religions of the world. The Khonds of Orissa may be counted its most perfect modern exponents in barbaric culture. To their supreme creative deity, Būra Pennu or Bella Pennu, Light-god or Sun-god, there stands opposed his evil consort Tari Pennu the Earth-goddess, and the history of good and evil in the world is the history of his work and her counterwork. He created a world paradisaic, happy, harmless; she rebelled against him, and to blast the lot of his new creature, man,

[1] Proyart, 'Loango,' in Pinkerton, vol. xvi. p. 504. Bastian, 'Mensch.' vol. ii. p. 109. See Kolbe, 'Kaap de Goede Hoop,' part i. xxix. ; Waitz, vol. ii. p. 342 (Hottentots).
[2] J. L. Wilson, 'W. Afr.' pp. 217, 387. Waitz, vol. ii. p. 173.
[3] Birch, in Bunsen, vol. v. p. 136. Wilkinson, 'Ancient Eg.' etc.

she brought in disease, and poison, and all disorder, "sowing the seeds of sin in mankind as in a ploughed field." Death became the divine punishment of wickedness, the spontaneously fertile earth went to jungle and rock and mud, plants and animals grew poisonous and fierce, throughout nature good and evil were commingled, and still the fight goes on between the two great powers. So far all Khonds agree, and it is on the practical relation of good and evil that they split into their two hostile sects of Būra and Tari. Būra's sect hold that he triumphed over Tari, in sign of her discomfiture imposed the cares of childbirth on her sex, and makes her still his subject instrument wherewith to punish; Tari's sect hold that she still maintains the struggle, and even practically disposes of the happiness of man, doing evil or good on her own account, and allowing or not allowing the Creator's blessings to reach mankind.[1]

Now that the sacred books of the Zend-Avesta are open to us, it is possible to compare the doctrines of savage tribes with those of the great faith through which of all others Dualism seems to have impressed itself on the higher nations. The religion of Zarathustra was a schism from that ancient Aryan nature-worship which is represented in a pure and early form in the Veda, and in depravity and decay in modern Hinduism. The leading thought of the Zarathustrian faith was the contest of Good and Evil in the world, a contrast typified and involved in that of Day and Night, Light and Darkness, and brought to personal shape in the warfare of Ahura-Mazda and Anra-Mainyu, the Good and Evil Deity, Ormuzd and Ahriman. The prophet Zarathustra said: "In the beginning there was a pair of twins, two spirits, each of a peculiar activity. These are the good and the base in thought, word, and deed. Choose one of these two spirits. Be good, not base!" The sacred Vendidad begins with the record of the primæval contest of the two principles. Ahura-mazda created the best of regions

[1] Macpherson, 'India,' p. 84.

and lands, the Aryan home, Sogdia, Bactria, and the rest; Anra-Mainyu against his work created snow and pestilence, buzzing insects and poisonous plants, poverty and sickness, sin and unbelief. The modern Parsi, in passages of his formularies of confession, still keeps alive the old antagonism. I repent, he says, of all kinds of sins which the evil Ahriman produced amongst the creatures of Ormazd in opposition. "That which was the wish of Ormazd the Creator, and I ought to have thought and have not thought, what I ought to have spoken and have not spoken, what I ought to have done and have not done; of these sins repent I with thoughts, words, and works, corporeal as well as spiritual, earthly as well as heavenly, with the three words: Pardon, O Lord, I repent of sin. That which was the wish of Ahriman, and I ought not to have thought and yet have thought, what I ought not to have spoken and yet have spoken, what I ought not to have done and yet have done; of these sins repent I with thoughts, words, and works, corporeal as well as spiritual, earthly as well as heavenly, with the three words: Pardon, O Lord, I repent of sin." ... "May Ahriman be broken, may Ormazd increase."[1] The Izedis or Yezidis, the so-called Devil-worshippers, still remain a numerous though oppressed people in Mesopotamia and adjacent countries. Their adoration of the sun and horror of defiling fire accord with the idea of a Persian origin of their religion (Persian ized = god), an origin underlying more superficial admixture of Christian and Moslem elements. This remarkable sect is distinguished by a special form of dualism. While recognizing the existence of a Supreme Being, their peculiar reverence is given to Satan, chief of the angelic host, who now has the means of doing evil to mankind, and in his restoration will have the power of rewarding them. "Will not Satan then reward the poor Izedis, who alone have never spoken ill of him, and have suffered so much for him?" Martyrdom for the rights

[1] Avesta (Spiegel and Bleeck) : Vendidad, i. ; 'Khordah-A.' xlv. xlvi. Max Müller, 'Lectures,' 1st Ser. p. 208.

of Satan! exclaims the German traveller to whom an old white-bearded devil-worshipper thus set forth the hopes of his religion.[1]

Direct worship of the Evil Principle, familiar as it is to low barbaric races, is scarcely to be found among people higher in civilization than these persecuted and stubborn sectaries of Western Asia. So far as such ideas extend in the development of religion, they seem fair evidence how far worship among low tribes turns rather on fear than love. That the adoration of a Good Deity should have more and more superseded the propitiation of an Evil Deity, is the sign of one of the great movements in the education of mankind, a result of happier experience of life, and of larger and more gladsome views of the system of the universe. It is not, however, through the inactive systems of modern Parsism and Izedism that the mighty Zoroastrian dualism has exerted its main influence on mankind. We must look back to long past ages for traces of its contact with Judaism and Christianity. It is often and reasonably thought that intercourse between Jews and ancient Persians was an effective agent in producing that theologic change which differences the later Jew of the Rabbinical books from the earlier Jew of the Pentateuch, a change in which one important part is the greater prominence of the dualistic scheme. So in later times (about the fourth century), the contact of Zoroastrism and Christianity appears to have been influential in producing Manichæism. We know Manichæism mostly on the testimony of its adversaries, but thus much seems clear, that it is based on the very doctrine of the two antagonistic principles of good and evil, of spirit and matter. It sets on the one hand God, original good and source of good alone, primal light and lord of the kingdom of light, and on the other hand the Prince of Darkness, with his kingdom of darkness, of matter, of confusion, and destruction. The theory of ceaseless conflict between these contending

[1] Layard, 'Nineveh,' vol. i. p. 297 ; Ainsworth, ' Izedis,' in 'Tr. Eth. Soc.' vol. i. p. 11.

powers becomes a key to the physical and moral nature and course of the universe.[1] Among Christian or semi-Christian sects, the Manichæans stand as representatives of dualism pushed to its utmost development. It need scarcely be said, however, that Christian dualism is not bounded by the limits of this or that special sect. In so far as the Evil Being, with his subordinate powers of darkness, is held to exist and act in any degree in independence of the Supreme Deity and his ministering spirits of light, so far theological schools admit, though in widely different grades of importance, a philosophy of nature and of life which has its basis rather in dualism than in monotheism.

We now turn to the last objects of our present survey, those theological beliefs of the lower tribes of mankind which point more or less distinctly toward a doctrine of Monotheism. Here it is by no means proposed to examine savage ideas from the point of view of doctrinal theology, an undertaking which would demand arguments quite beyond the present range. Their treatment is limited to classifying the actual beliefs of the lower races, with some ethnographic considerations as to their origin and their relation to higher religions. For this purpose it is desirable to distinguish the prevalent doctrines of the uncultured world from absolute monotheism. At the outset, care is needed to exclude an ambiguity of which the importance often goes unnoticed. How are the mighty but subordinate divinities, recognized in different religions, to be classed? Beings who in Christian or Moslem theology would be called angels, saints, demons, would under the same definitions be called deities in polytheistic systems. This is obvious, but we may realize it more distinctly from its actually having happened. The Chuwashes, a race of Turkish affinity, are stated to reverence a god of Death, who takes to himself the souls of the departed, and whom they call Esrel; it is curious that Castrén, in mentioning

[1] Beausobre, 'Hist. de Manichée,' etc. Neander, 'Hist. of Christian Religion,' vol. ii. p. 157, etc.

this, should fail to point out that this deity is no other than Azrael the angel of death, adopted under Moslem influence.[1] Again, in the mixed Pagan and Christian religion of the Circassians, which at least in its recently prevalent form would be reckoned polytheistic, there stand beneath the Supreme Being a number of mighty subordinate deities, of whom the principal are Iele the Thunder-god, Tleps the Fire-god, Seoseres the god of Wind and Water, Misitcha the Forest-god, and Mariam the Virgin Mary.[2] If the monotheistic criterion be simply made to consist in the Supreme Deity being held as creator of the universe and chief of the spiritual hierarchy, then its application to savage and barbaric theology will lead to perplexing consequences. Races of North and South America, of Africa, of Polynesia, recognizing a number of great deities, are usually and reasonably considered polytheists, yet under this definition their acknowledgment of a Supreme Creator, of which various cases will here be shown, would entitle them at the same time to the name of monotheists. To mark off the doctrines of the lower races, closer definition is required, assigning the distinctive attributes of deity to none save the Almighty Creator. It may be declared that, in this strict sense, no savage tribe of monotheists has been ever known. Nor are any fair representatives of the lower culture in a strict sense pantheists. The doctrine which they do widely hold, and which opens to them a course tending in one or other of these directions, is polytheism culminating in the rule of one supreme divinity. High above the doctrine of souls, of divine manes, of local nature-spirits, of the great deities of class and element, there are to be discerned in savage theology shadowings, quaint or majestic, of the conception of a Supreme Deity, henceforth to be traced onward in expanding power and brightening glory along the history of religion. It is no unimportant task, partial as it is, to select and group the typical data

[1] Castrén, 'Finn. Myth.' p. 155.
[2] Klemm, 'Cultur-Gesch.' vol. vi. p. 85.

which show the nature and position of the doctrine of supremacy, as it comes into view within the lower culture. On the threshold of the investigation, there meets us the same critical difficulty which obstructs the study of primitive dualism. Among low tribes who have been in contact with Christianity or Mohammedanism, how are we to tell to what extent, under this foreign influence, dim, uncouth ideas of divine supremacy may have been developed into more cultured forms, or wholly foreign ideas implanted? We know how the Jesuit missionaries caught and trained into their own theology the native Canadian thought of a Great Manitu, how they took up the native Brazilian name of the divine Thunder, Tupan, and adapted its meaning to convey in Christian teaching the idea of God. Thus, again, we find most distinctly-marked African ideas of a Supreme Deity in the West, where intercourse with Moslems has actually Islamized or semi-Islamized whole negro nations, and the name of Allah is in all men's mouths. The ethnographer must be ever on the look-out for traces of such foreign influence in the definition of the Supreme Deity acknowledged by any uncultured race, a divinity whose nature and even whose name may betray his adoption from abroad. Thus the supreme Iroquois deity, Neo or Hawaneu, the pre-existent creator, has been triumphantly adduced to show the monotheism underlying the native creeds of America. But Dr. Brinton considers this divinity as derived from Christian instruction, and his very name but a corruption of Dieu, le bon Dieu.[1] Among the list of supreme deities of the lower races who are also held to be first ancestors of man, we hear of Louquo, the uncreate first Carib, who descended from the eternal heaven, made the flat earth, and produced man from his own body. He lived long on earth among men, died and came to life again after three days, and returned to heaven.[2] It would be hardly reasonable to enumerate, among genuine deities of native

[1] Brinton, 'Myths of New World,' p. 53. Schoolcraft, 'Iroquois,' p. 33.
[2] De la Borde, 'Caraibes,' p. 524. J. G. Müller, 'Amer. Urrel.' p. 228.

West Indian religion, a being with characteristics thus on the face of them adopted from the religion of the white men. Yet even in such extreme cases, it does not necessarily follow that the definitions of these deities, vitiated as they are for ethnographical use by foreign influence, have not to some extent a native substratum. In criticising details, moreover, it must not be forgotten how largely the similarities in the religions of different races may be of independent origin, and how closely allied are many ideas in the rude native theologies of savages to ideas holding an immemorial place in the religions of their civilized invaders. For the present purpose, however, it is well to dwell especially on such evidence as by characteristic traits or early date is farthest removed from suspicion of being borrowed from a foreign source.

In surveying the peoples of the world, the ethnographer finds some who are not shown to have any definite conception of a supreme deity; and even where such a conception is placed on record, it is sometimes so vaguely asserted, or on such questionable authority, that he can but take note of it and pass on. In numerous cases, however, illustrated by the following collection from different regions, certain leading ideas, singly or blended, may be traced. There are many savage and barbaric religions which solve their highest problem by the simple process of raising to divine primacy one of the gods of polytheism itself. Even the system of the manes-worshipper has been stretched to reach the limit of supreme deity, in the person of the primæval ancestor. More frequently, it is the nature-worshipper's principle which has prevailed, giving to one of the great nature-deities the precedence of the rest. Here, by no recondite speculation, but by the plain teaching of nature, the choice has for the most part lain between two mighty visible divinities, the all-animating Sun and the all-encompassing Heaven. In the study of such schemes, we are on intellectual terra firma. There is among the religions of the lower races another notable group of systems, seemingly in close connection with the first. These display to us a heavenly pantheon arranged on the model of an earthly political constitution, where the commonalty are crowds of human souls and other tribes of world-pervading spirits, the aristocracy are great polytheistic gods, and the King is the Supreme Deity. To this comparatively intelligible side of the subject, a more perplexed and obscure side stands contrasted. Among men whose theory of the soul animating the body has already led them to suppose a divine spirit animating the huge mass of earth or sky, this idea needs but a last expansion to become a doctrine of the universe as animated by one greatest, all-pervading divinity, the World-Spirit. Moreover, where speculative philosophy, savage or cultured, grapples with the vast fundamental world-problem, the solution is attained by ascending from the Many to the One, by striving to discern through and beyond the Universe a First Cause. Let the basis of such reasoning be laid in theological ground, then the First Cause is realised as the Supreme Deity. In such ways, the result of carrying to their utmost limits the animistic conceptions which pervade the philosophy of religion, alike among low races and high, is to reach an idea of as it were a soul of the world, a shaper, animator, ruler of the universe, a Great Spirit. In no small measure, such definition answers to that of the highest deity adored by the lower races of mankind. As we enter these regions of transcendental theology, however, we are not to wonder that the comparative distinctness belonging to conceptions of lower spiritual beings here fades away. Human souls, subordinate nature-spirits, and huge polytheistic nature-gods, carry with the defined special functions they perform some defined character and figure, but beyond such limits form and function blend into the infinite and universal in the thought of supreme divinity. To realize this vast idea, two especial ways are open, and both are trodden even by uncultured men. The first way is to fuse the attributes of the great polytheistic powers into more or less of common personality, thus conceiving that, after all, it is the same Highest Being who holds up the heavens, shines in the sun, smites his foes in the thunder, stands first in the human pedigree as the divine ancestor. The second way is to remove the limit of theologic speculation into the region of the indefinite and the inane. An unshaped divine entity looming vast, shadowy, and calm beyond and over the material world, too benevolent or too exalted to need human worship, too huge, too remote, too indifferent, too supine, too merely existent, to concern himself with the petty race of men,—this is a mystic form or formlessness in which savage and barbaric tribes have not seldom pictured the Supreme.

Thus, then, it appears that the theology of the lower races already reaches its climax in conceptions of Supreme Deity, and that these conceptions in the savage and barbaric world are no copies stamped from one common type, but outlines widely varying among mankind. The degeneration-theory, in some instances no doubt with justice, may claim such beliefs as mutilated and perverted remnants of higher religions. Yet for the most part, the development-theory is competent to account for them without seeking their origin in grades of culture higher than those in which they are found existing. Looked upon as products of natural religion, such doctrines of divine supremacy seem in no way to transcend the powers of the low-cultured mind to reason out, nor of the low-cultured imagination to deck with mythic fancy. There have existed in times past, and do still exist, many savage or barbaric people who hold such views of a highest god as they may have attained to of themselves, without the aid of more cultured nations. Among these races, Animism has its distinct and consistent outcome, and Polytheism its distinct and consistent completion, in the doctrine of a Supreme Deity.

The native religions of South America and the West Indies display a well-marked series of types. The primacy of the Sun was long ago well stated by the Moluches when a Jesuit missionary preached to them, and they replied,

"Till this hour, we never knew nor acknowledged anything greater or better than the Sun."[1] So when a later missionary argued with the chief of the Tobas, "My god is good and punishes wicked people," the chief replied, "My God (the Sun) is good likewise; but he punishes nobody, satisfied to do good to all."[2] In various manifestations, moreover, there reigns in native faiths a supreme being whose characteristics are those of the Heaven-god. It is thus with the Tamoi of the Guaranis, that beneficent deity worshipped in his blended character of ancestor of mankind and ancient of heaven, lord of the celestial paradise."[3] It is so with the highest deity of the Araucanians, Pillan the Thunder or the Thunderer, called also Huenu-Pillan or Heaven-Thunder, and Vuta-gen or Great Being. "The universal government of Pillan," says Molina, "is a prototype of the Araucanian polity. He is the great Toqui (Governor) of the invisible world, and as such has his Apo-Ulmenes, and his Ulmenes, to whom he entrusts the administration of affairs of less importance. These ideas are certainly very rude, but it must be acknowledged that the Araucanians are not the only people who have regulated the things of heaven by those of the earth."[4] A different but not less characteristic type of the Supreme Deity is placed on record among the Caribs, a beneficent power dwelling in the skies, reposing in his own happiness, careless of mankind, and by them not honoured nor adored.[5]

The theological history of Peru, in ages before the Spanish conquest, has lately had new light thrown on it by the researches of Mr. Markham. Here the student comes into view of a rivalry full of interest in the history of barbaric religion, the rivalry between the Creator and

[1] Dobrizhoffer, 'Abipones,' vol. ii. p. 89.
[2] Hutchinson, 'Chaco Ind.' in 'Tr. Eth. Soc.' vol. iii. p. 327.
[3] D'Orbigny, 'L'Homme Américain,' vol. ii. p. 319.
[4] Molina, 'Hist. of Chili,' vol. ii. p. 84, etc. Compare Febres, 'Diccionario Chileño.'
[5] Rochefort, 'Iles Antilles,' p. 415. Musters, 'Patagonians,' p. 179.

the divine Sun. The Supreme Deity in the religion of the Incas was Uiracocha, whose titles were Pachayachachic, 'Teacher of the World,' and Pachacamac, 'Creator of the World.' The Sun (with whom was coupled his sister-wife the Moon) was the divine ancestor, the dawn or origin, the totem or lar, of the Inca family. The three great deities were the Creator, Sun, and Thunder; their images were brought out together at great festivals into the square of Cuzco, llamas were sacrificed to all three, and they could be addressed in prayer together, "O Creator, and Sun, and Thunder, be for ever young, multiply the people, and let them always be at peace." Yet the Thunder and Lightning was held to come by the command of the Creator, and the following prayer shows clearly that even "our father the Sun" was but his creature:—

"Uiracocha! Thou who gavest being to the Sun, and afterwards said let there be day and night. Raise it and cause it to shine, and preserve that which thou hast created, that it may give light to men. Grant this, Uiracocha!

"Sun! Thou who art in peace and safety, shine upon us, keep us from sickness, and keep us in health and safety."

Among the transitions of religion, however, it is not strange that a subordinate god, by virtue of his nearer intercourse and power, should usurp the place of the Supreme Deity. Among the various traces of this taking place under the Incas, are traditions of the great temple at Cuzco called "the Golden Place, the house of the Teacher of the World," where Manco Ccapac originally set up a flat oval golden plate to signify the Creator; Mayta Ccapac, it is said, renewed the Creator's symbol, but Huascar Inca took it down, and set up in its stead in the place of honour a round golden plate like the sun with rays. The famous temple itself, Ccuricancha the "Golden Place," was known to the Spaniards as the temple of the Sun; no wonder that the idea has come to be so generally accepted, that the Sun was the chief god of Peru. There is even on record a memorable protest made by one Inca, who dared to deny that the Sun could be the maker of all things, comparing him to a tethered beast that must make ever the same daily round, and to an arrow that must go whither it is sent, not whither it will. But what availed philosophic protest, even from the head of church and state himself, against a state church of which the world has seldom seen the equal for stiff and solid organization? The Sun reigned in Peru till Pizarro overthrew him, and his splendid golden likeness came down from the temple wall to be the booty of a Castilian soldier, who lost it in one night at play.[1]

Among rude tribes of the North American continent, evidence of the primacy of the divine Sun is not unknown. We may perhaps distrust Father Sagard's early identification of Atahocan the Creator with Iouskeha the Sun. Yet Father Hennepin's account of the Sioux worshipping the Sun as the Creator is explicit enough, and agrees with the argument of the modern Shawnees, that the Sun animates everything, and therefore must be the Master of Life or Great Spirit.[2] It is the widespread belief in this Great Spirit, whatever his precise nature and origin, that has long and deservedly drawn the attention of European thinkers to the native religions of the North American tribes. True, this is a district in which the native doctrine has been at times described by Europeans in exaggerated and mistaken terms, converting it into a rude analogue of theism, while also the ideas of the Indians themselves came to be remodelled under Christian influence. It has even

---

[1] 'Narratives of the Rites and Laws of the Yncas,' trans. from the original Spanish MSS., and ed. by C. R. Markham, Hakluyt Soc. 1873, p. ix. 5, 16, 30, 76, 84, 154, etc. The above remarks are based on the early evidence here printed for the first time, and on private suggestions for which I am also indebted to Mr. Markham. The title Pachacamac has been also considered to mean Animator or Soul of the World, camani = I create, camac = creator, cama = soul (note to 2nd ed.). Garcilaso de la Vega, lib. i., ii. c. 2, iii. c. 20; Herrera, dec. v. 4; Brinton, 'Myths of New World,' p. 177, see 142; Rivero and Tschudi, 'Peruvian Antiquities,' ch. vii. ; Waitz, vol. iv. p. 447; J. G. Müller, p. 317, etc.

[2] Sagard, 'Hist. du Canada,' p. 490. Hennepin, 'Voy dans l'Amérique," p. 302. Gregg, 'Commerce of Prairies,' vol. ii. p. 237.

been thought that the whole doctrine of the Great Spirit was borrowed by the savages from missionaries and colonists. But this view will not bear examination. After due allowance made for mis-rendering of savage answers and importation of white men's thoughts, it can hardly be judged that a divine being whose characteristics are often so unlike what European intercourse would have suggested, and who is heard of by such early explorers among such distant tribes, could be a deity of foreign origin. The Greenlanders' Torngarsuk or Great Spirit (his name is an augmentative of "torngak"—"spirit") seems no figure derived from the religion of Scandinavian colonists, ancient or modern. He is the oracular deity whom the angekoks go in spirit to consult about sickness and weather and sport, and to whose summer-land beneath the sea Greenland souls hope to descend at death. Imperfectly defined by native theologians, thought to be beneficent and therefore scarcely worshipped, he so clearly held his place as supreme deity in the native mind, that, as Cranz the missionary relates, many Greenlanders hearing of God and his almighty power were apt to fall on the idea that it was their Torngarsuk who was meant.[1] In like manner, Algonquin Indians, early in the 17th century, hearing of the white man's Deity, identified him with one known to their own native belief, Atahocan the Creator. When Le Jeune the missionary talked to them of an almighty creator of heaven and earth, they began to say to one another, "Atahocan, Atahocan, it is Atahocan!" The traditional idea of such a being seems indeed to have lain in utter mythic vagueness in their thoughts, for they had made his name into a verb, "Nitatahocan," meaning, "I tell a fable, an old fanciful story."[2]

The Great Spirit of the North American Indians is especially known to us in name and nature as the Kitchi

[1] Cranz, 'Grönland,' p. 263.
[2] Le Jeune in 'Rel. des Jésuites dans la Nouvelle France,' 1633, p. 16; 1634, p. 13.

Manitu of the Ojibwas and other Algonquin tribes. In late times, Schoolcraft represents this deity as a pantheistic Soul of the Universe, inhabiting and animating all things, recognized in rocks and trees, in cataracts and clouds, in thunder and lightning, in tempest and zephyr, becoming incarnate in birds and beasts as titular deities, existing in the world under every possible form, animate and inanimate.[1] Whether the Red Indian mind even in modern times really entertained this extreme pantheistic scheme, we may well doubt. In early times of American discovery, the records show a quite different and more usual conception of a supreme deity. Among the more noteworthy of these older documents are the following. Jacques Cartier, in his second Canadian voyage (1535) speaks of the people having no valid belief in God, for they believe in one whom they call Cudouagni, and say that he often speaks with them, and tells them what the weather will be; they say that when he is angry with them he casts earth in their eyes. Thevet's statement somewhat later is as follows: "As to their religion, they have no worship or prayer to God, except that they contemplate the new moon, called in their language Osannaha, saying that Andouagni calls it thus, sending it little by little to advance or retard the waters. For the rest, they fully believe that there is a Creator, greater than the Sun, the Moon, and the Stars, and who holds all in his power. He it is whom they call Andouagni, without however having any form or method of prayer to him.[2] In Virginia about 1586, we learn from Heriot that the natives believed in many gods, which they call "mantoac," but of different sorts and degrees, also that there is one chief god who first made other principal gods, and afterwards the sun, moon, and stars as petty gods. In New England, in 1622, Winslow says that they

[1] Schoolcraft, 'Indian Tribes,' part i. p. 15.
[2] Cartier, 'Relation;' Hakluyt, vol. iii. p. 212; Lescarbot, 'Nouvelle France,' p. 613. Thevet, 'Singularitez de la France Antarctique,' Paris, 1558, ch. 77. See also J. G. Müller, p. 102. Andouagni is perhaps a miscopied form of Cudouagni. Other forms, Cudruagni, etc., occur.

believe, as do the Virginians, in many divine powers, yet of one above all the rest; the Massachusetts call their great god Kiehtan, who made all the other gods; he dwells far westerly above the heavens, whither all good men go when they die; "They never saw *Kiehtan*, but they hold it a great charge and dutie, that one age teach another; and to him they make feasts, and cry and sing for plentie and victorie, or anything is good." Brinton's etymology is plausible, that this Kiehtan is simply the Great Spirit (Kittanitowit, Great Living Spirit, an Algonquin word compounded of "Kitta"—great; "manito"—spirit; "wit"—termination indicating life). Another famous native American name for the supreme deity is Oki. Captain John Smith, the hero of the colonization of Virginia in 1607, he who was befriended by Pocahontas, "La Belle Sauvage," thus describes the religion of the country, and especially of her tribe, the Powhatans: "There is yet in Virginia no place discovered to be so Savage in which they haue not a Religion, Deer, and Bow and Arrowes. All things that are able to doe them hurt beyond their prevention, they adore with their kinde of divine worship; as the fire, water, lightning, thunder, our Ordnance peeces, horses, &c. But their chiefe god they worship is the Devill. Him they call *Okee*, and serue him more of feare than loue. They say they haue conference with him, and fashion themselves as neare to his shape as they can imagine. In their Temples they haue his image evill favouredly carved, and then painted and adorned with chaines of copper, and beads, and covered with a skin in such manner as the deformities may well suit with such a God."[1] This quaint account deserves to be quoted at length as an example of the judgment which a half-educated and whole-prejudiced European is apt to pass on savage deities, which from his point of view seem of simply diabolic nature. It is known from other sources

[1] Smith, 'Hist. of Virginia,' London, 1632; in Pinkerton, vol. xiii. pp. 13, 39; New England, ibid. p. 244. Brinton, p. 58; Waitz, voL iii. p. 177, etc.; J. G. Müller, pp. 99, etc.; Loskiel, part i. pp. 33, 43.

that Oki, a word apparently meaning that which is "above," was in fact a general name for spirit or deity. We may judge the real belief of these Indians better from Father Brebeuf's description of the Heaven-god, cited here in a former chapter: they imagine in the heavens an Oki, that is, a Demon or power ruling the seasons of the year, and controlling the winds and waves, a being whose anger they fear, and whom they call on in making solemn treaties.[1] The longer rude tribes of America have been in contact with European belief, the less confidently can we ascribe to purely native sources the theologic schemes their religions have settled into. Yet the Creeks towards the end of the last century preserved some elements of native faith. They believed in the Great Spirit, the Master of Breath (a being whom Bartram represents as a soul and governor of the universe): to him they would address their frequent prayers and ejaculations, at the same time paying a kind of homage to the sun, moon, and stars, as the mediators or ministers of the Great Spirit, in dispensing his attributes for their comfort and well-being in this life.[2] In our own day, among the wild Comanches of the prairies, the Great Spirit, their creator and supreme deity, is above Sun and Moon and Earth; towards him is sent the first puff of tobacco-smoke before the Sun receives the second, and to him is offered the first morsel of the feast.[3]

Turning from the simple faiths of savage tribes of North America, to the complex religion of the half-civilized Mexican nation, we find what we might naturally expect, a cumbrous polytheism complicated by mixture of several national pantheons, and beside and beyond this, certain appearances of a doctrine of divine supremacy. But these

[1] Brebeuf in 'Rel. des Jés.' 1636, p. 107; see above, p. 255. Brinton, p. 47; Sagard, p. 494; J. G. Müller, p. 103. For other mention of a Supreme Deity among North American tribes see Joutel, 'Journal du Voyage, etc. Paris, 1713, p. 224 (Louisiana); Sproat in 'Tr. Eth. Soc.' vol. v. p. 253 (Vancouver's I.).
[2] Bartram in 'Tr. Amer. Eth. Soc.' vol. iii. pp. 20, 26.
[3] Schoolcraft, 'Ind. Tribes,' part ii. p. 127.

doctrines seem to have been spoken of more definitely than the evidence warrants. A remarkable native development of Mexican theism must be admitted, in so far as we may receive the native historian Ixtlilxochitl's account of the worship paid by Nezahualcoyotl, the poet-king of Tezcuco, to the invisible supreme Tloque Nahuaque, he who has all in him, the cause of causes, in whose star-roofed pyramid stood no idol, and who there received no bloody sacrifice, but only flowers and incense. Yet it would have been more satisfactory were the stories told by this Aztec panegyrist of his royal ancestor confirmed by other records. Traces of divine supremacy in Mexican religion are especially associated with Tezcatlipoca, "Shining Mirror," a deity who seems in his original nature the Sun-god, and thence by expansion to have become the soul of the world, creator of heaven and earth, lord of all things, Supreme Deity. Such conceptions may in more or less measure have arisen in native thought, but it should be pointed out that the remarkable Aztec religious formulas collected by Sahagun, in which the deity Tezcatlipoca is so prominent a figure, show traces of Christian admixture in their material, as well as of Christian influence in their style. For instance, all students of Mexican antiquities know the belief in Mictlan, the Hades of the dead. But when one of these Aztec prayer-formulas (concerning auricular confession, the washing away of sins, and a new birth) makes mention of sinners being plunged into a lake of intolerable misery and torment, the introduction of an idea so obviously European condemns the composition as not purely native. The question of the actual developments of ideas verging on pantheism or theism, among the priests and philosophers of native Mexico, is one to be left for further criticism.[1]

In the islands of the Pacific, the idea of Supreme Deity is especially manifested in that great mythologic divinity of

[1] Prescott, 'Mexico,' book i. ch. vi. Sahagun, 'Hist. de Nueva España,' lib. vi. in Kingsborough, vol. v. ; Torquemada, 'Monarq. Ind' lib. x. c. 14. Waitz, vol. iv. p. 136; J. G. Müller, p. 621, etc.

the Polynesian race, whom the New Zealanders call Tangaroa, the Hawaiians Kanaroa, the Tongans and Samoans Tangaloa, the Georgian and Society islanders Taaroa. Students of the science of religion who hold polytheism to be but the mis-development of a primal idea of divine unity, which in spite of corruption continues to pervade it, might well choose this South Sea Island divinity as their aptest illustration from the savage world. Taaroa, says Moerenhout, is their supreme or rather only god; for all the others, as in other known polytheisms, seem scarcely more than sensible figures and images of the infinite attributes united in his divine person. The following is given as a native poetic definition of the Creator. " He was; Taaroa was his name; he abode in the void. No earth, no sky, no men. Taaroa calls, but nought answers; and alone existing, he became the universe. The props are Taaroa; the rocks are Taaroa; the sands are Taaroa; it is thus he himself is named." According to Ellis, Taaroa is described in the Leeward Islands as the eternal parentless uncreate Creator, dwelling alone in the highest heaven, whose bodily form mortals cannot see, who after intervals of innumerable seasons casts off his body or shell and becomes renewed. It was he who created Hina his daughter, and with her aid formed the sky and earth and sea. He founded the world on a solid rock, which with all the creation he sustains by his invisible power. Then he created the ranks of lesser deities such as reign over sea and land and air, and govern peace and war, and preside over physic and husbandry, and canoe-building, and roofing, and theft. The version from the Windward Islands is that Taaroa's wife was the rock, the foundation of all things, and she gave birth to earth and sea. Now, fortunately for our understanding of this myth, the name of Taaroa's wife, with whom he begat the lesser deities, was taken down in Tahiti in Captain Cook's time. She was a rock called Papa, and her name plainly suggests her identity with Papa the Earth, the wife of Rangi the Heaven in the New Zealand myth of Heaven and Earth, the great first parents. If this inference be just, then it seems that Taaroa the Creator is no personification of a primæval theistic idea, but simply the divine personal Heaven transformed into the supreme Heaven-god. Thus, when Turner gives the Samoan myths of Tangaloa in heaven presiding over the production of the earth from beneath the waters, or throwing down from the sky rocks which are now islands, the classic name by which he calls him is that which rightly describes his nature and mythic origin— Tangaloa, the Polynesian Jupiter. Yet in island district after district, we find the name of the mighty heavenly creator given to other and lesser mythic beings. In Tahiti, the manes-worshipper's idea is applied not only to lesser deities, but to Taaroa the Creator himself, whom some maintained to be but a man deified after death. In the New Zealand mythology, Tangaroa figures on the one hand as Sea-god and father of fish and reptiles, on the other as the mischievous eaves-dropping god who reveals secrets. In Tonga, Tangaloa was god of artificers and arts, and his priests were carpenters; it was he who went forth to fish, and dragged up the Tonga islands from the bottom of the sea. Here, then, he corresponds with Maui, and indeed Tangaroa and Maui are found blending in Polynesia even to full identification. It is neither easy nor safe to fix to definite origin the Protean shapes of South Sea mythology, but on the whole the native myths are apt to embody cosmic ideas, and as the idea of the Sun preponderates in Maui, so the idea of the Heaven in Taaroa.[1] In the Fiji Islands, whose native mythology is on the whole distinct from that of Polynesia proper, a strange weird figure takes the supreme place among the gods. His name is Ndengei, the serpent is his shrine, some traditions represent him with a

[1] Moerenhout, 'Voy. aux Iles du Grand Océan,' vol. i. pp. 419, 437. Ellis, 'Polyn. Res.' vol. i. p. 321, etc. J. R. Forster, 'Voyage round the World,' pp. 540, 567. Grey, 'Polyn. Myth.' p. 6. Taylor, 'New Zealand,' p. 118; see above, vol. i. p. 290. Turner, 'Polynesia,' p. 244. Mariner, 'Tonga Is.' vol. ii. pp. 116, 121. Schirren, 'Wandersagen der Neuseeländer,' pp. 68, 89.

serpent's head and body and the rest of him stone. He passes a monotonous existence in his gloomy cavern, feeling no emotion nor sensation, nor any appetite but hunger; he takes no interest in any one but Uto, his attendant, and gives no sign of life beyond eating, answering his priest, and changing his position from one side to the other. No wonder Ndengei is less worshipped than most of the inferior gods. The natives have even made a comic song about him, where he talks with his attendant, Uto, who has been to attend the feast at Rakiraki, where Ndengei has especially his temple and worship.

Ndengei. " Have you been to the sharing of food to-day ? "
Uto. " Yes: and turtles formed a part; but only the undershell was shared to us two."
Ndengei. " Indeed, Uto! This is very bad. How is it? We made them men, placed them on the earth, gave them food, and yet they share to us only the under-shell. Uto, how is this?"[1]

The native religion of Africa, a land pervaded by the doctrines of divine hierarchy and divine supremacy, affords apt evidence for the problem before us. The capacity of the manes-worshipper's scheme to extend in this direction may be judged from the religious speculations of the Zulus, where we may trace the merging of the First Man, the Old-Old-One, Unkulunkulu, into the ideal of the Creator, Thunderer, and Heaven-god.[2] If we examine a collection of documents illustrating the doctrines of the West African races lying between the Hottentots on the south and the Berbers on the north, we may fairly judge their conceptions, influenced as these may have been by foreign intercourse, to be nevertheless for the most part based on native ideas of the personal Heaven.[3] Whether they think of their

[1] Williams, 'Fiji,' vol. i. p. 217.
[2] Callaway, ' Religion of Amazulu,' part i. See ante, pp. 116, 313.
[3] See especially Waitz, vol. ii. p. 167, etc.; J. L. Wilson, ' W. Afr.' pp. 209, 387 ; Bosman, Mungo Park, etc. Comp. Ellis, ' Madagascar,' vol. i. p. 390.

supreme deity as actively pervading and governing his universe, or as acting through his divine subordinates, or as retiring from his creation and leaving the lesser spirits to work their will, he is always to their minds the celestial ruler, the Heaven-god. Examples may be cited, each in its way full of instruction. In the mind of the Gold-coast negro, tendencies towards theistic religion seem to have been mainly developed through the idea of Nyongmo, the personal Heaven, or its animating personal deity. Heaven, wide-arching, rain-giving, light-giving, who has been and is and shall be, is to him the Supreme Deity. The sky is Nyongmo's creature, the clouds are his veil, the stars his face-ornaments. Creator of all things, and of their animating powers whose chief and elder he is, he sits in majestic rest surrounded by his children, the wongs, the spirits of the air who serve him and represent him on earth. Though men's worship is for the most part paid to these, reverence is also given to Nyongmo, the Eldest, the Highest. Every day, said a fetish-man, we see how the grass and corn and trees spring forth by the rain and sunshine that Nyongmo sends, how should he not be the Creator? Again, the mighty Heaven-god, far removed from man and seldom roused to interfere in earthly interests, is the type on which the Guinea negros may have modelled their thoughts of a Highest Deity who has abandoned the control of his world to lesser and evil spirits.[1] The religion of another district seems to show clearly the train of thought by which such ideas may be worked out. Among the Kimbunda race of Congo, Suku-Vakange is the highest being. He takes little interest in mankind, leaving the real government of the world to the good and evil kilulu or spirits, into whose ranks the souls of men pass at death. Now in that there are more bad spirits who torment, than good who favour living men, human misery would be unbearable, were it not that from

[1] Steinhauser, 'Religion des Negers,' in 'Mag. der Miss.' Basel, 1856. No. 2, p. 128. J. L. Wilson, 'W. Afr.' pp. 92, 209; Römer, 'Guinea,' p. 42. See also Waitz, vol. ii. pp. 171, 419.

time to time Suku-Vakange, enraged at the wickedness of the evil spirits, terrifies them with thunder, and punishes the more obstinate with his thunderbolts. Then he returns to rest, and lets the kilulu rule again.[1] Who, we may ask, is this divinity, calm and indifferent save when his wrath bursts forth in storm, but the Heaven himself? The relation of the Supreme Deity to the lesser gods of polytheism is graphically put in the following passage, where an American missionary among the Yorubas describes the relation of Olorung, the Lord of Heaven, to his lesser deities (orisa), among whom the chief are the androgynous Obatala, representing the reproductive power of nature, and Shango the Thunder-god. "The doctrine of idolatry prevalent in Yoruba appears to be derived by analogy from the form and customs of the civil government. There is but one king in the nation, and one God over the universe. Petitioners to the king approach him through the intervention of his servants, courtiers, and nobles: and the petitioner conciliates the courtier whom he employs by good words and presents. In like manner no man can directly approach God; but the Almighty himself, they say, has appointed various kinds of orisas, who are mediators and intercessors between himself and mankind. No sacrifices are made to God, because he needs nothing; but the orisas, being much like men, are pleased with offerings of sheep, pigeons, and other things. They conciliate the orisa or mediator that he may bless them, not in his own power, but in the power of God."[2]

Rooted as they are in the depths of nature-worship, the doctrines of the supreme Sun and Heaven both come to the surface again in the native religions of Asia. The divine Sun holds his primacy distinctly enough among the rude indigenous tribes of India. Although one sect of the Khonds of Orissa especially direct their worship to Tari

[1] Magyar, 'Reisen in Süd-Afrika,' pp. 125, 335.
[2] Bowen, 'Gr. and Dic. of Yoruba,' p. xvi. in 'Smithsonian Contr.' vol. i.

Pennu the Earth-goddess, yet even they agree theoretically with the sect who worship Būra Pennu or Bella Pennu, Light-god or Sun-god, in giving to him supremacy above the manes-gods and nature-gods, and all spiritual powers.[1] Among the Kol tribes of Bengal, the acknowledged primate of all classes of divinities is the beneficent supreme deity, Sing-bonga, Sun-god. Among some Munda tribes his authority is so real that they will appeal to him for help where recourse to minor deities has failed; while among the Santals his cultus has so dwindled away that he receives less practical worship than his malevolent inferiors, and is scarce honoured with more than nominal dignity and an occasional feast.[2] These are rude tribes who, so far as we know, have never been other than rude tribes. The Japanese are a comparatively civilized nation, one of those so instructive to the student of culture from the stubborn conservatism with which they have consecrated by traditional reverence, and kept up by state authority, the religion of their former barbarism. This is the Kami-religion, Spirit-religion, the remotely ancient faith of divine spirits of ancestors, nature-spirits, and polytheistic gods, which still holds official place by the side of the imported Buddhism and Confucianism. In this ancient faith the Sun-god is supreme. He is "Ama-terasu oho Kami," the "heaven-enlightening great Spirit." Below him stand all lesser kamis or spirits, through whom, as mediators, guardians, and protectors, worship is paid by men. The Sun-god's race, as in Peru, is the royal family, and his spirit animates the reigning ruler, the Son of Heaven. Kempfer, in his 'History of Japan,' written early in the 18th century, showed how absolutely the divine Tensio Dai Sin was looked upon as ruler of the minor powers, by his mention of the Japanese tenth month, called the "god-less month," because then the lesser gods are considered to be away from their temples, gone to pay their annual

[1] Macpherson, 'India,' p. 84, etc.
Dalton, 'Kols,' in 'Tr. Eth. Soc.' vol. vi. p. 32. Hunter, 'Rural Bengal,' p. 184.

homage to their celestial Dairi. He describes, as it was in his time, the great Japanese place of pilgrimage, Ysse, the home of Tensio Dai Sin. There may be seen the small cavern in a hill near the sea, where he once hid himself, depriving the world, sun, and stars of their light, and thus showing himself to be lord of light and supreme above all gods. Within his small ancient temple hard by, there are to be seen round the walls pieces of cut white paper, symbols of purity, and in the midst nothing but a polished metal mirror, emblem of the all-seeing eye of this great god.[1]

Over the vast range of the Tatar races, it is the type of the supreme Heaven that comes prominently into view. Nature-worshippers in the extreme sense, these rude tribes conceived their ghosts and elves and demons and great powers of the earth and air to be, like men themselves, within the domain of the divine Heaven, almighty and all-encompassing. To trace the Samoyed's thought of Num the personal Sky passing into vague conceptions of pervading deity; to see with the Tunguz how Boa the Heaven-god, unseen but all-knowing, kindly but indifferent, has divided the business of his world among such lesser powers as sun and moon, earth and fire; to discern the meaning of the Mongol Tengri, shading from Heaven into Heaven-god, and thence into god or spirit in general; to follow the records of Heaven-worship among the ancient Turks and Hiong-nu; to compare the supremacy among the Lapps of Tiermes, the Thunderer, with the supremacy among the Finns of Jumala and Ukko, the Heaven-god and heavenly Grandfather—such evidence seems good ground for Castrén's argument, that the doctrine of the divine Sky underlay the first Turanian conceptions, not merely of a Heaven-god, but of a highest deity who in after ages of Christian conversion blended into the Christian God.[2] Here, again, we may have

[1] Siebold, 'Nippon,' part v. p. 9. Kempfer, 'Japan,' ch. xi. in Pinkerton, vol. vii. Wuttke, 'Gesch. d. Heidenthums,' part ii. p. 220.

[2] Castrén, 'Finn. Myth.' p. 1, etc. Klemm, 'Cultur-Gesch.' vol. iii. p. 101. Samoiedia in 'Pinkerton,' vol. i. p. 531. Georgi, Reise im Russ. Reich.' vol. i. p. 275.

the advantage of studying among a cultured race† he survival of religion from ruder ancient times, kept up by official ordinance. The state religion of China is in its dominant doctrine the worship of Tien, Heaven, identified with Shang-ti, the Emperor-above, next to whom stands Tu, Earth; while below them are worshipped great nature-spirits and ancestors. It is possible that this faith, as Professor Max Müller argues, may be ethnologically and even linguistically part and parcel of the general Heaven-worship of the Turanian tribes of Siberia. At any rate, it is identical with it in its primary idea, the adoration of the supreme Heaven. Dr. Legge charges Confucius with an inclination to substitute in his religious teaching the name of Tien, Heaven, for that known to more ancient religion and used in more ancient books, Shang-ti, the personal ruling Deity. But it seems rather that the sage was in fact upholding the traditions of the ancient faith, thus acting according to the character on which he prided himself, that of a transmitter and not a maker, a preserver of old knowledge, not a new revealer. It is in accordance with the usual course of theologic development, for the divine Heaven to reign in rude mythologic religion over the lesser spirits of the world, before the childlike poetic thought passes into the statesman's conception of a Celestial Emperor. As Plath well remarks, "It belongs to the Chinese system that all nature is animated by spirits, and that all these follow one order. As the Chinese cannot think of a Chinese Empire with an Emperor only, and without the host of vassal-princes and officials, so he cannot think of the Upper Emperor without the host of spirits." Developed in a different line, the idea of a supreme Heaven comes to pervade Chinese philosophy and ethics as a general expression of fate, ordinance, duty. " Heaven's order is nature "—" The wise man readily awaits Heaven's command "—" Man must first do his own part; when he has done all, then he can wait for Heaven to complete it "—" All state officers are Heaven's workmen, and represent him "—" How does Heaven speak? The four seasons have their course, the hundred things arise, what speaks he?"—"No, Heaven speaks not; by the course of events he makes himself understood, no more."[1]

These stray scraps from old Chinese literature are intelligible to European ears, for our Aryan race has indeed worked out religious ideas from the like source and almost in the like directions. The Samoyed or Tunguz Heaven-God had his analogue in Dyu, Heaven, of the Vedic hymns. Once meaning the sky, and the sky personified, this Zeus came to mean far more than mere heaven in the minds of Greek poets and philosophers, when it rose toward "that conception which in sublimity, brightness, and infinity transcended all others as much as the bright blue sky transcended all other things visible upon earth." At the lower level of mythic religion, the ideal process of shaping the divine world into a monarchic constitution was worked out by the ancient Greeks, on the same simple plan as among such barbarians as the Kols of Chota-Nagpur or the Gallas of Abyssinia; Zeus is King over Olympian gods, and below these again are marshalled the crowded ranks of demigods, heroes, demons, nymphs, ghosts. At the higher level of theologic speculation, exalted thoughts of universal cause and being, of physical and moral law, took personality under the name of Zeus. It is in direct derivation along this historic line, that the classical heaven-cultus still asserts itself in song and pageant among us, in that quaintest of quaint survivals, the factitious religion of the Italian Opera, where such worship as artistic ends require is still addressed to the divine Cielo. Even in our daily talk, colloquial expressions call up before the mind of the ethnographer outlines of remotest religious history. Heaven grants, forbids, blesses still in phrase, as heretofore in fact.

Vast and difficult as is the research into the full scope and history of the doctrine of supremacy among the higher

[1] Plath, 'Rel. der Alten Chinesen,' part i. p. 18, etc. See Max Müller, 'Lectures on Science of Religion, No. III. in 'Fraser's Mag.' 1870. Legge, 'Confucius,' p. 100.

nations, it may be at least seen that helpful clues exist to lead the explorer. The doctrine of mighty nature-spirits, inhabiting and controlling sky and earth and sea, seems to expand in Asia into such ideas as that of Mahâtman the Great Spirit, Paramâtman the Highest Spirit, taking personality as Brahma the all-pervading universal soul[1]—in Europe into philosophic conceptions of which a grand type stands out in Kepler's words, that the universe is a harmonious whole, whose soul is God. There is a saying of Comte's that throws strong light upon this track of speculative theology: he declares that the conception among the ancients of the Soul of the Universe, the notion that the earth is a vast living animal, and in our own time, the obscure pantheism which is so rife among German metaphysicians, are only fetishism generalized and made systematic.[2] Polytheism, in its inextricable confusion of the persons and functions of the great divinities, and in its assignment of the sovereignty of the world to a supreme being who combines in himself the attributes of several such minor deities, tends toward the doctrine of fundamental unity. Max Müller, in a lecture on the Veda, has given the name of kathenotheism to the doctrine of divine unity in diversity which comes into view in these instructive lines :—

"Indram Mitram Varunam Agnim âhur atho
divyah sa suparno Garutmân:
Ekam sad viprâ bahudha vadanti Agnim
Yamam Mâtariçvânam âhuh."

"They call him Indra, Mitra, Varuna, Agni; then he is the beautiful-winged heavenly Garutmat: That which is One the wise call it in divers manners; they call it Agni, Yama, Mâtariçvan."[3]

[1] See Colebrooke, 'Essays,' vol. ii. Wuttke, 'Heidenthum,' part i. p. 254. Ward, 'Hindoos,' vol. i. p. xxi. vol. ii. p. 1.
[2] Comte, 'Philosophie Positive.' Cf. Bp. Berkeley's 'Siris'; and for a modern dissertation on the universal æther as the divine soul of the world, see Phil. Spiller, 'Gott im Lichte der Naturwissenschaften,' Berlin, 1873 (note to 2nd ed.).
[3] 'Rig-Veda,' i. 164, 46. Max Müller, 'Chips,' vol. i. pp. 27, 241.

ANIMISM. 355

The figure of the supreme deity, be he Heaven-god, Sun-god, Great Spirit, beginning already in savage thought to take the form and function of a divine ruler of the world, represents a conception which it becomes the age-long work of systematic theology to develope and to define. Thus in Greece arises Zeus the highest, greatest, best, "who was and is and shall be," "beginning and chief of all things," "who rules over all mortals and immortals," "Zeus the god of gods."[1] Such is Ahura Mazda in the Persian faith, among whose seventy-two names of might are these: Creator, Protector, Nourisher, Holiest Heavenly One, Healing, Priest, Most Pure, Most Majestic, Most Knowing, Most Ruling at Will.[2] There may be truth in the assertion that the esoteric religion of ancient Egypt centred in a doctrine of divine unity, manifested through the heterogeneous crowd of popular deities.[3] It may be a hopeless task to disentangle the confused personalities of Baal, Bel, and Moloch, and no antiquary may ever fully solve the enigma how far the divine name of El carried in its wide range among the Jewish and other Semitic nations a doctrine of divine supremacy.[4] The great Syro-Phœnician kingdoms and religions have long since passed away into darkness, leaving but antiquarian relics to vouch for their former might. Far other has been the history of their Jewish kindred, still standing fast to their ancient nationality, still upholding to this day their patriarchal religion, in the midst of nations who inherit from the faith of Israel the belief in one God, highest, almighty, who in the beginning made the heavens and the earth, whose throne is established of old, who is from everlasting to everlasting.

Before now bringing these researches to a close, it will be well to state compactly the reasons for treating the animism

[1] See Welcker, 'Griech. Götterlehre,' pp. 143, 175.
[2] Avesta; trans. by Spiegel, 'Ormazd-Yasht.' 12.
[3] Wilkinson, 'Ancient Eg.' vol. iv. ch. xii.; Bunsen, 'Egypt,' vol. iv. p. 325.
[4] Movers, 'Phönizier,' vol. i. p. 169, etc. See Max Müller, 'Lecture,' iii. l. c.

of the modern savage world as more or less representing the animism of remotely ancient races of mankind. Savage animism, founded on a doctrine of souls carried to an extent far beyond its limits in the cultivated world, and thence expanding to a yet wider doctrine of spiritual beings animating and controlling the universe in all its parts, becomes a theory of personal causes developed into a general philosophy of man and nature. As such, it may be reasonably accounted for as the direct product of natural religion, using this term according to the sense of its definition by Wilkins: "I call that natural religion, which men might know, and should be obliged unto, by the meer principles of reason, improved by consideration and experience, without the help of revelation."[1] It will scarcely be argued by theologians familiar with the religions of savage tribes, that they are direct or nearly direct products of revelation, for the theology of our time would abolish or modify their details till scarce one was left intact. The main issue of the problem is this, whether savage animism is a primary formation belonging to the lower culture, or whether it consists, mostly or entirely, of beliefs originating in some higher culture, and conveyed by adoption or degradation into the lower. The evidence for the first alternative, though not amounting to complete demonstration, seems reasonably strong, and not met by contrary evidence approaching it in force. The animism of the lower tribes, self-contained and self-supporting, maintained in close contact with that direct evidence of the senses on which it appears to be originally based, is a system which might quite reasonably exist among mankind, had they never any-

[1] Cited in Johnson's Dictionary. The term "natural religion" is used in various and even incompatible senses. Thus Butler in his 'Analogy of Religion, Natural and Revealed, to the Constitution and Course of Nature,' signifies by "natural religion" a primæval system which he expressly argues to have been not reasoned out, but taught first by revelation. This system, of which the main tenets are the belief in one God, the Creator and Moral Governor of the World, and in a future state of moral retribution, differs in the extreme from the actual religions of the lower races.

where risen above the savage condition. Now it does not seem that the animism of the higher nations stands in a connexion so direct and complete with their mental state. It is by no means so closely limited to doctrines evidenced by simple contemplation of nature. The doctrines of the lower animism appear in the higher often more and more modified, to bring them into accordance with an advancing intellectual condition, to adapt them at once to the limits of stricter science and the needs of higher faith ; and in the higher animism these doctrines are retained side by side with other and special beliefs, of which the religions of the lower world show scarce a germ. In tracing the course of animistic thought from stage to stage of history, instruction is to be gained alike from the immensity of change and from the intensity of permanence. Savage animism, both by what it has and by what it wants, seems to represent the earlier system in which began the age-long course of the education of the world. Especially it is to be noticed that various beliefs and practices, which in the lower animism stand firm upon their grounds as if they grew there, in the higher animism belong rather to peasants than philosophers, exist rather as ancestral relics than as products belonging to their age, are falling from full life into survival. Thus it is that savage religion can frequently explain doctrines and rites of civilized religion. The converse is far less often the case. Now this is a state of things which seems to carry a historical as well as a practical meaning. The degradation-theory would expect savages to hold beliefs and customs intelligible as broken-down relics of former higher civilization. The development-theory would expect civilized men to keep up beliefs and customs which have their reasonable meaning in less cultured states of society. So far as the study of survival enables us to judge between the two theories, it seems that what is intelligible religion in the lower culture is often meaningless superstition in the higher, and thus the development-theory has the upper hand. Moreover, this evidence fits with the teaching of prehistoric archæology. Savage life, carrying on into our own day the life of the Stone Age, may be legitimately claimed as representing remotely ancient conditions of mankind, intellectual and moral as well as material. If so, a low but progressive state of animistic religion occupies a like ground in savage and in primitive culture.

Lastly, a few words of explanation may be offered as to the topics which this survey has included and excluded. To those who have been accustomed to find theological subjects dealt with on a dogmatic, emotional, and ethical, rather than an ethnographic scheme, the present investigation may seem misleading, because one-sided. This one-sided treatment, however, has been adopted with full consideration. Thus, though the doctrines here examined bear not only on the development but the actual truth of religious systems, I have felt neither able nor willing to enter into this great argument fully and satisfactorily, while experience has shown that to dispose of such questions by an occasional dictatorial phrase is one of the most serious of errors. The scientific value of descriptions of savage and barbarous religions, drawn up by travellers and especially by missionaries, is often lowered by their controversial tone, and by the affectation of infallibility with which their relation to the absolutely true is settled. There is something pathetic in the simplicity with which a narrow student will judge the doctrines of a foreign religion by their antagonism or conformity to his own orthodoxy, on points where utter difference of opinion exists among the most learned and enlightened scholars. The systematization of the lower religions, the reduction of their multifarious details to the few and simple ideas of primitive philosophy which form the common groundwork of them all, appeared to me an urgently needed contribution to the science of religion. This work I have carried out to the utmost of my power, and can now only leave the result in the hands of other students, whose province it is to deal with such evidence in wider schemes of argument. Again, the intellectual rather than the emotional side of religion has here been kept in view. Even in the life of the rudest savage, religious belief is associated with intense emotion, with awful reverence, with agonizing terror, with rapt ecstasy when sense and thought utterly transcend the common level of daily life. How much the more in faiths where not only does the believer experience such enthusiasm, but where his utmost feelings of love and hope, of justice and mercy, of fortitude and tenderness and self-sacrificing devotion, of unutterable misery and dazzling happiness, twine and clasp round the fabric of religion. Language, dropping at times from such words as soul and spirit their mere philosophic meaning, can use them in full conformity with this tendency of the religious mind, as phrases to convey a mystic sense of transcendent emotion. Yet of all this religion, the religion of vision and of passion, little indeed has been said in these pages, and even that little rather in incidental touches than with purpose. Those to whom religion means above all things religious feeling, may say of my argument that I have written soullessly of the soul, and unspiritually of spiritual things. Be it so : I accept the phrase not as needing an apology, but as expressing a plan. Scientific progress is at times most furthered by working along a distinct intellectual line, without being tempted to diverge from the main object to what lies beyond, in however intimate connexion. The anatomist does well to discuss bodily structure independently of the world of happiness and misery which depends upon it. It would be thought a mere impertinence for a strategist to preface a dissertation on the science of war, by an enquiry how far it is lawful for a Christian man to bear weapons and serve in the wars. My task has been here not to discuss Religion in all its bearings, but to pourtray in outline the great doctrine of Animism, as found in what I conceive to be its earliest stages among the lower races of mankind, and to show its transmission along the lines of religious thought.

The almost entire exclusion of ethical questions from

this investigation has more than a mere reason of arrangement. It is due to the very nature of the subject. To some the statement may seem startling, yet the evidence seems to justify it, that the relation of morality to religion is one that only belongs in its rudiments, or not at all, to rudimentary civilization. The comparison of savage and civilized religions brings into view, by the side of deep-lying resemblance in their philosophy, a deep-lying contrast in their practical action on human life. So far as savage religion can stand as representing natural religion, the popular idea that the moral government of the universe is an essential tenet of natural religion simply falls to the ground. Savage animism is almost devoid of that ethical element which to the educated modern mind is the very mainspring of practical religion. Not, as I have said, that morality is absent from the life of the lower races. Without a code of morals, the very existence of the rudest tribe would be impossible; and indeed the moral standards of even savage races are to no small extent well-defined and praiseworthy. But these ethical laws stand on their own ground of tradition and public opinion, comparatively independent of the animistic beliefs and rites which exist beside them. The lower animism is not immoral, it is unmoral. For this plain reason, it has seemed desirable to keep the discussion of animism, as far as might be, separate from that of ethics. The general problem of the relation of morality to religion is difficult, intricate, and requiring immense array of evidence, and may be perhaps more profitably discussed in connexion with the ethnography of morals. To justify their present separation, it will be enough to refer in general terms to the accounts of savage tribes whose ideas have been little affected by civilized intercourse; proper caution being used not to trust vague statements about good and evil, but to ascertain whether these are what philosophic moralists would call virtue and vice, righteousness and wickedness, or whether they are mere personal advantage and disadvantage. The essential con-

ANIMISM. 361

nexion of theology and morality is a fixed idea in many minds. But it is one of the lessons of history that subjects may maintain themselves independently for ages, till the event of coalescence takes place. In the course of history, religion has in various ways attached to itself matters small and great outside its central scheme, such as prohibition of special meats, observance of special days, regulation of marriage as to kinship, division of society into castes, ordinance of social law and civil government. Looking at religion from a political point of view, as a practical influence on human society, it is clear that among its greatest powers has been its divine sanction of ethical laws, its theological enforcement of morality, its teaching of moral government of the universe, its supplanting the "continuance-doctrine" of a future life by the "retribution-doctrine" supplying moral motive in the present. But such alliance belongs almost wholly to religions above the savage level, not to the earlier and lower creeds. It will aid us to see how much more the fruit of religion belongs to ethical influence than to philosophical dogma, if we consider how the introduction of the moral element separates the religions of the world, united as they are throughout by one animistic principle, into two great classes, those lower systems whose best result is to supply a crude childlike natural philosophy, and those higher faiths which implant on this the law of righteousness and of holiness, the inspiration of duty and of love.

CHAPTER XVIII.

RITES AND CEREMONIES.

Religious Rites: their purpose practical or symbolic—Prayer: its continuity from low to high levels of Culture; its lower phases Unethical; its higher phases Ethical—Sacrifice: its original Gift-theory passes into the Homage-theory and the Abnegation-theory—Manner of reception of Sacrifice by Deity—Material Transfer to elements, fetish-animals, priests; consumption of substance by deity or idol; offering of blood; transmission by fire; incense—Essential Transfer: consumption of essence, savour, etc. —Spiritual Transfer: consumption or transmission of soul of offering— Motive of Sacrificer—Transition from Gift-theory to Homage-theory: insignificant and formal offerings; sacrificial banquets—Abnegation theory; sacrifice of children, etc.—Sacrifice of Substitutes; part given for whole; inferior life for superior; effigies—Modern survival of Sacrifice in folklore and religion—Fasting, as a means of producing ecstatic vision; its course from lower to higher Culture—Drugs used to produce ecstasy— Swoons and fits induced for religious purposes—Orientation: its relation to Sun-myth and Sun-Worship; rules of East and West as to burial of dead, position of worship, and structure of temple—Lustration by Water and Fire: its transition from material to symbolic purification; its connexion with special events of life; its appearance among the lower races—Lustration of new-born children; of women; of those polluted by bloodshed or the dead—Lustration continued at higher levels of Culture —Conclusion.

RELIGIOUS rites fall theoretically into two divisions, though these blend in practice. In part, they are expressive and symbolic performances, the dramatic utterance of religious thought, the gesture-language of theology. In part, they are means of intercourse with and influence on spiritual beings, and as such, their intention is as directly practical as any chemical or mechanical process, for doctrine and worship correlate as theory and practice. In the science of religion, the study of ceremony has its strong and weak sides. On the one hand, it is generally easier to obtain accurate accounts of ceremonies by eyewitnesses, than anything like trustworthy and intelligible statements of doctrine; so that very much of our knowledge of religion in the savage and barbaric world consists in acquaintance with its ceremonies. It is also true that some religious ceremonies are marvels of permanence, holding substantially the same form and meaning through age after age, and far beyond the range of historic record. On the other hand, the signification of ceremonies is not to be rashly decided on by mere inspection. In the long and varied course in which religion has adapted itself to new intellectual and moral conditions, one of the most marked processes has affected time-honoured religious customs, whose form has been faithfully and even servilely kept up, while their nature has often undergone transformation. In the religions of the great nations, the natural difficulty of following these changes has been added to by the sacerdotal tendency to ignore and obliterate traces of the inevitable change of religion from age to age, and to convert into mysteries ancient rites whose real barbaric meaning is too far out of harmony with the spirit of a later time. The embarrassments, however, which beset the inquirer into the ceremonies of a single religion, diminish in a larger comparative study. The ethnographer who brings together examples of a ceremony from different stages of culture can often give a more rational account of it, than the priest, to whom a special signification, sometimes very unlike the original one, has become matter of orthodoxy. As a contribution to the theory of religion, with especial view to its lower phases as explanatory of the higher, I have here selected for ethnographic discussion a group of sacred rites, each in its way full of instruction, different as these ways are. All have early place and rudimentary meaning in savage culture, all belong to barbaric ages, all have their representatives within the limits of modern Christendom. They are the rites of Prayer, Sacrifice,

Fasting and other methods of Artificial Ecstasy, Orientation, Lustration.

Prayer, "the soul's sincere desire, uttered or unexpressed," is the address of personal spirit to personal spirit. So far as it is actually addressed to disembodied or deified human souls, it is simply an extension of the daily intercourse between man and man; while the worshipper who looks up to other divine beings, spiritual after the nature of his own spirit, though of place and power in the universe far beyond his own, still has his mind in a state where prayer is a reasonable and practical act. So simple and familiar indeed is the nature of prayer, that its study does not demand that detail of fact and argument which must be given to rites in comparison practically insignificant. It is not indeed to be claimed as an immediate or necessary outcome of animistic belief, for especially at low levels of civilization there are many races who distinctly admit the existence of spirits, but are not certainly known to pray to them even in thought. Beyond this lower level, however, animism and prayer become more and more nearly conterminous; and a view of their relation in their earlier stages may be easiest and best gained from a selection of actual prayers taken down word for word, within the limits of savage and barbaric life. These agree with an opinion that prayer appeared in the religion of the lower culture, but that in this its earlier stage it was unethical. The accomplishment of desire is asked for, but desire is as yet limited to personal advantage. It is at later and higher moral levels, that the worshipper begins to add to his entreaty for prosperity the claim for help toward virtue and against vice, and prayer becomes an instrument of morality.

In the Papuan Island of Tanna, where the gods are the spirits of departed ancestors, and preside over the growth of fruits, a prayer after the offering of first-fruits is spoken aloud by the chief who acts as high priest to the silent assembly: "Compassionate father! Here is some food for you; eat it; be kind to us on account of it!" Then all shout together.[1] In the Samoan Islands, when the libation of ava was poured out at the evening meal, the head of the family prayed thus :—

"Here is ava for you, O Gods! Look kindly towards this family: let it prosper and increase ; and let us all be kept in health. Let our plantations be productive; let food grow ; and may there be abundance of food for us, your creatures. Here is ava for you, our war gods! Let there be a strong and numerous people for you in this land.
"Here is ava for you, O sailing gods (gods who come in Tongan canoes and foreign vessels). Do not come on shore at this place ; but be pleased to depart along the ocean to some other land."[2]

Among the Indians of North America, the Sioux will say, " Spirits of the dead, have mercy on me!" then they will add what they want, if good weather they say so, if good luck in hunting, they say so.[3] Among the Osages, prayers used not long since to be offered at daybreak to Wohkonda, the Master of Life. The devotee retired a little from the camp or company, and with affected or real weeping, in loud uncouth voice of plaintive piteous tone, howled such prayers as these :—" Wohkonda, pity me, I am very poor; give me what I need ; give me success against mine enemies, that I may avenge the death of my friends. May I be able to take scalps, to take horses! &c." Such prayers might or might not have allusion to some deceased relative or friend.[4] How an Algonquin Indian undertakes a dangerous voyage, we may judge from John Tanner's account of a fleet of frail Indian bark canoes setting out at dawn one calm morning on Lake Superior. We had proceeded, he writes, about two hundred yards into the lake, when the canoes all stopped together, and the chief, in a very loud voice, addressed a prayer to the Great Spirit, entreating him to give us a good look to cross the lake. "You," said he,

[1] Turner, 'Polynesia,' p. 88 ; see p. 427.
[2] Ibid. p. 200 ; see 174. See also Ellis, 'Polyn. Res.' vol. i. p. 343. Mariner, 'Tonga Is.' vol. ii. p. 235.
[3] Schoolcraft, 'Ind. Tribes,' part iii. p. 237.
[4] M'Coy, 'Baptist Indian Missions,' p. 359.

"have made this lake, and you have made us, your children; you can now cause that the water shall remain smooth while we pass over in safety." In this manner he continued praying for five or ten minutes; he then threw into the lake a small quantity of tobacco, in which each of the canoes followed his example.[1] A Nootka Indian, preparing for war, prayed thus: "Great Quahootze, let me live, not be sick, find the enemy, not fear him, find him asleep, and kill a great many of him."[2] There is more pathos in these lines from the war-song of a Delaware:—

> "O Great Spirit there above
> Have pity on my children
> And my wife!
> Prevent that they shall mourn for me!
> Let me succeed in this undertaking,
> That I may slay my enemy
> And bring home the tokens of victory
> To my dear family and my friends
> That we may rejoice together . . .
> Have pity on me and protect my life,
> And I will bring thee an offering."[3]

The following two prayers are among those recorded by Molina, from the memory of aged men who described to him the religion of Peru under the Incas, in whose rites they had themselves borne part. The first is addressed to the Sun, the second to the World-creator:—

"O Sun! Thou who hast said, let there be Cuzcos and Tampus, grant that these thy children may conquer all other people. We beseech thee that thy children the Yncas may be conquerors always, for this hast thou created them."

"O conquering Uiracocha! Ever present Uiracocha! Thou who art in the ends of the earth without equal! Thou who gavest life and valour to men, saying, 'Let this be a man!' and to women, saying, 'Let this be a woman!' Thou who madest them and gavest them being! Watch over them that they may live in health and peace.

---

[1] Tanner, 'Narrative,' p. 46.
[2] Brinton, 'Myths of New World,' p. 297.
[3] Heckewelder, 'Ind. Völkerschaften,' p. 354.

RITES AND CEREMONIES. 367

Thou who art in the high heavens, and among the clouds of the tempest, grant this with long life, and accept this sacrifice, O Uiracocha!"[1]

In Africa, the Zulus, addressing the spirits of their ancestors, think it even enough to call upon them without saying what they want, taking it for granted that the spirits know, so that the mere utterance "People of our house!" is a prayer. When a Zulu sneezes, and is thus for the moment in close relation to the divine spirits, it is enough for him to mention what he wants ("to wish a wish," as our own folklore has it), and thus the words "A cow!" "Children!" are prayers. Fuller forms are such as these: "People of our house! Cattle!"—"People of our house! Good luck and health!"—"People of our house! Children!" On occasions of ancestral cattle-sacrifice the prayers extend to actual harangues, as when, after the feast is over, the headman speaks thus amid dead silence: "Yes, yes, our people, who did such and such noble acts, I pray to you—I pray for prosperity after having sacrificed this bullock of yours. I say, I cannot refuse to give you food, for these cattle which are here you gave me. And if you ask food of me which you have given me, is it not proper that I should give it to you? I pray for cattle, that they may fill this pen. I pray for corn, that many people may come to this village of yours, and make a noise, and glorify you. I ask also for children, that this village may have a large population, and that your name may never come to an end." So he finishes.[2] From among the negro races near the equator, the following prayers may be cited, addressed to that Supreme Deity whose nature is, as we have seen, more or less that of the Heaven-god. The Gold Coast negro would raise his eyes to Heaven and thus address him: "God, give me to-day rice and yams, gold and agries, give me

[1] 'Narratives of Rites and Laws of Yncas,' tr. and ed. by C. R. Markham, p. 31, 33. See also Brinton, p. 298.
[2] Callaway, 'Religion of Amazulu,' pp. 124, 141, 174, 182. 'Remarks on Zulu lang.' Pietermaritzburg, 1870, p. 22.

slaves, riches, and health, and that I may be brisk and swift!" the fetish-man will often in the morning take water in his mouth and say, "Heaven! grant that I may have something to eat to-day;" and when giving medicine shown him by the fetish, he will hold it up to heaven first, and say, "Ata Nyongmo! (Father Heaven!) bless this medicine that I now give." The Yebu would say, "God in heaven, protect me from sickness and death. God give me happiness and wisdom!"[1] When the Manganja of Lake Nyassa were offering to the Supreme Deity a basketful of meal and a pot of native beer, that he might give them rain, the priestess dropped the meal handful by handful on the ground, each time calling, in a high-pitched voice, "Hear thou, O God, and send rain!" and the assembled people responded, clapping their hands softly and intoning (they always intone their prayers) "Hear thou, O God!"[2]

Typical forms of prayer may be selected in Asia near the junction-line of savage and barbaric culture. Among the Karens of Birma, the Harvest-goddess has offerings made to her in a little house in the paddy-field, in which two strings are put for her to bind the spirits of any persons who may enter her field. Then they entreat her on this wise: "Grandmother, thou guardest my field, thou watchest over my plantation. Look out for men entering; look sharp for people coming in. If they come, bind them with this string, tie them with this rope, do not let them go!" And at the threshing of the rice they say: "Shake thyself, Grandmother, shake thyself. Let the paddy ascend till it equals a hill, equals a mountain. Shake thyself, Grandmother, shake thyself!"[3] The following are extracts from the long-drawn prayers of the Khonds of Orissa: "O Boora Pennu! and O Tari Pennu, and all other gods! (naming them). You, O Boora Pennu! created us, giving us the attribute of hunger; thence corn food was necessary to us,

---

[1] Waitz, vol. ii. p. 169. Steinhauser, l. c. p. 129.
[2] Rowley, 'Universities' Mission to Central Africa,' p. 226.
[3] Mason, 'Karens,' l. c. p. 215.

and thence were necessary producing fields. You gave us every seed, and ordered us to use bullocks, and to make ploughs, and to plough. Had we not received this art, we might still indeed have existed upon the natural fruits of the jungle and the plain, but, in our destitution, we could not have performed your worship. Do you, remembering this—the connexion betwixt our wealth and your honour—grant the prayers which we now offer. In the morning, we rise before the light to our labour, carrying the seed. Save us from the tiger, and the snake, and from stumblingblocks. Let the seed appear earth to the eating birds, and stones to the eating animals of the earth. Let the grain spring up suddenly like a dry stream that is swelled in a night. Let the earth yield to our ploughshares as wax melts before hot iron. Let the baked clods melt like hailstones. Let our ploughs spring through the furrows with a force like the recoil of a bent tree. Let there be such a return from our seed, that so much shall fall and be neglected in the fields, and so much on the roads in carrying it home, that, when we shall go out next year to sow, the paths and the fields shall look like a young corn-field. From the first times we have lived by your favour. Let us continue to receive it. Remember that the increase of our produce is the increase of your worship, and that its diminution must be the diminution of your rites." The following is the conclusion of a prayer to the Earth-goddess: " Let our herds be so numerous that they cannot be housed; let children so abound that the care of them shall overcome their parents—as shall be seen by their burned hands; let our heads ever strike against brass pots innumerable hanging from our roofs; let the rats form their nests of shreds of scarlet cloth and silk; let all the kites in the country be seen in the trees of our village, from beasts being killed there every day. We are ignorant of what it is good to ask for. You know what is good for us. Give it to us!"[1]

[1] Macpherson, 'India,' pp. 110, 128. See also Hunter, 'Rural Bengal,' p. 182 (Santals).

Such are types of prayer in the lower levels of culture, and in no small degree they remain characteristic of the higher nations. If, in long-past ages, the Chinese raised themselves from the condition of rude Siberian tribes to their peculiar culture, at any rate their conservative religion has scarce changed the matter-of-fact prayers for rain and good harvest, wealth and long life, addressed to manes and nature-spirits and merciful Heaven.[1] In other great national religions of the world, not the whole of prayer, but a smaller or larger part of it, holds closely to the savage definition. This is a Vedic prayer: "What, Indra, has not yet been given me by thee, Lightning-hurler, all good things bring us hither with both hands . . . . with mighty riches fill me, with wealth of cattle, for thou art great!"[2] This is Moslem: "O Allah! unloose the captivity of the captives, and annul the debts of the debtors: and make this town to be safe and secure, and blessed with wealth and plenty, and all the towns of the Moslems, O Lord of all creatures! and decree safety and health to us and to all travellers, and pilgrims, and warriors, and wanderers, upon thy earth, and upon thy sea, such as are Moslems, O Lord of all creatures!"[3] Thus also, throughout the rituals of Christendom, stand an endless array of supplications unaltered in principle from savage times—that the weather may be adjusted to our local needs, that we may have the victory over all our enemies, that life and health and wealth and happiness may be ours.

So far, then, is permanence in culture: but now let us glance at the not less marked lines of modification and new formation. The vast political effect of a common faith in developing the idea of exclusive nationality, a process scarcely expanding beyond the germ among savage tribes, but reaching its full growth in the barbaric world, is apt to have its outward manifestation in hostility to those of another

[1] Plath, 'Religion der Chinesen,' part ii. p. 2; Doolittle, vol. ii. p. 116.
[2] 'Sama-Veda,' i. 4, 2. Wuttke, 'Gesch. des Heidenthums,' part ii. p. 342.
[3] Lane, 'Modern Egyptians,' vol. i. p. 128.

creed, a sentiment which finds vent in characteristic prayers. Such are these from the Rig-Veda: "Take away our calamities. By sacred verses may we overcome those who employ no holy hymns! Distinguish between the Aryas and those who are Dasyus: chastising those who observe no sacred rites, subject them to the sacrificer . . . Indra subjects the impious to the pious, and destroys the irreligious by the religious."[1] The following is from the closing prayer which the boys in many schools in Cairo used to repeat some years ago, and very likely do still: "I seek refuge with Allah from Satan the accursed. In the name of Allah, the Compassionate, the Merciful . . . O Lord of all creatures! O Allah! destroy the infidels and polytheists, thine enemies, the enemies of the religion! O Allah! make their children orphans, and defile their abodes, and cause their feet to slip, and give them and their families and their households and their women and their children and their relations by marriage and their brothers and their friends and their possessions and their race and their wealth and their lands as booty to the Moslems! O Lord of all creatures!"[2] Another powerful tendency of civilization, that of regulating human affairs by fixed ordinance, has since early ages been at work to arrange worship into mechanical routine. Here, so to speak, religion deposits itself in sharply defined shape from a supersaturated solution, and crystallizes into formalism. Thus prayers, from being at first utterances as free and flexible as requests to a living patriarch or chief, stiffened into traditional formulas, whose repetition required verbal accuracy, and whose nature practically assimilated more or less to that of charms. Liturgies, especially in those three quarters of the world where the ancient liturgical language has become at once unintelligible and sacred, are crowded with examples of this historical process. Its extremest development in Europe is connected with the use of the rosary. This devotional

[1] 'Rig-Veda,' i. 51, 8, x. 105, 8. Muir, 'Sanskrit Texts,' part ii. ch. iii.
[2] Lane, 'Modern Egyptians,' vol. ii. p. 383.

calculating-machine is of Asiatic invention; it had if not its origin at least its special development among the ancient Buddhists, and its 108 balls still slide through the modern Buddhist's hands as of old, measuring out the sacred formulas whose reiteration occupies so large a fraction of a pious life. It was not till toward the middle ages that the rosary passed into Mohammedan and Christian lands, and finding there conceptions of prayer which it was suited to accompany, has flourished ever since. How far the Buddhist devotional formulas themselves partake of the nature of prayer, is a question opening into instructive considerations, which need only be suggested here. By its derivation from Brahmanism and its fusion with the beliefs of rude spirit-worshipping populations, Buddhism practically retains in no small measure a prayerful temper and even practice. Yet, according to strict and special Buddhist philosophy, where personal divinity has faded into metaphysical idea, even devotional utterances of desire are not prayers; as Köppen says, there is no "Thou!" in them. It must be only with reservation that we class the rosary in Buddhist hands as an instrument of actual prayer. The same is true of the still more extreme development of mechanical religion, the prayer-mill of the Tibetan Buddhists. This was perhaps originally a symbolic "chakra" or wheel of the law, but has become a cylinder mounted on an axis, which by each rotation is considered to repeat the sentences written on the papers it is filled with, usually the "Om mani padme hûm!" Prayer-mills vary in size, from the little wooden toys held in the hand, to the great drums turned by wind or water-power, which repeat their sentences by the million.[1] The Buddhist idea, that "merit" is produced by the recitation of these sentences, may perhaps lead us to form an opinion of large application in the study of religion and superstition, namely, that the theory of prayers may explain the origin of charms. Charm-formulas are in very many cases actual

[1] See Köppen, 'Religion des Buddha,' vol. i. pp. 345, 556; vol. ii. pp. 303, 319. Compare Fergusson, 'Tree and Serpent Worship,' pl. xlii.

prayers, and as such are intelligible. Where they are mere verbal forms, producing their effect on nature and man by some unexplained process, may not they or the types they were modelled on have been originally prayers, since dwindled into mystic sentences?

The worshipper cannot always ask wisely what is for his good, therefore it may be well for him to pray that the greater power of the deity may be guided by his greater wisdom—this is a thought which expands and strengthens in the theology of the higher nations. The simple prayer of Sokrates, that the gods would give such things as are good, for they know best what are good,[1] raises a strain of supplication which has echoed through Christendom from its earliest ages. Greatest of all changes which difference the prayers of lower from those of higher nations, is the working out of the general principle that the ethical element, so scanty and rudimentary in the lower forms of religion, becomes in the higher its most vital point; while it scarcely appears as though any savage prayer, authentically native in its origin, were ever directed to obtain moral goodness or to ask pardon for moral sin. Among the semi-civilized Aztecs, in the elaborate ritual which from its early record and its original characteristics seems to have at least a partial authenticity, we mark the appearance of ethical prayer. Such is the supplication concerning the newly-elect ruler: "Make him, Lord, as your true image, and permit him not to be proud and haughty in your throne and court; but vouchsafe, Lord, that he may calmly and carefully rule and govern them whom he has in charge, the people, and permit not, Lord, that he may injure or vex his subjects, nor without reason and justice cause loss to any; and permit not, Lord, that he may spot or soil your throne or court with any injustice or wrong, etc."[2] Moral prayer, sometimes appearing in rudiment, sometimes shrunk into

[1] Xenoph. Memorabilia Socrat. i. 3, 2.
[2] Sahagun, 'Retorica, etc., de la Gente Mexicana,' lib. vi. c. 4, in Kingsborough; 'Antiquities of Mexico,' vol. v.

insignificance, sometimes overlaid by formalism, sometimes maintained firm and vigorous in the inmost life, has its place without as well as within the Jewish-Christian scheme. The ancient Aryan prayed: "Through want of strength, thou strong and bright god, have I gone wrong; have mercy, almighty, have mercy! . . . . Whenever we men, O Varuna, commit an offence before the heavenly host, whenever we break the law through thoughtlessness, have mercy, almighty, have mercy!"[1] The modern Parsi prays: "Of my sins which I have committed against the ruler Ormazd, against men, and the different kinds of men. . . . . Deceit, contempt, idol-worship, lies, I repent of. . . . . All and every kind of sin which men have committed because of me, or which I have committed because of men; pardon, I repent with confession!"[2] As a general rule it would be misleading to judge utterances of this kind in the religions of classic Greece and Rome as betokening the intense habitual prayerfulness which pervades the records of Judaism, Mahommedanism, Christianity. Moralists admit that prayer can be made an instrument of evil, that it may give comfort and hope to the superstitious robber, that it may strengthen the heart of the soldier to slay his foes in an unrighteous war, that it may uphold the tyrant and the bigot in their persecution of freedom in life and thought. Philosophers dwell on the subjective operation of prayer, as acting not directly on outward events, but on the mind and will of the worshipper himself, which it influences and confirms. The one argument tends to guide prayer, the other to suppress it. Looking on prayer in its effect on man himself through the course of history, both must recognize it as even in savage religion a means of strengthening emotion, of sustaining courage and exciting hope, while in higher faiths it becomes a great motive power of the ethical system, controlling and enforcing, under an ever-present sense of supernatural intercourse and aid, the emotions and energies of moral life.

[1] 'Rig-Veda,' vii. 89. 3. Max Müller, 'Chips,' vol. i. p. 39.
[2] 'Avesta,' tr. by Spiegel; 'Khordah-Avesta,' Patet Qod.

Sacrifice has its apparent origin in the same early period of culture and its place in the same animistic scheme as prayer, with which through so long a range of history it has been carried on in the closest connexion. As prayer is a request made to a deity as if he were a man, so sacrifice is a gift made to a deity as if he were a man. The human types of both may be studied unchanged in social life to this day. The suppliant who bows before his chief, laying a gift at his feet and making his humble petition, displays the anthropomorphic model and origin at once of sacrifice and prayer. But sacrifice, though in its early stages as intelligible as prayer is in early and late stages alike, has passed in the course of religious history into transformed conditions, not only of the rite itself but of the intention with which the worshipper performs it. And theologians, having particularly turned their attention to the rite as it appears in the higher religions, have been apt to gloss over with mysticism ceremonies which, when traced ethnographically up from their savage forms, seem open to simply rational interpretation. Many details of sacrifice have already been given incidentally here, as a means of elucidating the nature of the deities they are offered to. Moreover, a main part of the doctrine of sacrifice has been anticipated in examining the offerings to spirits of the dead, and indeed the ideal distinction between soul and deity breaks down among the lower races, when it appears how often the deities receiving sacrifice are themselves divine human souls. In now attempting to classify sacrifice in its course through the religions of the world, it seems a satisfactory plan to group the evidence as far as may be according to the manner in which the offering is given by the worshipper, and received by the deity. At the same time, the examples may be so arranged as to bring into view the principal lines along which the rite has undergone alteration. The ruder conception that the deity takes and values the offering for itself, gives place on the one hand to the idea of mere homage expressed by a gift, and on the other to the negative view that the virtue lies in the worshipper depriving himself of something prized. These ideas may be broadly distinguished as the gift-theory, the homage-theory, and the abnegation-theory. Along all three the usual ritualistic change may be traced, from practical reality to formal ceremony. The originally valuable offering is compromised for a smaller tribute or a cheaper substitute, dwindling at last to a mere trifling token or symbol.

The gift-theory, as standing on its own independent basis, properly takes the first place. That most childlike kind of offering, the giving of a gift with as yet no definite thought how the receiver can take and use it, may be the most primitive as it is the most rudimentary sacrifice. Moreover, in tracing the history of the rite from level to level of culture, the same simple unshaped intention may still largely prevail, and much of the reason why it is often found difficult to ascertain what savages and barbarians suppose to become of the food and valuables they offer to the gods, may be simply due to ancient sacrificers knowing as little about it as modern ethnologists do, and caring less. Yet rude races begin and civilized races continue to furnish with the details of their sacrificial ceremonies the key also to their meaning, the explanation of the manner in which the offering is supposed to pass into the possession of the deity.

Beginning with cases in which this transmission is performed bodily, it appears that when the deity is the personal Water, Earth, Fire, Air, or a fetish-spirit animating or inhabiting such element, he can receive and sometimes actually consume the offerings given over to this material medium. How such notions may take shape is not ill shown in the quaintly rational thought noticed in old Peru, that the Sun drinks the libations poured out before him; and in modern Madagascar, that the Angatra drinks the arrack left for him in the leaf-cup. Do not they see the liquids diminish from day to day?[1] The sacrifice to Water

[1] Garcilaso de la Vega, 'Commentarios Reales,' v. 19. Ellis, 'Madagascar,' vol. i. p. 421.

is exemplified by Indians caught in a storm on the North American lakes, who would appease the angry tempest-raising deity by tying the feet of a dog and throwing it overboard.[1] The following case from Guinea well shows the principle of such offerings. Once in 1693, the sea being unusually rough, the headmen complained to the king, who desired them to be easy, and he would make the sea quiet next day. Accordingly he sent his fetishman with a jar of palm oil, a bag of rice and corn, a jar of pitto, a bottle of brandy, a piece of painted calico, and several other things to present to the sea. Being come to the seaside, he made a speech to it, assuring it that his king was its friend, and loved the white men; that they were honest fellows and came to trade with him for what he wanted; and that he requested the sea not to be angry, nor hinder them to land their goods; he told it, that if it wanted palm oil, his king had sent it some; and so threw the jar with the oil into the sea, as he did, with the same compliment, the rice, corn, pitto, brandy, calico, &c.[2] Among the North American Indians the Earth also receives offerings buried in it. The distinctness of idea with which such objects may be given is well shown in a Sioux legend. The Spirit of the Earth, it seems, requires an offering from those who perform extraordinary achievements, and accordingly the prairie gapes open with an earthquake before the victorious hero of the tale; he casts a partridge into the crevice, and springs over.[3] One of the most explicit recorded instances of the offering to the Earth, is the hideous sacrifice to the Earth-goddess among the Khonds of Orissa, tearing the flesh of the human victim from the bones, the priest burying half of it in a hole in the earth behind his back without

[1] Charlevoix, 'Nouv. Fr.' vol. i. p. 394. See also Smith, 'Virginia,' in 'Pinkerton,' vol. xiii. p. 41.
[2] Phillips in Astley's 'Voyages,' vol. ii. p. 411; Lubbock, 'Origin of Civilization,' p. 216. Bosman, 'Guinea,' in 'Pinkerton,' vol. xvi. p. 500. Bastian in 'Ztschr. für Ethnologie,' 1869, p. 315.
[3] Schoolcraft, 'Algic Res.' vol. ii. p. 75. See also Tanner, 'Narr.' p. 193, and above, p. 270.

looking round, and each householder carrying off a particle to bury in like manner in his favourite field.[1] For offerings to the Fire, we may take for an example the Yakuts, who not only give him the first spoonful of food, but instead of washing their earthen pots allow him to clean out the remains.[2] Here is a New Zealand charm called Wangaihau, *i.e.*, feeding the Wind:—

" Lift up his offering,
To Uenga a te Rangi his offering,
Eat, O invisible one, listen to me,
Let that food bring you down from the sky." [3]

Beside this may be set the quaint description of the Fanti negroes assisting at the sacrifice of men and cattle to the local fetish; the victims were considered to be carried up in a whirlwind out of the midst of the small inner ring of priests and priestesses; this whirlwind was, however, not perceptible to the senses of the surrounding worshippers.[4] These series of details collected from the lower civilization throw light on curious problems as to sacrificial ideas in the religions of the classic world; such questions as what Xerxes meant when he threw the golden goblet and the sword into the Hellespont, which he had before chained and scourged; why Hannibal cast animals into the sea as victims to Poseidon; what religious significance underlay the patriotic Roman legend of the leap of Marcus Curtius.[5]

Sacred animals, in their various characters of divine beings, incarnations, representatives, agents, symbols, naturally receive meat and drink offerings, and sometimes other gifts. For examples, may be mentioned the sun-birds (tonatzuli), for which the Apalaches of Florida set out

[1] Macpherson, 'India,' p. 129.
[2] Billings, 'Exp. to Northern Russia,' p. 125. Chinese sacrifices buried for earth spirits, see ante, vol. i. p. 107; Plath, part ii. p. 50.
[3] Taylor, 'New Zealand,' p. 182.
[4] Römer, 'Guinea,' p. 67.
[5] Herod. vii. 35, 54. Liv. vii. 6. Grote, 'Hist. of Greece,' vol. x. p. 589, see 715.

crushed maize and seed;[1] the Polynesian deities coming incarnate in the bodies of birds to feed on the meat-offerings and carcases of human victims set out upon the altar-scaffolds;[2] the well-fed sacred snakes of West Africa, and local fetish animals like the alligator at Dix Cove which will come up at a whistle, and follow a man half a mile if he carries a white fowl in his hands, or the shark at Bonny that comes to the river bank every day to see if a human victim has been provided for his repast;[3] in modern India the cows reverently fed with fresh grass, Durga's meat-offerings laid out on stones for the jackals, the famous alligators in their temple-tanks.[4] The definition of sacred animal from this point of view distinctly includes man. Such in Mexico was the captive youth adored as living representative of Tezcatlipoca, and to whom banquets were made during the luxurious twelvemonth which preceded his sacrifice at the festival of the deity whom he personated: such still more definitely was Cortes himself, when Montezuma supposed him to be the incarnate Quetzalcoatl come back into the land, and sent human victims accordingly to be slaughtered before him, should he seem to lust for blood.[5] Such in modern India is the woman who as representative of Radha eats and drinks the offerings at the shameless orgies of the Saktas.[6] More usually it is the priest who as minister of the deities has the lion's share of the offerings or the sole privilege of consuming them, from the Fijian priest who watches for the turtle and puddings apportioned to his god,[7] and the West African priest who carries the allowances of food sent to the local spirits of mountain, or

---

[1] Rochefort, 'Iles Antilles,' p. 367.
[2] Ellis, 'Polyn. Res.' vol. i. pp. 336, 358. Williams, 'Fiji,' vol. L p. 220.
[3] Bosman, 'Guinea,' in Pinkerton, vol. xvi. p. 494; J. L. Wilson, 'W. Afr.' p. 218; Burton, 'W. & W. fr. W. Afr.' p. 331.
[4] Ward, 'Hindoos,' vol. ii. p. 195, etc.
[5] Clavigero, 'Messico,' vol. ii. p. 69. J. G. Müller, p. 631.
[6] Ward, vol. ii. p. 194; 'Mem. Anthrop. Soc.' vol. i. p. 332.
[7] Williams, 'Fiji,' vol. i. p. 226.

river, or grove, which food he eats himself as the spirit's proxy,[1] to the Brahmans who receive for the divine ancestors the oblation of a worshipper who has no sacred fire to consume it, "for there is no difference between the Fire and a Brahman, such is the judgment declared by them who know the Veda."[2] It is needless to collect details of a practice so usual in the great systematic religions of the world, where priests have become professional ministers and agents of deity, as for them to partake of the sacrificial meats. It by no means follows from this usage that the priest is necessarily supposed to consume the food as representative of his divinity; in the absence of express statement to such effect, the matter can only be treated as one of ceremonial ordinance. Indeed, the case shows the caution needed in interpreting religious rites, which in particular districts may have meanings attached to them quite foreign to their general intent.

The feeding of an idol, as when Ostyaks would pour daily broth into the dish at the image's mouth,[3] or when the Aztecs would pour the blood and put the heart of the slaughtered human victim into the monstrous idol's mouth,[4] seems ceremonial make-believe, but shows that in each case the deity was somehow considered to devour the meal. The conception among the lower races of deity, as in disembodied spiritual form, is even less compatible with the notion that such a being should consume solid matter. It is true that the notion does occur. In old times it appears in the legend of Bel and the Dragon, where the footprints in the strewn ashes betray the knavish priests who come by secret doors to eat up the banquet set before Bel's image.[5] In modern centuries, it may be exemplified by the negroes of Labode, who could hear the noise of their god Jimawong emptying one after another the bottles of brandy handed in

[1] J. L. Wilson, 'W. Afr.' p. 218.
[2] Manu, iii. 212. See also 'Avesta,' tr. by Spiegel and Bleek, vol. ii. p. 2.
[3] Ysbrants Ides, 'Reize naar China,' p. 38. Meiners, vol. i. p. 162.
[4] Clavigero, vol. ii. p. 46. J. G. Müller, p. 631.
[5] Bel and the Dragon.

RITES AND CEREMONIES. 381

at the door of his straw-roofed temple;[1] or among the Ostyaks, who, as Pallas relates, used to leave a horn of snuff for their god, with a shaving of willow bark to stop his nostrils with after the country fashion; the traveller describes their astonishment when sometimes an unbelieving Russian has emptied it in the night, leaving the simple folk to conclude that the deity must have gone out hunting to have snuffed so much.[2] But these cases turn on fraud, whereas absurdities in which low races largely agree are apt to have their origin rather in genuine error. Indeed, their dominant theories of the manner in which deities receive sacrifice are in accordance not with fraud but with facts, and must be treated as strictly rational and honest developments of the lower animism. The clearest and most general of these theories are as follows.

When the deity is considered to take actual possession of the food or other objects offered, this may be conceived to happen by abstraction of their life, savour, essence, quality, and in yet more definite conception their spirit or soul. The solid part may die, decay, be taken away or consumed or destroyed, or may simply remain untouched. Among this group of conceptions, the most materialized is that which carries out the obvious primitive world-wide doctrine that the life is the blood. Accordingly, the blood is offered to the deity, and even disembodied spirits are thought capable of consuming it, like the ghosts for whom Odysseus entering Hades poured into the trench the blood of the sacrificed ram and black ewe, and the pale shades drank and spoke;[3] or the evil spirits which the Mintira of the Malay Peninsula keep away from the wife in childbirth by placing her near the fire, for the demons are believed to drink human blood when they can find it.[4] Thus in Virginia the Indians (in pretence or reality) sacrificed children, whose

[1] Römer, 'Guinea,' p. 47.
[2] Bastian, 'Mensch,' part ii. p. 210.
[3] Homer, Odyss. xi. xii.
[4] 'Journ. Ind. Archip.' vol. i. p. 270.

blood the oki or spirit was said to suck from their left breast.[1] The Kayans of Borneo used to offer human sacrifice when a great chief took possession of a newly built house; in one late case, about 1847, a Malay slave girl was bought for the purpose and bled to death, the blood, which alone is efficacious, being sprinkled on the pillars and under the house, and the body being thrown into the river.[2] The same ideas appear among the indigenes of India, alike in North Bengal and in the Deccan, where the blood alone of the sacrificed animal is for the deities, and the votary retains the meat.[3] Thus, in West Africa, the negroes of Benin are described as offering a cock to the idol, but it receives only the blood, for they like the flesh very well themselves;[4] while in the Yoruba country, when a beast is sacrificed for a sick man, the blood is sprinkled on the wall and smeared on the patient's forehead, with the idea, it is said, of thus transferring to him the victim's life.[5] The Jewish law of sacrifice marks clearly the distinction between shedding the blood as life, and offering it as food. As the Israelites themselves might not eat with the flesh the blood which is the life, but must pour it on the earth as water, so the rule applies to sacrifice. The blood must be sprinkled before the sanctuary, put upon the horns of the altar, and there sprinkled or poured out, but not presented as a drink offering—"their drink-offerings of blood will I not offer."[6]

Spirit being considered in the lower animism as somewhat of the ethereal nature of smoke or mist, there is an

[1] Smith, 'Virginia,' in Pinkerton, vol. xiii. p. 41; see J. G. Müller, p. 143; Waitz, vol. iii. p. 207. Comp. Meiners, vol. ii. p. 89. See also Bollaert in 'Mem. Anthrop. Soc.' vol. ii. p. 96.
[2] 'Journ. Ind. Archip.' vol. iii. p. 145. See also St. John, 'Far East,' vol. i. p. 160.
[3] Hodgson, 'Abor. of India,' p. 147; Hunter, 'Rural Bengal,' p. 181; Forbes Leslie, 'Early Races of Scotland,' vol. ii. p. 458.
[4] Bosman, 'Guinea,' letter xxi.; in 'Pinkerton,' vol. xvi. p. 531. See also Waitz, vol. ii. p. 192.
[5] Bastian, 'Psychologie,' p. 96.
[6] Levit. i., etc.; Deuteron. xii. 23; Psalm xvi. 4.

obvious reasonableness in the idea that offerings reduced to this condition are fit to be consumed by, or transmitted to, spiritual beings towards whom the vapour rises in the air. This idea is well shown in the case of incense, and especially a peculiar kind of incense offered among the native tribes of America. The habit of smoking tobacco is not suggestive of religious rites among ourselves, but in its native country, where it is so widely diffused as to be perhaps the best point assignable in favour of a connexion in the culture of the northern and southern continent, its place in worship is very important. The Osages would begin an undertaking by smoking a pipe, with such a prayer as this: " Great Spirit, come down to smoke with me as a friend! Fire and Earth, smoke with me and help me to overthrow my foes!" The Sioux in Hennepin's time would look toward the Sun when they smoked, and when the calumet was lighted, they presented it to him, saying: "Smoke, Sun!" The Natchez chief at sunrise smoked first to the east and then to the other quarters; and so on. It is not merely, however, that puffs from the tobacco-pipe are thus offered to deities as drops of drink or morsels of food might be. The calumet is a special gift of the Sun or the Great Spirit, tobacco is a sacred herb, and smoking is an acceptable sacrifice ascending into the air to the abode of gods and spirits.[1] Among the Caribs, the native sorcerer evoking a demon would puff tobacco-smoke into the air as an agreeable perfume to attract the spirit; while among Brazilian tribes the sorcerers smoked round upon the bystanders and on the patient to be cured.[2] How thoroughly incense and burnt-offering are of the same nature, the Zulus well show, burning incense together with the fat of the caul of the slaughtered beast, to give the spirits of the people a sweet

[1] Waitz, vol. iii. p. 181. Hennepin, 'Voyage,' p. 302. Charlevoix, 'Nouvelle France,' vol. v. p. 311, vi. p. 178. Schoolcraft, 'Ind. Tribes,' part i. p. 49, part ii. p. 127. Catlin, vol. i. pp. 181, 229. Morgan, 'Iroquois,' p. 164. J. G. Müller, p. 58.

[2] Rochefort, 'Iles Antilles,' pp. 418, 507. Lery, 'Voy. en Brésil,' p. 268. See also Musters in 'Journ. Anthrop. Inst.' vol. i. p. 202 (Patagonians).

savour.[1] As to incense more precisely of the sort we are familiar with, it was in daily use in the temples of Mexico, where among the commonest antiquarian relics are the earthen incense-pots in which "copalli" (whence our word copal) and bitumen were burnt.[2] Though incense was hardly usual in the ancient religion of China, yet in modern Chinese houses and temples the "joss-stick" and censer do honour to all divine beings, from the ancestral manes to the great gods and Heaven and Earth.[3] The history of incense in the religion of Greece and Rome points the contrast between old thrift and new extravagance, where the early fumigations with herbs and chips of fragrant wood are contrasted with the later oriental perfumes, myrrh and cassia and frankincense.[4] In the temples of ancient Egypt, numberless representations of sacrificial ceremony show the burning of the incense-pellets in censers before the images of the gods; and Plutarch speaks of the incense burnt thrice daily to the Sun, resin at his rising, myrrh at his meridian, kuphi at his setting.[5] The ordinance held as prominent a place among the Semitic nations. At the yearly festival of Bel in Babylon, the Chaldæans are declared by Herodotus to have burned a thousand talents of incense on the large altar in the temple where sat his golden image.[6] In the records of ancient Israel, there has come down to us the very recipe for compounding incense after the art of the apothecary. The priests carried every man his censer, and on the altar of incense, overlaid with gold, standing before the vail in the tabernacle, sweet spices

[1] Callaway, 'Religion of Amazulu,' pp. 11, 141, 177. See also Casalis, 'Basutos,' p. 258.
[2] Clavigero, 'Messico,' vol. ii. p. 39. See also Piedrahita, part i. lib. i. c. 3 (Muyscas).
[3] Plath, 'Religion der alten Chinesen,' part ii. p. 81. Doolittle, 'Chinese.'
[4] Porphyr. de Abstinentia, ii. 5. Arnob. contra Gentes. vii. 26. Meiners, vol. ii. p. 14.
[5] Wilkinson, 'Ancient Egyptians,' vol. v. pp. 315, 338. Plutarch de Is. et Osir.
[6] Herodot. i. 183.

were burned morn and even, a perpetual incense before the Lord.[1] The sacrifice by fire is familiar to the religion of North American tribes. Thus the Algonquins knew the practice of casting into the fire the first morsel of the feast; and throwing fat into the flames for the spirits, they would pray to them "make us find food." Catlin has described and sketched the Mandans dancing round the fire where the first kettleful of the green-corn is being burned, an offering to the Great Spirit before the feast begins.[2] The Peruvians burnt llamas as offerings to the Creator, Sun, Moon, and Thunder, and other lesser deities. As to the operation of sacrifice, an idea of theirs comes well into view in the legend of Manco Ccapac ordering the sacrifice of the most beautiful of his sons, "cutting off his head, and sprinkling the blood over the fire, that the smoke might reach the Maker of heaven and earth."[3] In Siberia the sacrifices of the Tunguz and Buraets, in the course of which bits of meat and liver and fat are cast into the fire, carry on the same idea.[4] Chinese sacrifices to sun and moon, stars and constellations, show their purpose in most definite fashion; beasts and even silks and precious stones are burned, that their vapour may ascend to these heavenly spirits.[5] No less significant, though in a different sense, is the Siamese offering to the household deity, incense and arrack and rice steaming hot; he does not eat it all, not always any part of it, it is the fragrant steam which he loves to inhale.[6] Looking now to the records of Aryan sacrifice, views similar to these are not obscurely expressed. When the Brahman burns the offerings on the altar-fire, they are received by

---

[1] Exod. xxx., xxxvii. Lev. x. 1, xvi. 12, etc.
[2] Smith, 'Virginia,' in 'Pinkerton,' vol. xiii. p. 41. Le Jeune in 'Rel. des Jes.' 1634, p. 16. Catlin, 'N. A. Ind.' vol. i. p. 189.
[3] 'Rites and Laws of Incas,' p. 16, etc., 79; see 'Ollanta, an ancient Ynca Drama,' tr. by C. R. Markham, p. 81. Garcilaso de la Vega, lib. i. ii. vi.
[4] Klemm, 'Cultur-Gesch.' vol. iii. pp. 106, 114.
[5] Plath, part ii. p. 65.
[6] Latham, 'Descr. Eth.' vol. i. p. 191.

Agni the divine Fire, mouth of the gods, messenger of the All-knowing, to whom is chanted the Vedic strophe, "Agni! the sacrifice which thou encompassest whole, it goes unto the gods!"[1] The Homeric poems show the plain meaning of the hecatombs of old barbaric Greece, where the savour of the burnt offering went up in wreathing smoke to heaven, "Κνίσση δ' οὐρανὸν ἷκεν ἑλισσομένη περὶ καπνῷ."[2] Passed into a far other stage of history, men's minds had not lost sight of the archaic thought even in Porphyry's time, for he knows how the demons who desire to be gods rejoice in the libations and fumes of sacrifice, whereby their spiritual and bodily substance fattens, for this lives on the steam and vapours and is strengthened by the fumes of the blood and flesh.[3]

The view of commentators that sacrifice, as a religious rite of remote antiquity and world-wide prevalence, was adopted, regulated, and sanctioned in the Jewish law, is in agreement with the general ethnography of the subject. Here sacrifice appears not with the lower conception of a gift acceptable and even beneficial to deity, but with the higher significance of devout homage or expiation for sin. As is so usual in the history of religion, the offering consisted in general of food, and the consummation of the sacrifice was by fire. To the ceremonial details of the sacrificial rites of Israel, whether prescribing the burning of the carcases of oxen and sheep or of the bloodless gifts of flour mingled with oil, there is appended again and again the explanation of the intent of the rite; it is "an offering made by fire, of a sweet savour unto the Lord." The copious records of sacrifice in the Old Testament enable us to follow its expansion from the simple patriarchal forms of a pastoral tribe, to the huge and complex system organized to carry on the ancient service in a now populous and settled kingdom. Among writers on the Jewish religion,

[1] 'Rig-Veda,' i. 1, 4.
[2] Homer, Il. i. 317.
[3] Porphyr. De Abstinentia, ii. 42; see 58.

Dean Stanley has vividly pourtrayed the aspect of the Temple, with the flocks of sheep and droves of cattle crowding its courts, the vast apparatus of slaughter, the huge altar of burnt-offering towering above the people, where the carcases were laid, the drain beneath to carry off the streams of blood. To this historian, in sympathy rather with the spirit of the prophet than the ceremony of the priest, it is a congenial task to dwell upon the great movement in later Judaism to maintain the place of ethical above ceremonial religion.[1] In these times of Hebrew history, the prophets turned with stern rebuke on those who ranked ceremonial ordinance above weightier matters of the law. "I desired mercy and not sacrifice, and the knowledge of God more than burnt offerings." "I delight not in the blood of bullocks, or of lambs, or of he goats . . . Wash you, make you clean; put away the evil of your doings from before mine eyes. Cease to do evil, learn to do well."

Continuing the enquiry into the physical operation ascribed to sacrifice, we turn to a different conception. It is an idea well vouched for in the lower culture, that the deity, while leaving apparently untouched the offering set out before him, may nevertheless partake of or abstract what in a loose way may be described as its essence. The Zulus leave the flesh of the sacrificed bullock all night, and the divine ancestral spirits come and eat, yet next morning everything remains just as it was. Describing this practice, a native Zulu thus naively comments on it: "But when we ask, 'What do the Amadhlozi eat? for in the morning we still see all the meat,' the old men say, 'The Amatongo lick it.' And we are unable to contradict them, but are silent, for they are older than we, and tell us all things and we listen; for we are told all things, and assent without seeing clearly whether they are true or not."[2] Such imagination

---

[1] Stanley, 'Jewish Church,' 2d Ser. pp. 410, 424. See Kalisch on Leviticus; Barry in Smith's 'Dictionary of the Bible,' art. 'sacrifice.'

[2] Callaway, 'Religion of Amazulu,' p. 11 (amadhlozi or amatongo = ancestral spirits).

was familiar to the native religion of the West Indian islands. In Columbus' time, and with particular reference to Hispaniola, Roman Pane describes the native mode of sacrifice. Upon any solemn day, when they provide much to eat, whether fish, flesh, or any other thing, they put it all into the house of the cemis, that the idol may feed on it. The next day they carry all home, after the cemi has eaten. And God so help them (says the friar), as the cemi eats of that or anything else, they being inanimate stocks or stones. A century and a half later, a similar notion still prevailed in these islands. Nothing could show it more neatly than the fancy of the Caribs that they could hear the spirits in the night moving the vessels and champing the food set out for them, yet next morning there was nothing touched; it was held that the viands thus partaken of by the spirits had become holy, so that only the old men and considerable people might taste them, and even these required a certain bodily purity.[1] Islanders of Pulo Aur, though admitting that their banished disease-spirits did not actually consume the grains of rice set out for them, nevertheless believed them to appropriate its essence.[2] In India, among the indigenes of the Garo hills, we hear of the head and blood of the sacrificed animal being placed with some rice under a bamboo arch covered with a white cloth; the god comes and takes what he wants, and after a time this special offering is dressed for the company with the rest of the animal.[3] The Khond deities live on the flavours and essences drawn from the offerings of their votaries, or from animals or grain which they cause to die or disappear.[4] When the Buraets of Siberia have sacrificed a sheep and boiled the mutton, they set it up on a scaffold for the gods while the shaman is

[1] Roman Pane, ch. xvi. in 'Life of Colon,' in Pinkerton, vol. xii. p. 86. Rochefort, 'Iles Antilles,' p. 418; see Meiners, vol. ii. p. 516; J. G. Müller, p. 212.
[2] 'Journ. Ind. Archip.' vol. iv. p. 194.
[3] Eliot in 'As. Res.' vol. iii. p. 30.
[4] Macpherson, 'India,' pp. 88, 100.

chanting his song, and then themselves fall to.[1] And thus, in the folklore of mediæval Europe, Domina Abundia would come with her dames into the houses at night, and eat and drink from the vessels left uncovered for their increase-giving visit, yet nothing was consumed.[2]

The extreme animistic view of sacrifice is that the soul of the offered animal or thing is abstracted by or transmitted to the deity. This notion of spirits taking souls is in a somewhat different way exemplified among the Binua of Johore, who hold that the evil River-spirits inflict diseases on man by feeding on the 'semangat,' or unsubstantial body (in ordinary parlancé the spirit) in which his life resides,[3] while the Karen demon devours not the body but the "la," spirit or vital principle; thus when it eats a man's eyes, their material part remains, but they are blind.[4] Now an idea similar to this furnished the Polynesians with a theory of sacrifice. The priest might send commissions by the sacrificed human victim; spirits of the dead are eaten by the gods or demons; the spiritual part of the sacrifices is eaten by the spirit of the idol (i. e., the deity dwelling or embodied in the idol) before whom it is presented.[5] Of the Fijians it is observed that of the great offerings of food native belief apportions merely the soul to the gods, who are described as being enormous eaters; the substance is consumed by the worshippers. As in various other districts of the world, human sacrifice is here in fact a meat-offering; cannibalism is a part of the Fijian religion, and the gods are described as delighting in human flesh.[6] Such ideas are explicit among Indian tribes of the American lakes, who consider that offerings, whether abandoned or consumed by the worshippers, go in a spiritual form to the

[1] Klemm, 'Cultur-Gesch.' vol. iii. p. 114.
[2] Grimm, 'Deutsche Myth.' p. 264.
[3] 'Journ. Ind. Archip.' vol. i. p. 27.
[4] Mason, 'Karens,' l. c. p. 208.
[5] Bastian, 'Mensch,' vol. ii. p. 407. Ellis, 'Polyn. Res.' vol. i. p. 358. Taylor, 'New Zealand,' pp. 104, 220.
[6] Williams, 'Fiji,' vol. i. p. 231.

spirit they are devoted to. Native legends afford the clearest illustrations. The following is a passage from an Ottawa tale which recounts the adventures of Wassamo, he who was conveyed by the spirit-maiden to the lodge of her father, the Spirit of the Sand Downs, down below the waters of Lake Superior. "Son-in-law," said the Old Spirit, "I am in want of tobacco. You shall return to visit your parents, and can make known my wishes. For it is very seldom that those few who pass these Sand Hills, offer a piece of tobacco. When they do it, it immediately comes to me. Just so," he added, putting his hand out of the side of the lodge, and drawing in several pieces of tobacco, which some one at that moment happened to offer to the Spirit, for a smooth lake and prosperous voyage. "You see," he said, "every thing offered me on earth, comes immediately to the side of my lodge." Wassamo saw the women also putting their hands to the side of the lodge, and then handing round something, of which all partook. This he found to be offerings of food made by mortals on earth. The distinctly spiritual nature of this transmission is shown immediately after, for Wassamo cannot eat such mere spirit-food, wherefore his spirit-wife puts out her hand from the lodge and takes in a material fish out of the lake to cook for him.[1] Another Ottawa legend, the already cited nature-myth of the Sun and Moon, is of much interest not only for its display of this special thought, but as showing clearly the motives with which savage animists offer sacrifices to their deities, and consider these deities to accept them. Onowuttokwutto, the Ojibwa youth who has followed the Moon up to the lovely heaven-prairies to be her husband, is taken one day by her brother the Sun to see how he gets his dinner. The two look down together through the hole in the sky upon the earth below, the Sun points out a group of children playing beside a lodge, at the same time throwing a tiny stone to hit a beautiful boy. The child falls, they see him carried into the lodge, they

[1] Schoolcraft, 'Algic Researches,' vol. ii. p. 140; see 190.

hear the sound of the sheesheegwun (the rattle), and the song and prayer of the medicine-man that the child's life may be spared. To this entreaty of the medicine-man, the Sun makes answer, "Send me up the white dog." Then the two spectators above could distinguish on the earth the hurry and bustle of preparation for a feast, a white dog killed and singed, and the people who were called assembling at the lodge. While these things were passing, the Sun addressed himself to Onowuttokwutto, saying, "There are among you in the lower world some whom you call great medicine-men; but it is because their ears are open, and they hear my voice, when I have struck any one, that they are able to give relief to the sick. They direct the people to send me whatever I call for, and when they have sent it, I remove my hand from those I had made sick." When he had said this, the white dog was parcelled out in dishes for those that were at the feast; then the medicine-man when they were about to begin to eat, said, "We send thee this, Great Manito." Immediately the Sun and his Ojibwa companion saw the dog, cooked and ready to be eaten, rising to them through the air—and then and there they dined upon it.[1] How such ideas bear on the meaning of human sacrifice, we may perhaps judge from this prayer of the Iroquois, offering a human victim to the War-god: "To thee, O Spirit Arieskoi, we slay this sacrifice, that thou mayst feed upon the flesh, and be moved to give us henceforth luck and victory over our enemies!"[2] So among the Aztec prayers, there occurs this one addressed to Tezcatlipoca-Yautl in time of war: "Lord of battles; it is a very certain and sure thing, that a great war is beginning to make, ordain, form, and concert itself; the War-god opens his mouth, hungry to swallow the blood of many who shall die in this war; it seems that the Sun and the Earth-God Tlatecutli desire to rejoice; they desire to give meat and drink to the gods of Heaven and Hades, making them a

[1] Tanner's 'Narrative,' pp. 286, 318. See also Waitz, vol. iii. p. 207.
[2] J. G. Müller, p. 142; see 282.

banquet of the flesh and blood of the men who are to die in this war," &c.[1] There is remarkable definiteness in the Peruvian idea that the souls of human victims are transmitted to another life in divine as in funeral sacrifice; at one great ceremony, where children of each tribe were sacrificed to propitiate the gods, "they strangled the children, first giving them to eat and drink, that they might not enter the presence of the Creator discontented and hungry."[2] Similar ideas of spiritual sacrifice appear in other regions of the world. Thus in West Africa we read of the tree-fetish enjoying the spirit of the food-offering, but leaving its substance, and an account of the religion of the Gold Coast mentions how each great wong or deity has his house, and his priest and priestess to clean the room and give him daily bread kneaded with palm-oil, "of which, as of all gifts of this kind, the wong eats the invisible soul."[3] So, in India, the Limbus of Darjeeling make small offerings of grain, vegetables, and sugar-cane, and sacrifice cows, pigs, fowls, &c., on the declared principle "the life breath to the gods, the flesh to ourselves."[4] It seems likely that such meaning may largely explain the sacrificial practices of other religions. In conjunction with these accounts, the unequivocal meaning of funeral sacrifices, whereby offerings are conveyed spiritually into the possession of spirits of the dead, may perhaps justify us in inferring that similar ideas of spiritual transmission prevail extensively among the many nations whose sacrificial rites we know in fact, but cannot trace with certainty to their original significance.

Having thus examined the manner in which the operation of sacrifice is considered to take physical effect, whether indefinitely or definitely, and having distinguished its actual transmission as either substantial, essential, or spiritual,

[1] Sahagun, lib. vi. in Kingsborough, vol. v.
[2] 'Rites and Laws of Yncas,' tr. and ed. by C. R. Markham, pp. 55, 58, 166. See ante, p. 385 (possible connexion of smoke with soul).
[3] Waitz, vol. ii. pp. 188, 196. Steinhauser, l. c. p. 136. See also Schlegel, 'Ewe-Sprache,' p. xv.; Magyar, 'Süd-Afrika,' p. 273.
[4] A. Campbell in 'Tr. Eth. Soc.' vol. vii. p. 153.

let us now follow the question of the sacrificer's motive in presenting the sacrifice. Important and complex as this problem is, its key is so obvious that it may be almost throughout treated by mere statement of general principle. If the main proposition of animistic natural religion be granted, that the idea of the human soul is the model of the idea of deity, then the analogy of man's dealings with man ought, *inter alia*, to explain his motives in sacrifice. It does so, and very fully. The proposition may be maintained in wide generality, that the common man's present to the great man, to gain good or avert evil, to ask aid or to condone offence, needs only substitution of deity for chief, and proper adaptation of the means of conveying the gift to him, to produce a logical doctrine of sacrificial rites, in great measure explaining their purpose directly as they stand, and elsewhere suggesting what was the original meaning which has passed into changed shape in the course of ages. Instead of offering a special collection of evidence here on this proposition, it may be enough to ask attentive reference to any extensive general collection of accounts of sacrifice, such for instance as those cited for various purposes in these volumes. It will be noticed that offerings to divinities may be classed in the same way as earthly gifts. The occasional gift made to meet some present emergency, the periodical tribute brought by subject to lord, the royalty paid to secure possession or protection of acquired wealth, all these have their evident and well-marked analogues in the sacrificial systems of the world. It may impress some minds with a stronger sense of the sufficiency of this theory of sacrifice, to consider how the transition is made in the same imperceptible way from the idea of substantial value received, to that of ceremonial homage rendered, whether the recipient be man or god. We do not find it easy to analyse the impression which a gift makes on our own feelings, and to separate the actual value of the object from the sense of gratification in the giver's good-will or respect, and thus we may well scruple to define closely how uncultured men work out this very same distinction in their dealings with their deities. In a general way it may be held that the idea of practical acceptableness of the food or valuables presented to the deity, begins early to shade into the sentiment of divine gratification or propitiation by a reverent offering, though in itself of not much account to so mighty a divine personage. These two stages of the sacrificial idea may be fairly contrasted, the one among the Karens who offer to a demon arrack or grain or a portion of the game they kill, considering invocation of no avail without a gift,[1] the other among the negroes of Sierra Leone, who sacrifice an ox "to make God glad very much, and do Kroomen good."[2]

Hopeless as it may be in hundreds of accounts of sacrifice to guess whether the worshipper means to benefit or merely to gratify the deity, there are also numbers of cases in which the thought in the sacrificer's mind can scarcely be more than an idea of ceremonial homage. One of the best-marked sacrificial rites of the world is that of offering by fire or otherwise morsels or libations at meals. This ranges from the religion of the North American Indian to that of the classic Greek and the ancient Chinese, and still holds its place in peasant custom in Europe.[3] Other groups of cases pass into yet more absolute formality of reverence. See the Guinea negro passing in silence by the sacred tree or cavern, and dropping a leaf or a sea-shell as an offering to the local spirit;[4] the Talein of Birma holding up the dish at his meal to offer it to the nat, before the company fall to;[5] the Hindu holding up a little of his rice in his

---

[1] O'Riley, in 'Journ. Ind. Archip.' vol. iv. p. 592. Bastian, 'Oestl. Asien,' vol. ii. p. 12.
[2] R. Clarke, 'Sierra Leone,' p. 43.
[3] Smith, 'Virginia,' in 'Pinkerton,' vol. xiii. p. 41. Welcker, 'Griech. Götterlehre,' vol. ii. p. 693. Legge, 'Confucius,' p. 179. Grohmann, 'Aberglauben aus Böhmen,' p. 41, etc.
[4] J. L. Wilson, 'W. Afr.' p. 218; Bosman, 'Guinea,' in 'Pinkerton,' vol. xvi. p. 400.
[5] Bastian, 'Oestl. Asien,' vol. ii. p. 387.

fingers to the height of his forehead, and offering it in thought to Siva or Vishnu before he eats it.[1] The same argument applies to the cases ranging far and wide through religion, where, whatever may have been the original intent of the sacrifice, it has practically passed into a feast. A banquet where the deity has but the pretence and the worshippers the reality, may seem to us a mere mockery of sacrifice. Yet how sincerely men regard it as a religious ceremony, the following anecdote of a North American Indian tribe will show. A travelling party of Potawatomis, for three days finding no game, were in great distress for want of food. On the third night, a chief, named Saugana, had a dream, wherein a person appearing to him showed him that they were suffering because they had set out without a sacrificial feast. He had started on this important journey, the dreamer said, " as a white man would," without making any religious preparation. Therefore the Great Spirit had punished them with scarcity. Now, however, twelve men were to go and kill four deer before the sun was thus high (about nine o'clock). The chief in his dream had seen these four deer lying dead, the hunters duly killed them, and the sacrificial feast was held.[2] Further illustrative examples of such sacred banquets may be chosen through the long range of culture. The Zulus propitiate the Heaven-god above with a sacrifice of black cattle, that they may have rain; the village chiefs select the oxen, one is killed, the rest are merely mentioned; the flesh of the slaughtered ox is eaten in the house in perfect silence, a token of humble submission; the bones are burnt outside the village; and after the feast they chant in musical sounds, a song without words.[3] The Serwatty Islanders sacrifice buffalos, pigs, goats, and fowls to the idols when an individual or the community undertakes an affair or expedition of importance, and as the carcases are

[1] Roberts, 'Oriental Illustrations,' p. 545.
[2] M'Coy, 'Baptist Indian Missions,' p. 305.
[3] Callaway, ' Religion of Amazulu,' p. 59. See Casalis, p. 252.

devoured by the devotees, this ensures a respectable attendance when the offerings are numerous.[1] Thus among rude tribes of Northern India, sacrifices of beasts are accompanied by libations of fermented liquor, and in fact sacrifice and feast are convertible words.[2] Among the Aztecs, prisoners of war furnished first an acceptable sacrifice to the deity, and then the staple of a feast for the captors and their friends;[3] while in ancient Peru whole flocks of sacrificed llamas were eaten by the people.[4] The history of Greek religion plainly records the transition from the early holocausts devoted by fire to the gods, to the great festivals where the sacrifices provided meat for the public banquets held to honour them in ceremonial homage.[5]

Beside this development from gift to homage, there arises also a doctrine that the gist of sacrifice is rather in the worshipper giving something precious to himself, than in the deity receiving benefit. This may be called the abnegation-theory, and its origin may be fairly explained by considering it as derived from the original gift-theory. Taking our own feelings again for a guide, we know how it satisfies us to have done our part in giving, even if the gift be ineffectual, and how we scruple to take it back if not received, but rather get rid of it in some other way—it is corban. Thus we may enter into the feelings of the Assinaboin Indians, who considered that the blankets and pieces of cloth and brass kettles and such valuables abandoned in the woods as a medicine-sacrifice, might be carried off by any friendly party who chanced to discover them;[6] or of the Ava Buddhists bringing to the temples offerings of boiled rice and sweetmeats and cocoa-nut fried

---

[1] Earl in 'Journ. Ind. Archip.' vol. iv. p. 174.
[2] Hodgson, 'Abor. of India,' p. 170, see 146; Hooker, 'Himalayan Journals,' vol. ii. p. 276.
[3] Prescott, 'Mexico,' book i. ch. iii.
[4] 'Rites and Laws of Yncas,' p. 33, etc.
[5] Welcker, 'Griech. Götterlehre,' vol. ii. p. 50; Pauly, 'Real-Encyclopedie.' s. v. 'Sacrificia.'
[6] Tanner's 'Nar.' p. 154; see also Waitz, vol. iii. p. 167.

in oil, and never attempting to disturb the crows and wild dogs who devoured it before their eyes;[1] of the modern Moslems sacrificing sheep, oxen, and camels in the valley of Muna on their return from Mekka, it being a meritorious act to give away a victim without eating any of it, while parties of Takruri watch around like vultures, ready to pounce upon the carcases.[2] If the offering to the deity be continued in ceremonial survival, in spite of a growing conviction that after all the deity does not need and cannot profit by it, sacrifice will be thus kept up in spite of having become practically unreasonable, and the worshipper may still continue to measure its efficacy by what it costs him. But to take this abnegation-theory as representing the primitive intention of sacrifice would be, I think, to turn history upside down. The mere fact of sacrifices to deities, from the lowest to the highest levels of culture, consisting to the extent of nine-tenths or more of gifts of food and sacred banquets, tells forcibly against the originality of the abnegation-theory. If the primary motive had been to give up valuable property, we should find the sacrifice of weapons, garments, ornaments, as prevalent in the lower culture as in fact it is unusual. Looking at the subject in a general view, to suppose men to have started by devoting to their deities what they considered practically useless to them, in order that they themselves might suffer a loss which none is to gain, is to undervalue the practical sense of savages, who are indeed apt to keep up old rites after their meaning has fallen away, but seldom introduce new ones without a rational motive. In studying the religion of the lower races, we find men dealing with their gods in as practical and straightforward a way as with their neighbours, and where plain original purpose is found, it may well be accepted as sufficient explanation. Of the way in which gift can pass into abnegation, an instructive example is forth-

[1] Symes, 'Ava,' in 'Pinkerton,' vol. ix. p. 440.
[2] Burton, 'Medinah,' etc. vol. iii. p. 302; Lane, 'Modern Egyptians,' vol. i. p. 132.

coming in Buddhism. It is held that sinful men are liable to be re-born in course of transmigration as wandering, burning, miserable demons (preta). Now those demons may receive offerings of food and drink from their relatives, who can further benefit them by acts of merit done in their name, as giving food to priests, unless the wretched spirits be so low in merit that this cannot profit them. Yet even in this case it is held that though the act does not benefit the spirit whom it is directed to, it does benefit the person who performs it.[1] Unequivocal examples of abnegation in sacrifice may be best found among those offerings of which the value to the offerer utterly exceeds the value they can be supposed to have to the deity. The most striking of these, found among nations somewhat advanced in general culture, appear in the history of human sacrifice among Semitic nations. The king of Moab, when the battle was too sore for him, offered up his eldest son for a burnt-offering on the wall. The Phœnicians sacrificed the dearest children to propitiate the angry gods, they enhanced their value by choosing them of noble families, and there was not wanting among them even the utmost proof that the efficacy of the sacrifice lay in the sacrificer's grievous loss, for they must have for yearly sacrifice only-begotten sons of their parents (Κρόνῳ γὰρ Φοίνικες καθ' ἕκαστον ἔτος ἔθυον τὰ ἀγαπητὰ καὶ μονογενῆ τῶν τέκνων). Heliogabalus brought the hideous Oriental rite into Italy, choosing for victims to his solar divinity high-born lads throughout the land. Of all such cases, the breaking of the sacred law of hospitality by sacrificing the guest to Jupiter hospitalis, Ζεὺς ξένιος, shows in the strongest light in Semitic regions how the value to the offerer might become the measure of acceptableness to the god.[2] In such ways, slightly within the range of the lower culture, but strongly in the religion of the higher

[1] Hardy, 'Manual of Buddhism,' p. 59.
[2] II. Kings, iii. 27. Euseb. Præp. Evang. i. 10, iv. 156. Laud. Constant. xiii. Porphyr. De Abstin. ii. 56, etc. Lamprid. Heliogabal. vii. Movers, 'Phönizier,' vol. i. p. 300, etc.

nations, the transition from the gift-theory to the abnegation-theory seems to have come about. Our language displays it in a word, if we do but compare the sense of presentation and acceptance which " sacrificium " had in a Roman temple, with the sense of mere giving up and loss which "sacrifice" conveys in an English market.

Through the history of sacrifice, it has occurred to many nations that cost may be economized without impairing efficiency. The result is seen in ingenious devices to lighten the burden on the worshipper by substituting something less valuable than what he ought to offer, or pretends to. Even in such a matter as this, the innate correspondence in the minds of men is enough to produce in distant and independent races so much uniformity of development, that three or four headings will serve to class the chief divisions of sacrificial substitution among mankind.

To give part for the whole is a proceeding so closely conformed to ordinary tribute by subject to lord, that in great measure it comes directly under the gift-theory, and as such has already had its examples here. It is only when the part given to the gods is of contemptible value in proportion to the whole, that full sacrifice passes gradually into substitution. This is the case when in Madagascar the head of the sacrificed beast is set up on a pole, and the blood and fat are rubbed on the stones of the altar, but the sacrificers and their friends and the officiating priest devour the whole carcase;[1] when rich Guinea negroes sacrifice a sheep or goat to the fetish, and feast on it with their friends, only leaving for the deity himself part of the entrails[2]; when Tunguz, sacrificing cattle, would give a bit of liver and fat and perhaps hang up the hide in the woods as the god's share, or Mongols would set the heart of the beast before the idol till next day.[3] Thus the most ancient whole

[1] Ellis, 'Madagascar,' vol. i. p. 419.
[2] Römer, 'Guinea,' p. 59. Bosman in Pinkerton, vol. xvi. p. 399.
[3] Klemm, 'Cultur-Gesch.' vol. iii. p. 106 ; Castrén, ' Finn. Myth.' p. 232.

burnt-offering of the Greeks dwindled to burning for the gods only the bones and fat of the slaughtered ox, while the worshippers feasted themselves on the meat, an economic rite which takes mythic shape in the legend of the sly Prometheus giving Zeus the choice of the two parts of the sacrificed ox he had divided for gods and mortals, on the one side bones covered seemly with white fat, on the other the joints hidden under repulsive hide and entrails.[1] With a different motive, not that of parsimony, but of keeping up in survival an ancient rite, the Zarathustrian religion performed by substitution the old Aryan sacrifice by fire. The Vedic sacrifice Agnishtoma required that animals should be slain, and their flesh partly committed to the gods by fire, partly eaten by sacrificers and priests. The Parsi ceremony Izeshne, formal successor of this bloody rite, requires no animal to be killed, but it suffices to place the hair of an ox in a vessel, and show it to the fire.[2]

The offering of a part of the worshipper's own body is a most usual rite, whether its intention is simply that of gift or tribute, or whether it is considered as a *pars pro toto* representing the whole man, either in danger and requiring to be ransomed, or destined to actual sacrifice for another and requiring to be redeemed. How a finger-joint may thus represent a whole body, is perfectly shown in the funeral sacrifices of the Nicobar islanders; they bury the dead man's property with him, and his wife has a finger-joint cut off (obviously a substitute for herself), and if she refuses even this, a deep notch is cut in a pillar of the house.[3] We are now concerned, however, with the finger-offering, not as a sacrifice to the dead, but as addressed to other deities. This idea is apparently worked out in the Tongan custom of tutu-nima, the chopping off a portion of the little finger with a hatchet or sharp stone as a sacrifice to the gods, for the recovery of a sick relation of higher rank; Mariner saw

[1] Hesiod. Theog. 537. Welcker, vol. i. p. 764; vol. ii. p. 51.
[2] Haug, 'Parsis,' Bombay, 1862, p. 238.
[3] Hamilton in 'As. Res.' vol. ii. p. 342.

children of five years old quarrelling for the honour of having it done to them.[1] In the Mandan ceremonies of initiation into manhood, when the youth at last hung senseless and (as they called it) lifeless by the cords made fast to splints through his flesh, he was let down, and coming to himself crawled on hands and feet round the medicine-lodge to where an old Indian sat with a hatchet in his hand and a buffalo skull before him; then the youth, holding up the little finger of his left hand to the Great Spirit, offered it as a sacrifice, and it was chopped off, and sometimes the forefinger afterwards, upon the skull.[2] In India, probably as a Dravidian rather than Aryan rite, the practice with full meaning comes into view; as Siva cut off his finger to appease the wrath of Kali, so in the southern provinces mothers will cut off their own fingers as sacrifices lest they lose their children, and we hear of a golden finger being allowed instead, the substitute of a substitute.[3] The New Zealanders hang locks of hair on branches of trees in the burying-ground, a recognized place for offerings.[4] That hair may be a substitute for its owner is well shown in Malabar, where we read of the demon being expelled from the possessed patient and flogged by the exorcist to a tree; there the sick man's hair is nailed fast, cut away, and left for a propitiation to the demon.[5] Thus there is some ground for interpreting the consecration of the boy's cut hair in Europe as a representative sacrifice.[6] As for the formal shedding of blood, it may represent fatal bloodshed, as when

[1] Mariner's 'Tonga. Is.' vol. i. p. 454; vol. ii. p. 222. Cook's '3rd Voy.' vol. i. p. 403. Details from S. Africa in Bastian, 'Mensch,' vol. iii. pp. 4, 24; Scherzer, 'Voy. of Novara,' vol. i. p. 212.
[2] Catlin, 'N. A. Ind.' vol. i. p. 172; Klemm, 'Cultur-Gesch.' vol. ii. p. 170. See also Venegas, 'Noticia de la California,' vol. i. p. 117; Garcilaso de la Vega, lib. ii. c. 8 (Peru).
[3] Buchanan, 'Mysore,' etc., in 'Pinkerton,' vol. viii. p. 661; Meiners, vol. ii. p. 472; Bastian, l. c. See also Dubois, 'India,' vol. i. p. 5.
[4] Polack, 'New Zealand,' vol. i. p. 264.
[5] Bastian, 'Psychologie,' p. 184.
[6] Theodoret. in Levit. xix.; Hanusch, 'Slaw. Myth.' Details in Bastian, 'Mensch,' vol. ii. p. 229, etc.

the Jagas or priests in Quilombo only marked with spears the children brought in, instead of running them through;[1] or when in Greece a few drops of human blood had come to stand instead of the earlier and more barbaric human sacrifice;[2] or when in our own time and under our own rule a Vishnuite who has inadvertently killed a monkey, a garuda, or a cobra, may expiate his offence by a mock sacrifice, in which a human victim is wounded in the thigh, pretends to die, and goes through the farce of resuscitation, his drawn blood serving as substitute for his life.[3] One of the most noteworthy cases of the survival of such formal bloodshed within modern memory in Europe must be classed as not Aryan but Turanian, belonging as it does to the folklore of Esthonia. The sacrificer had to draw drops of blood from his forefinger, and therewith to pray this prayer, which was taken down verbatim from one who remembered it :—" I name thee with my blood and betroth thee with my blood, and point thee out my buildings to be blessed, stables and cattle-pens and hen-roosts ; let them be blessed through my blood and thy might!" "Be my joy, thou Almighty, upholder of my forefathers, my protector and guardian of my life! I beseech thee by strength of flesh and blood ; receive the food that I bring thee to thy sustenance and the joy of my body; keep me as thy good child, and I will thank and praise thee. By the help of the Almighty, my own God, hearken to me! What through negligence I have done imperfectly toward thee, do thou forget! But keep it truly in remembrance, that I have honestly paid my gifts to my parent's honour and joy and requital. Moreover falling down I thrice kiss the earth. Be with me quick in doing, and peace be with thee hitherto!"[4] These various rites of finger-cutting, hair-cutting, and blood-letting, have required mention here from the special point of view of their

[1] Bastian, 'Mensch,' vol. iii. p. 113 (see other details).
[2] Pausan. viii. 23 ; ix. 8.
[3] 'Encyc. Brit.' art. 'Brahma.' See 'Asiat. Res.' vol. ix. p. 387.
[4] Boecler, 'Ehsten Aberglaübische Gebraüche,' etc., p. 4.

connexion with sacrifice. They belong to an extensive series of practices, due to various and often obscure motives, which come under the general heading of ceremonial mutilations. When a life is given for a life, it is still possible to offer a life less valued than the life in danger. When in Peru the Inca or some great lord fell sick, he would offer to the deity one of his sons, imploring him to take this victim in his stead.[1] The Greeks found it sufficient to offer to the gods criminals or captives;[2] and the like was the practice of the heathen tribes of northern Europe, to whom indeed Christian dealers were accused of selling slaves for sacrificial purposes.[3] Among such accounts, the typical story belongs to Punic history. The Carthaginians had been overcome and hard pressed in the war with Agathokles, and they set down the defeat to divine wrath. Kronos (Moloch) had in former times received his sacrifice of the chosen of their sons, but of late they had put him off with children bought and nourished for the purpose. In fact they had obeyed the sacrificer's natural tendency to substitution, but now in time of misfortune the reaction set in. To balance the account and condone the parsimonious fraud, a monstrous sacrifice was celebrated. Two hundred children, of the noblest of the land, were brought to the idol of Moloch. "For there was among them a brazen statue of Kronos, holding out his hands sloping downward, so that the child placed on them rolled off and fell into a certain chasm full of fire."[4] Next, it will help us to realize how the sacrifice of an animal may atone for a human life, if we notice in South Africa how a Zulu will redeem a lost child from the finder by a bullock, or a Kimbunda will expiate the blood of a slave by the offering of an ox, whose blood will wash away

[1] Rivero and Tschudi, p. 196. See 'Rites of Yncas,' p. 79.
[2] Bastian, p. 112, etc.; Smith's 'Dic. of Gr. and Rom. Ant.' art. 'Sacrificium.'
[3] Grimm, 'Deutsche Myth.' p. 40.
[4] Diodor. Sic. xx. 14.

the other.[1] For instances of the animal substituted for man in sacrifice the following may serve. Among the Khonds of Orissa, when Colonel Macpherson was engaged in putting down the sacrifice of human victims by the sect of the Earth-goddess, they at once began to discuss the plan of sacrificing cattle by way of substitutes. Now there is some reason to think that this same course of ceremonial change may account for the following sacrificial practice in the other Khond sect. It appears that those who worship the Light-god hold a festival in his honour, when they slaughter a buffalo in commemoration of the time when, as they say, the Earth-goddess was prevailing on men to offer human sacrifices to her, but the Light-god sent a tribe-deity who crushed the bloody-minded Earth-Goddess under a mountain, and dragged a buffalo out of the jungle, saying, " Liberate the man, and sacrifice the buffalo!"[2] It looks as though this legend, divested of its mythic garb, may really record a historical substitution of animal for human sacrifice. In Ceylon, the exorcist will demand the name of the demon possessing a demoniac, and the patient in frenzy answers, giving the demon's name, " I am So-and-so, I demand a human sacrifice and will not go out without!" The victim is promised, the patient comes to from the fit, and a few weeks later the sacrifice is made, but instead of a man they offer a fowl.[3] Classic examples of substitution of this sort may be found in the sacrifice of a doe for a virgin to Artemis in Laodicæa, a goat for a boy to Dionysos at Potniæ. There appears to be Semitic connexion here, as there clearly is in the story of the Æolians of Tenedos sacrificing to Melikertes (Melkarth) instead of a new-born child a new-born calf, shoeing it with buskins and tending the mother-cow as if a human mother.[4]

One step more in the course of substitution leads the

[1] Callaway, 'Zulu Tales,' vol. i. p. 88; Magyar, 'Süd-Afrika,' p. 256.
[2] Macpherson, 'India,' pp. 108, 187.
[3] De Silva in Bastian, 'Psychologie,' p. 181.
[4] Details in Pauly, ' Real-Encyclop.' s. v. 'Sacrificia' ; Bastian, 'Mensch,' vol. iii. p. 114 ; Movers, 'Phönizier,' vol. i. p. 300.

worshipper to make his sacrifice by effigy. An instructive example of the way in which this kind of substitution arises may be found in the rites of ancient Mexico. At the yearly festival of the water-gods and mountain-gods, certain actual sacrifices of human victims took place in the temples. At the same time, in the houses of the people, there was celebrated an unequivocal but harmless imitation of this bloody rite. They made paste images, adored them, and in due pretence of sacrifice cut them open at the breast, took out their hearts, cut off their heads, divided and devoured their limbs.[1] In the classic religions of Greece and Rome, the desire to keep up the consecrated rites of ages more barbaric, more bloodthirsty, or more profuse, worked itself out in many a compromise of this class, such as the brazen statues offered for human victims, the cakes of dough or wax in the figure of the beasts for which they were presented as symbolic substitutes.[2] Not for economy, but to avoid taking life, Brahmanic sacrifice has been known to be brought down to offering models of the victim-animals in meal and butter.[3] The modern Chinese, whose satisfaction in this kind of make-believe is so well shown by their dispatching paper figures to serve as attendants for the dead, work out in the same fanciful way the idea of the sacrificial effigy, in propitiating the presiding deity of the year for the cure of a sick man. The rude figure of a man is drawn on or cut out of a piece of paper, pasted on a slip of bamboo, and stuck upright in a packet of mock money. With proper exorcism, this representative is carried out into the street with the disease, the priest squirts water from his mouth over patient, image, and mock-money, the two latter are burnt, and the company eat up the little feast

---

[1] Clavigero, 'Messico,' vol. ii. p. 82 ; Torquemada, 'Monarquia Indiana,' x. c. 29 ; J. G. Müller, pp. 502, 640, See also *ibid.* p. 379 (Peru) ; 'Rites and Laws of Yncas,' pp. 46, 54.
[2] Grote, vol. v. p. 366. Schmidt in Smith's ' Dic. of Gr. and Rom. Ant.' art. 'Sacrificium.' Bastian, 1. c.
[3] Bastian, 'Oestl. Asien,' vol. iii. p. 501.

laid out for the year-deity.[1] There is curious historical significance in the custom at the inundation of the Nile at Cairo, of setting up a conical pillar of earth which the flood washes away as it rises. This is called the arûseh or bride, and appears to be a substitute introduced under humaner Moslem influence, for the young virgin in gay apparel who in older time was thrown into the river, a sacrifice to obtain a plentiful inundation.[2] Again, the patient's offering the model of his diseased limb is distinctly of the nature of a sacrifice, whether it be propitiatory offering before cure, or thank-offering after. On the one hand, the ex-voto models of arms and ears dedicated in ancient Egyptian temples are thought to be grateful memorials,[3] as seems to have been the case with metal models of faces, breasts, hands, &c. in Bœotian temples.[4] On the other hand, there are cases where the model and, as it were, substitute of the diseased part is given to obtain a cure; thus in early Christian times in Germany protest was made against the heathen custom of hanging up carved wooden limbs to a helpful idol for relief,[5] and in modern India the pilgrim coming for cure will deposit in the temple the image of his diseased limb, in gold or silver or copper according to his means.[6]

If now we look for the sacrificial idea within the range of modern Christendom, we shall find it in two ways not obscurely manifest. It survives in traditional folklore, and it holds a place in established religion. One of its most remarkable survivals may be seen in Bulgaria, where sacrifice of live victims is to this day one of the accepted rites of the land. They sacrifice a lamb on St. George's day, telling to account for the custom a legend which combines the episodes of the offering of Isaac and the miracle of the Three Children.

[1] Doolittle, 'Chinese,' vol. i. p. 152.
[2] Lane, 'Modern Eg.' vol. ii. p. 262. Meiners, vol. ii. p. 85.
[3] Wilkinson, 'Ancient Eg.' vol. iii. p. 395; and in Rawlinson's Herodotus, vol. ii. p. 137. See 1. Sam. vi. 4.
[4] Grimm, 'Deutsche Myth.' p. 1131.
[5] Ibid.
[6] Bastian, vol. iii. p. 116.

On the feast of the Panagia (Virgin Mary) sacrifices of lambs, kids, honey, wine, &c., are offered in order that the children of the house may enjoy good health throughout the year. A little child divines by touching one of three saints' candles to which the offering is to be dedicated; when the choice is thus made, the bystanders each drink a cup of wine, saying "Saint So-and-so, to thee is the offering." Then they cut the throat of the lamb, or smother the bees, and in the evening the whole village assembles to eat the various sacrifices, and the men end the ceremony with the usual drunken bout.[1] Within the borders of Russia, many and various sacrifices are still offered; such is the horse with head smeared with honey and mane decked with ribbons, cast into the river with two millstones to its neck to appease the water-spirit, the Vodyany, at his spiteful flood-time in early spring; and such is the portion of supper left out for the house-demon, the domovoy, who if not thus fed is apt to turn spirit-rapper, and knock the tables and benches about at night.[2] In many another district of Europe, the tenacious memory of the tiller of the soil has kept up in wondrous perfection heirlooms from præ-Christian faiths. In Franconia, people will pour on the ground a libation before drinking; entering a forest they will put offerings of bread and fruit on a stone, to avert the attacks of the demon of the woods, the "bilberry-man;" the bakers will throw white rolls into the oven flue for luck, and say, "Here, devil, they are thine!" The Carinthian peasant will fodder the wind by setting up a dish of food in a tree before his house, and the fire by casting in lard and dripping, in order that gale and conflagration may not hurt him. At least up to the end of last century this most direct elemental sacrifice might be seen in Germany at the midsummer festival in the most perfect form; some of the porridge

---

[1] St. Clair and Brophy, 'Bulgaria,' p. 43. Compare modern Circassian sacrifice of animal before cross, as substitute for child, in Bell, 'Circassia,' vol. ii.
[2] Ralston, 'Songs of Russian People,' pp. 123, 153, etc.

from the table was thrown into the fire, and some into running water, some was buried in the earth, and some smeared on leaves and put on the chimney-top for the winds.[1] Relics of such ancient sacrifice may be found in Scandinavia to this day; to give but one example, the old country altars, rough earth-fast stones with cup-like hollows, are still visited by mothers whose children have been smitten with sickness by the trolls, and who smear lard into the hollows and leave rag-dolls as offerings.[2] France may be represented by the country-women's custom of beginning a meal by throwing down a spoonful of milk or bouillon; and by the record of the custom of Andrieux in Dauphiny, where at the solstice the villagers went out upon the bridge when the sun rose, and offered him an omelet.[3] The custom of burning alive the finest calf, to save a murrain-struck herd, had its last examples in Cornwall in the present century; the records of bealtuinn sacrifices in Scotland continue in the Highlands within a century ago; and Scotchmen still living remember the corner of a field being left untilled for the Goodman's Croft (*i.e.*, the Devil's), but the principle of "cheating the devil" was already in vogue, and the piece of land allotted was but a worthless scrap.[4] It is a remnant of old sacrificial rite, when the Swedes still bake at yule-tide a cake in the shape of a boar, representing the boar sacrificed of old to Freyr, and Oxford to this day commemorates the same ancestral ceremony, when the boar's head is carried in to the Christmas feast at Queen's College, with its appointed carol, "Caput apri defero, Reddens laudes Domino."[5] With a lingering recollection of the old

[1] Wuttke, 'Deutsche Volksaberglaube,' p. 86. See also Grimm, 'Deutsche Myth.' pp. 417, 602.
[2] Hyltén-Cavallius, 'Wärend och Wirdarne,' part i. pp. 131, 146, 157, etc.
[3] Monnier, 'Traditions Populaires,' pp. 187, 666.
[4] R. Hunt, 'Pop. Rom. of W. of England,' 1st Ser. p. 237. Pennant, 'Tour in Scotland,' in Pinkerton, vol. iii. p. 49. J. Y. Simpson, Address to Soc. Antiq. Scotland, 1861, p. 33; Brand, 'Pop. Ant.' vol. iii. pp. 74, 317.
[5] Brand, vol. i. p. 484. Grimm, 'D. M.' pp. 45, 194, 1188, see 250; 'Deutsche Rechtsalterthümer,' p. 900; Hyltén-Cavallius, part i. p. 175.

libations, the German toper's saying still runs that heeltaps are a devil's offering.[1]

As for sacrificial rites most fully and officially existing in modern Christendom, the presentation of ex-votos is one. The ecclesiastical opposition to the continuance of these classic thank-offerings was but temporary and partial. In the 5th century it seems to have been usual to offer silver and gold eyes, feet, &c., to saints in acknowledgment of cures they had effected. At the beginning of the 16th century, Polydore Vergil, describing the classic custom, goes on to say: "In the same manner do we now offer up in our churches sigillaria, that is, little images of wax, and oscilla. As oft as any part of the body is hurt, as the hand, foot, breast, we presently make a vow to God, and his saints, to whom upon our recovery we make an offering of that hand or foot or breast shaped in wax, which custom has so far obtained that this kind of images have passed to the other animals. Wherefore so for an ox, so for a horse, so for a sheep, we place puppets in the temples. In which thing any modestly scrupulous person may perhaps say he knows not whether we are rivalling the religion or the superstition of the ancients."[2] In modern Europe the custom prevails largely, but has perhaps somewhat subsided into low levels of society, to judge by the general use of mock silver and such like worthless materials for the dedicated effigies. In Christian as in præ-Christian temples, clouds of incense rise as of old. Above all, though the ceremony of sacrifice did not form an original part of Christian worship, its prominent place in the ritual was obtained in early centuries. In that Christianity was recruited among nations to whom the conception of sacrifice was among the deepest of religious ideas, and the ceremony of sacrifice among the sincerest efforts of worship, there arose an observance suited to supply the vacant place.

[1] Grimm, 'D. M.' p. 962.
[2] Beausobre, vol. ii. p. 667. Polydorus Vergilius, De Inventoribus Reru (Basel, 1521), lib. v. 1.

This result was obtained not by new introduction, but by transmutation. The solemn eucharistic meal of the primitive Christians in time assumed the name of the sacrifice of the mass, and was adapted to a ceremonial in which an offering of food and drink is set out by a priest on an altar in a temple, and consumed by priest and worshippers. The natural conclusion of an ethnographic survey of sacrifice, is to point to the controversy between Protestants and Catholics, for centuries past one of the keenest which have divided the Christian world, on this express question whether sacrifice is or is not a Christian rite.

The next group of rites to be considered comprises Fasting and certain other means of producing ecstasy and other morbid exaltation for religious ends. In the foregoing researches on animism, it is frequently observed or implied that the religious beliefs of the lower races are in no small measure based on the evidence of visions and dreams, regarded as actual intercourse with spiritual beings. From the earliest phases of culture upward, we find religion in close alliance with ecstatic physical conditions. These are brought on by various means of interference with the healthy action of body and mind, and it is scarcely needful to remind the reader that, according to philosophic theories antecedent to those of modern medicine, such morbid disturbances are explained as symptoms of divine visitation, or at least of superhuman spirituality. Among the strongest means of disturbing the functions of the mind so as to produce ecstatic vision, is fasting, accompanied as it so usually is with other privations, and with prolonged solitary contemplation in the desert or the forest. Among the ordinary vicissitudes of savage life, the wild hunter has many a time to try involuntarily the effects of such a life for days and weeks together, and under these circumstances he soon comes to see and talk with phantoms which are to him visible personal spirits. The secret of spiritual intercourse thus learnt, he has thenceforth but to reproduce the cause in order to renew the effects.

The rite of fasting, and the utter objective reality ascribed to what we call its morbid symptoms, are shown in striking details among the savage tribes of North America. Among the Indians (the accounts mostly refer to the Algonquin tribes), long and rigorous fasting is enjoined among boys and girls from a very early age; to be able to fast long is an enviable distinction, and they will abstain from food three to seven days, or even more, taking only a little water. During these fasts, especial attention is paid to dreams. Thus Tanner tells the story of a certain Net-no-kwa, who at twelve years old fasted ten successive days, till in a dream a man came and stood before her, and after speaking of many things gave her two sticks, saying, "I give you these to walk upon, and your hair I give it to be like snow;" this assurance of extreme old age was through life a support to her in times of danger and distress. At manhood the Indian lad, retiring to a solitary place to fast and meditate and pray, receives visionary impressions which stamp his character for life, and especially he waits till there appears to him in a dream some animal or thing which will be henceforth his "medicine," the fetish-representative of his manitu or protecting genius. For instance, an aged warrior who had thus in his youth dreamed of a bat coming to him, wore the skin of a bat on the crown of his head henceforth, and was all his life invulnerable to his enemies as a bat on the wing. In after life, an Indian who wants anything will fast till he has a dream that his manitu will grant it him. While the men are away hunting, the children are sometimes made to fast, that in their dreams they may obtain omens of the chase. Hunters fasting before an expedition are informed in dreams of the haunts of the game, and the means of appeasing the wrath of the bad spirits; if the dreamer fancies he sees an Indian who has been long dead, and hears him say, "If thou wilt sacrifice to me thou shalt shoot deer at pleasure," he will prepare a sacrifice, and burn the whole or part of a deer, in honour of the apparition. Especially the "meda" or "medicine-man" receives in fasts much of his qualification for his sacred office. The Ojibwa prophetess, known in after life as Catherine Wabose, in telling the story of her early years, relates how at the age of womanhood she fasted in her secluded lodge till she went up into the heavens and saw the spirit at the entrance, the Bright Blue Sky; this was the first supernatural communication of her prophetic career. The account given to Schoolcraft by Chingwauk, an Algonquin chief deeply versed in the mystic lore and picture-writing of his people, is as follows: "Chingwauk began by saying that the ancient Indians made a great merit of fasting. They fasted sometimes six or seven days, till both their bodies and minds became free and light, which prepared them to dream. The object of the ancient seers was to dream of the sun; as it was believed that such a dream would enable them to see everything on the earth. And by fasting long and thinking much on the subject, they generally succeeded. Fasts and dreams were at first attempted at an early age. What a young man sees and experiences during these dreams and fasts, is adopted by him as truth, and it becomes a principle to regulate his future life. He relies for success on these revelations. If he has been much favoured in his fasts, and the people believe that he has the art of looking into futurity, the path is open to the highest honours. The prophet, he continued, begins to try his power in secret, with only one assistant, whose testimony is necessary should he succeed. As he goes on, he puts down the figures of his dreams and revelations, by symbols, on bark or other material, till a whole winter is sometimes passed in pursuing the subject, and he thus has a record of his principal revelations. If what he predicts is verified, the assistant mentions it, and the record is then appealed to as proof of his prophetic power and skill. Time increases his fame. His *kee-keé-wins*, or records, are finally shown to the old people, who meet together and consult upon them, for the whole nation believe in these revelations. They in the end give their approval, and declare that he is gifted as a prophet—is inspired with wisdom, and is fit to lead the opinions of the nation. Such, he concluded, was the ancient custom, and the celebrated old war-captains rose to their power in this manner." It remains to say that among these American tribes, the "jossakeed" or soothsayer prepares himself by fasting and the use of the sweating-bath for the state of convulsive ecstasy in which he utters the dictates of his familiar spirits.[1]

The practice of fasting is described in other districts of the uncultured world as carried on to produce similar ecstasy and supernatural converse. The account by Roman Pane in the Life of Colon describes the practice in Hayti of fasting to obtain knowledge of future events from the spirits (cemi); and a century or two later, rigorous fasting formed part of the apprentice's preparation for the craft of "boyé" or sorcerer, evoker, consulter, propitiator, and exorciser of spirits.[2] The "keebèt" or conjurors of the Abipones were believed by the natives to be able to inflict disease and death, cure all disorders, make known distant and future events, cause rain, hail, and tempests, call up the shades of the dead, put on the form of tigers, handle serpents unharmed, etc. These powers were imparted by diabolical assistance, and Father Dobrizhoffer thus describes the manner of obtaining them:—" Those who aspire to the office of juggler are said to sit upon an aged willow, overhanging some lake, and to abstain from food for several days, till they begin to see into futurity. It always appeared probable to me that these rogues, from long fasting, contract a weakness of brain, a giddiness, and kind

---

[1] Tanner's 'Narrative,' p. 288. Loskiel, 'N. A. Ind.' part i. p. 76. Schoolcraft, 'Ind. Tribes,' part i. pp. 34, 113, 360, 391; part iii. p. 227. Catlin, 'N. A. Ind.' vol. i. p. 36. Charlevoix, 'Nouv. Fr.' vol. ii. p. 170; vol. vi. p. 67. Klemm, 'Cultur-Gesch.' vol. ii. p. 170. Waitz, 'Anthropologie,' vol. iii. pp. 206, 217.

[2] Colombo, 'Vita,' ch. xxv. Rochefort, 'Iles Antilles,' p. 501. See also Meiners, vol. ii. p. 143 (Guyana).

of delirium, which makes them imagine that they are gifted with superior wisdom, and give themselves out for magicians. They impose upon themselves first, and afterwards upon others."[1] The Malay, to make himself invulnerable, retires for three days to solitude and scanty food in the jungle, and if on the third day he dreams of a beautiful spirit descending to speak to him, the charm is worked.[2] The Zulu doctor qualifies himself for intercourse with the "amadhlozi," or ghosts, from whom he is to obtain direction in his craft, by spare abstemious diet, want, suffering, castigation, and solitary wandering, till fainting fits or coma bring him into direct intercourse with the spirits. These native diviners fast often, and are worn out by fastings, sometimes of several days' duration, when they become partially or wholly ecstatic, and see visions. So thoroughly is the connexion between fasting and spiritual intercourse acknowledged by the Zulus, that it has become a saying among them, "The continually stuffed body cannot see secret things." They have no faith in a fat prophet.[3]

The effects thus looked for and attained by fasting among uncultured tribes continue into the midst of advanced civilization. No wonder that, in the Hindu tale, king Vasavadatta and his queen after a solemn penance and a three days' fast should see Siva in a dream and receive his gracious tidings; no wonder that, in the actual experience of to-day, the Hindu yogi should bring on by fasting a state in which he can with bodily eyes behold the gods.[4] The Greek oracle-priests recognized fasting as a means of bringing on prophetic dreams and visions; the Pythia of Delphi herself fasted for inspiration; Galen remarks that fasting dreams are the clearer.[5] Through after ages, both cause

[1] Dobrizhoffer, 'Abipones,' vol. ii. p. 68.
[2] St. John, 'Far East,' vol. i. p. 144.
[3] Döhne, 'Zulu Dic.' s. v. 'nyanga;' Grout, 'Zulu-land,' p. 158; Callaway, 'Religion of Amazulu,' p. 387.
Somadeva Bhatta, tr. Brockhaus, vol. ii. p. 81. Meiners, vol. ii. p. 147.
[5] Maury, 'Magie,' etc., p. 237; Pausan. i. 34; Philostrat. Apollon. Tyan. i.; Galen. Comment. in Hippocrat. i.

and consequence have held their places in Christendom. Thus Michael the Archangel, with sword in right hand and scales in left, appears to a certain priest of Siponte, who during a twelvemonth's course of prayer and fasting had been asking if he would have a temple built in his honour:—

"precibus jejunia longis
Addiderat, totoque orans se afflixerat anno."[1]

Reading the narratives of the wondrous sights seen by St. Theresa and her companions, how the saint went in spirit into hell and saw the darkness and fire and unutterable despair, how she saw often by her side her good patrons Peter and Paul, how when she was raised in rapture above the grate at the nunnery where she was to take the sacrament, Sister Mary Baptist and others being present, they saw an angel by her with a golden fiery dart at the end whereof was a little fire, and he thrust it through her heart and bowels and pulled them out with it, leaving her wholly inflamed with a great love of God—the modern reader naturally looks for details of physical condition and habit of life among the sisterhood, and as naturally finds that St. Theresa was of morbid constitution and subject to trances from her childhood, in after life subduing her flesh by long watchings and religious discipline, and keeping severe fast during eight months of the year.[2] It is needless to multiply such mediæval records of fasts which have produced their natural effects in beatific vision—are they not written page after page in the huge folios of the Bollandists? So long as fasting is continued as a religious rite, so long its consequences in morbid mental exaltation will continue the old and savage doctrine that morbid phantasy is supernatural experience. Bread and meat would have robbed the ascetic of many an angel's visit; the opening of the refectory door must many a time have closed the gates of heaven to his gaze.

[1] Baptist. Mantuan. Fast. ix. 350.
[2] 'Acta Sanctorum Bolland.' S. Theresa.

It is indeed not the complete theory of fasting as a religious rite, but only an important and perhaps original part of it, that here comes into view. Abstinence from food has a principal place among acts of self-mortification or penance, a province of religious ordinance into which the present argument scarcely enters. Looking at the practice of fasting here from an animistic point of view, as a process of bringing on dreams and visions, it will be well to mention with it certain other means by which ecstatic phenomena are habitually induced.

One of these means is the use of drugs. In the West India Islands at the time of the discovery, Columbus describes the religious ceremony of placing a platter containing "cohoba" powder on the head of the idol, the worshippers then snuffing up this powder through a cane with two branches put to the nose. Pane further describes how the native priest, when brought to a sick man, would put himself in communication with the spirits by thus snuffing cohoba, "which makes him drunk, that he knows not what he does, and so says many extraordinary things, wherein they affirm that they are talking with the cemis, and that from them it is told them that the infirmity came." On the Amazons, the Omaguas have continued to modern times the use of narcotic plants, producing an intoxication lasting twenty-four hours, during which they are subject to extraordinary visions; from one of these plants they obtain the "curupa" powder which they snuff into their nostrils with a Y-shaped reed.[1] Here the similar names and uses of the drug plainly show historical connexion between the Omaguas and the Antilles islanders. The Californian Indians would give children narcotic potions, in order to gain from the ensuing visions information about their enemies; and thus the Mundrucus

[1] Colombo, 'Vita,' ch. lxii.; Roman Pane, ibid. ch. xv.; and in Pinkerton, vol. xii. Condamine, 'Travels,' in Pinkerton, vol. xiv. p. 226; Martius, 'Ethnog. Amer.' vol. i. pp. 441, 631 (details of snuff-powders among Omaguas, Otomacs, etc.; native names curupá, paricá, niopo, nupa; made from seeds of Mimosa acacioides, Acacia niopo).

of North Brazil, desiring to discover murderers, would administer such drinks to seers, in whose dreams the criminals appeared.[1] The Darien Indians used the seeds of the Datura sanguinea to bring on in children prophetic delirium, in which they revealed hidden treasure. In Peru the priests who talked with the "huaca" or fetishes used to throw themselves into an ecstatic condition by a narcotic drink called "tonca," made from the same plant, whence its name of "huacacacha" or fetish-herb.[2] The Mexican priests also appear to have used an ointment or drink made with seeds of "ololiuhqui," which produced delirium and visions.[3] In both Americas tobacco was used for such purposes. It must be noticed that smoking is more or less used among native races to produce full intoxication, the smoke being swallowed for the purpose. By smoking tobacco, the sorcerers of Brazilian tribes raised themselves to ecstasy in their convulsive orgies, and saw spirits; no wonder tobacco came to be called the "holy herb."[4] So North American Indians held intoxication by tobacco to be supernatural ecstasy, and the dreams of men in this state to be inspired.[5] This idea may explain a remarkable proceeding of the Delaware Indians. At their festival in honour of the Fire-god with his twelve attendant manitus, inside the house of sacrifice a small oven-hut was set up, consisting of twelve poles tied together at the top and covered with blankets, high enough for a man to stand nearly upright within it. After the feast this oven was heated with twelve red-hot stones, and twelve men crept inside. An old man threw twelve pipefulls of tobacco on these stones, and when the patients had borne to the utmost

[1] Maury, 'Magie,' etc. p. 425.
[2] Seemann, 'Voy. of Herald,' vol. i. p. 256. Rivero and Tschudi, 'Peruvian Antiquities,' p. 184. J. G. Müller, p. 397.
[3] Brasseur, 'Mexique,' vol. iii. p. 558; Clavigero, vol. ii. p. 40; J. G. Müller, p. 656.
[4] J. G. Müller, 'Amer. Urrelig.' p. 277; Hernandez, 'Historia Mexicana,' lib. v. c. 51; Purchas, vol. iv. p. 1292.
[5] D. Wilson, 'Prehistoric Man,' vol. i. p. 487.

the heat and suffocating smoke, they were taken out, generally falling in a swoon.[1] This practice, which was carried on in the last century, is remarkable for its coincidence with the Scythian mode of purification after a funeral, as described by Herodotus. He relates that they make their hut with three stakes sloping together at the top and covered in with woollen felts; then they cast red-hot stones into a trough placed within and throw hemp-seed on them, which sends forth fumes such as no Greek vapour-bath could exceed, and the Scyths in their sweating-hut roar with delight.[2]

Not to dwell on the ancient Aryan deification of an intoxicating drink, the original of the divine Soma of the Hindus and the divine Haoma of the Parsis, nor on the drunken orgies of the worship of Dionysos in ancient Greece, we find more exact Old World analogues of the ecstatic medicaments used in the lower culture. Such are the decoctions of thalassægle which Pliny speaks of as drunk to produce delirium and visions; the drugs mentioned by Hesychius, whereby Hekate was evoked; the mediæval witch-ointments which brought visionary beings into the presence of the patient, transported him to the witches' sabbath, enabled him to turn into a beast.[3] The survival of such practices is most thorough among the Persian dervishes of our own day. These mystics are not only opium-eaters, like so large a proportion of their countrymen; they are hashish-smokers, and the effect of this drug is to bring them into a state of exaltation passing into utter hallucination. To a patient in this condition, says Dr. Polak, a little stone in the road will seem a great block that he must stride over; a gutter becomes a wide stream to his eyes, and he calls for a boat to ferry him

---

[1] Loskiel, 'Ind. of N. A.' part i. p. 42.
[2] Herodot. iv. 73–5.
[3] Maury, 'Magie,' etc. l. c.; Plin. xxiv. 102; Hesych. s. v. 'ὠπήτειρα.' See also Bastian, 'Mensch,' vol. ii. p. 152, etc.; Baring-Gould, 'Werewolves,' p. 149.

RITES AND CEREMONIES. 419

across; men's voices sound like thunder in his ears; he fancies he has wings and can rise from the ground. These ecstatic effects, in which miracle is matter of hourly experience, are considered in Persia as high religious developments; the visionaries and their rites are looked on as holy, and they make converts.[1]

Many details of the production of ecstasy and swoon by bodily exercises, chanting and screaming, etc., have been incidentally given in describing the doctrine of demoniacal possession. I will only further cite a few typical cases to show that the practice of bringing on swoons or fits by religious exercises, in reality or pretence, is one belonging originally to savagery, whence it has been continued into higher grades of civilization. We may judge of the mental and bodily condition of the priest or sorcerer in Guyana, by his preparation for his sacred office. This consisted in the first place in fasting and flagellation of extreme severity; at the end of his fast he had to dance till he fell senseless, and was revived by a potion of tobacco-juice causing violent nausea and vomiting of blood; day after day this treatment was continued till the candidate, brought into or confirmed in the condition of a " convulsionary," was ready to pass from patient into doctor.[2] Again, at the Winnebago medicine-feast, members of the fraternity assemble in a long arched booth, and with them the candidates for initiation, whose preparation is a three days' fast, with severe sweating and steaming with herbs, under the direction of the old medicine-men. The initiation is performed in the assembly by a number of medicine-men. These advance in line, as many abreast as there are candidates; holding their medicine-bags before them with both hands, they dance forward slowly at first, uttering low guttural sounds as they approach the candidates, their step and voice increasing in energy, until with a violent " Ough!" they thrust their medicine-

[1] Polak, 'Persien,' vol. ii. p. 245 ; Vambéry in ' Mem. Anthrop. Soc.' vol. ii. p. 20 ; Meiners, vol. ii. p. 216.
[2] Meiners, vol. ii. p. 162.

bags at their breasts. Instantly, as if struck with an electric shock, the candidates fall prostrate on their faces, their limbs extended, their muscles rigid and quivering. Blankets are now thrown over them, and they are suffered to lie thus a few moments; as soon as they show signs of recovering from the shock, they are assisted to their feet and led forward. Medicine-bags are then put in their hands, and medicine-stones in their mouths; they are now medicine men or women, as the case may be, in full communion and fellowship; and they now go round the bower in company with the old members, knocking others down promiscuously by thrusting their medicine-bags at them. A feast and dance to the music of drum and rattle carry on the festival.[1] Another instance may be taken from among the Alfurus of Celebes, inviting Empong Lembej to descend into their midst. The priests chant, the chief priest with twitching and trembling limbs turns his eyes towards heaven; Lembej descends into him, and with horrible gestures he springs upon a board, beats about with a bundle of leaves, leaps and dances, chanting legends of an ancient deity. After some hours another priest relieves him, and sings of another deity. So it goes on day and night till the fifth day, and then the chief priest's tongue is cut, he falls into a swoon like death, and they cover him up. They fumigate with benzoin the piece taken from his tongue, and swing a censer over his body, calling back his soul; he revives and dances about, lively but speechless, till they give him back the rest of his tongue, and with it his power of speech.[2] Thus, in the religion of uncultured races, the phenomenon of being "struck" holds so recognized a position that impostors will even counterfeit it. In its morbid nature, its genuine cases at least plainly correspond with the fits which history records among the convulsionnaires of St. Medard and the enthusiasts of the Cevennes. Nor need we go even a gene-

[1] Schoolcraft, 'Indian Tribes,' part iii. p. 286.
[2] Bastian, 'Mensch,' vol. ii. p. 145. Compare 'Oestl. Asien,' vol ii. p. 247 (Aracan).

ration back to see symptoms of the same type accepted as signs of grace among ourselves. Medical descriptions of the scenes brought on by fanatical preachers at "revivals" in England, Ireland, and America, are full of interest to students of the history of religious rites. I will but quote a single case. "A young woman is described as lying extended at full length; her eyes closed, her hands clasped and elevated, and her body curved in a spasm so violent that it appeared to rest arch-like upon her heels and the back portion of her head. In that position she lay without speech or motion for several minutes. Suddenly she uttered a terrific scream, and tore handfuls of hair from her uncovered head. Extending her open hands in a repelling attitude of the most appalling terror, she exclaimed, 'Oh, that fearful pit!' During this paroxysm three strong men were hardly able to restrain her. She extended her arms on either side, clutching spasmodically at the grass, shuddering with terror, and shrinking from some fearful inward vision; but she ultimately fell back exhausted, nerveless, and apparently insensible."[1] Such descriptions carry us far back in the history of the human mind, showing modern men still in ignorant sincerity producing the very fits and swoons to which for untold ages savage tribes have given religious import. These manifestations in modern Europe indeed form part of a revival of religion, the religion of mental disease.

From this series of rites, practical with often harmful practicality, we turn to a group of ceremonies whose characteristic is picturesque symbolism. In discussing sun-myth and sun-worship, it has come into view how deeply the association in men's minds of the east with light and warmth, life and happiness and glory, of the west with darkness and chill, death and decay, has from remote ages rooted itself in religious belief. It will illustrate and confirm this view to observe how the same symbolism of east and west has taken shape in actual ceremony, giving rise to a series of practices

[1] D. H. Tuke in 'Journal of Mental Science,' Oct. 1870, p. 368.

concerning the posture of the dead in their graves and the living in their temples, practices which may be classed under the general heading of Orientation.

While the setting sun has shown to men, from savage ages onward, the western region of death, the rising sun has displayed a scene more hopeful, an eastern home of deity. It seems to be the working out of the solar analogy, on the one hand in death as sunset, on the other in new life as sunrise, that has produced two contrasted rules of burial, which agree in placing the dead in the sun's path, the line of east and west. Thus the natives of Australia have in some districts well-marked thoughts of the western land of the dead, yet the custom of burying the dead sitting with face to the east is also known among them.[1] The Samoans and Fijians, agreeing that the land of the departed lies in the far west, bury the corpse lying with head east and feet west;[2] the body would but have to rise and walk straight onward to follow its soul home. This idea is stated explicitly among the Winnebagos of North America; they will sometimes bury a dead man sitting up to the breast in a hole in the ground, looking westward; or graves are dug east and west, and the bodies laid in them with the head eastward, with the motive "that they may look towards the happy land in the west."[3] With these customs may be compared those of certain South American tribes. The Yumanas bury their dead bent double with faces looking toward the heavenly region of the sunrise, the home of their great good deity, who they trust will take their souls with him to his dwelling;[4] the Guarayos bury the corpses with heads turned to the east, for it is in the eastern sky that their god Tamoi, the Ancient of Heaven, has his happy hunting-grounds where the dead will meet again.[5]

[1] Grey, 'Australia,' vol. ii. p. 327.
[2] Turner, 'Polynesia,' p. 230. Seemann, 'Viti,' p. 151.
[3] Schoolcraft, 'Indian Tribes,' part iv. p. 54.
[4] Martius, 'Ethnog. Amer.' vol. i. p. 485.
[5] D'Orbigny, 'L'Homme Americain,' vol. ii. pp. 319, 330.

RITES AND CEREMONIES. 423

On the other hand the Peruvian custom was to place the dead huddled up in a sitting posture and with faces turned to the west.[1] Barbaric Asia may be represented by the modern Ainos of Yesso, burying the dead lying robed in white with the head to the east, " because that is where the sun rises ;" or by the Tunguz who bury with the head to the west; or by the mediæval Tatars, raising a great mound over the dead, and setting up thereon a statue with face turned toward the east, holding a drinking-cup in his hand before his navel; or by the modern Siamese, who do not sleep with their heads to the west, because it is in this significant position that the dead are burned.[2] The burial of the dead among the ancient Greeks in the line of east and west, whether according to Athenian custom of the head toward the sunset, or the converse, is another link in the chain of custom.[3] Thus it is not to late and isolated fancy, but to the carrying on of ancient and widespread solar ideas, that we trace the well-known legend that the body of Christ was laid with the head toward the west, thus looking eastward, and the Christian usage of digging graves east and west, which prevailed through mediæval times and is not yet forgotten. The rule of laying the head to the west, and its meaning that the dead shall rise looking toward the east, are perfectly stated in the following passage from an ecclesiastical treatise of the 16th century: " Debet autem quis sic sepeliri, ut capite ad occidentem posito, pedes dirigat ad orientem, in quo quasi ipsa positione orat: et innuit quod promptus est, ut de occasu festinet ad ortum : de mundo ad seculum."[4]

---

[1] Rivero and Tschudi, 'Peruvian Antiquities,' p. 202. See also Arbousset and Daumas, 'Voyage,' p. 277 (Kafirs).
[2] Bickmore, in 'Tr. Eth. Soc.' vol. vii. p. 20. Georgi, ' Reise,' vol. i. p. 266. Gul. de Rubruquis in Hakluyt, vol. i. p. 78. Bastian, 'Oestl. Asien,' vol. iii. p. 228.
[3] Ælian. Var. Hist. v. 14, vii. 19 ; Plutarch. Solon, x. ; Diog. Laert. Solon ; Welcker, vol. i. p. 404.
[4] Beda in Die S. Paschæ. Durand, Rationale Divinorum Officiorum, lib. vii. c. 35-9. Brand, 'Popular Antiquities,' vol. ii. pp. 295, 318.

Where among the lower races sun-worship begins to consolidate itself in systematic ritual, the orientation of the worshipper and the temple becomes usual and distinct. The sun-worshipping Comanches, preparing for the war-path, will place their weapons betimes on the east side of the lodge to receive the sun's first rays; it is a remnant of old solar rite, that the Christianized Pueblo Indians of New Mexico turn to the sun at his rising.[1] It has been already noticed how in old times each morning at sunrise the Sun-chief of the Natchez of Louisiana stood facing the east at the door of his house, and smoked toward the sun first, before he turned to the other three quarters of the world.[2] The cave-temple of the sun-worshipping Apalaches of Florida had its opening looking east, and within stood the priests on festival days at dawn, waiting till the first rays entered to begin the appointed rites of chant and incense and offering.[3] In old Mexico, where sun-worship was the central doctrine of the complex religion, men knelt in prayer towards the east, and the doors of the sanctuaries looked mostly westward.[4] It was characteristic of the solar worship of Peru that even the villages were habitually built on slopes toward the east, that the people might see and greet the national deity at his rising. In the temple of the sun at Cuzco, his splendid golden disc on the western wall looked out through the eastern door, so that as he rose his first beams fell upon it, reflected thence to light up the sanctuary.[5]

In Asia, the ancient Aryan religion of the sun manifests itself not less plainly in rites of orientation. They have their place in the weary ceremonial routine which the Brah-

[1] Gregg, 'Commerce of Prairies,' vol. i. pp. 270, 273; vol. ii. p. 318.
[2] Charlevoix, 'Nouvelle France,' vol. vi. p. 178.
[3] Rochefort, 'Iles Antilles,' p. 365.
[4] Clavigero, 'Messico,' vol. ii. p. 24; J. G. Müller, p. 641. See Oviedo, 'Nicaragua,' p. 29.
[5] J. G. Müller, p. 363; Prescott, 'Peru,' book i. ch. 3. Garcilaso de la Vega, 'Commentarios Reales,' lib. iii. c. 20, says it was at the east end; cf. lib. vi. c. 21 (llama sacrificed with head to east).

man must daily accomplish. When he has performed the dawn ablution, and meditated on the effulgent sun-light which is Brahma, the supreme soul, he proceeds to worship the sun, standing on one foot and resting the other against his ankle or heel, looking toward the east, and holding his hands open before him in a hollow form. At noon, when he has again adored the sun, it is sitting with his face to the east that he must read his daily portion of the Veda; it is looking toward the east that his offering of barley and water must be first presented to the gods, before he turns to north and south; it is with first and principal direction to the east that the consecration of the fire and the sacrificial implements, a ceremony which is the groundwork of all his religious acts, has to be performed.[1] The significance of such reverence paid by adorers of the sun to the glorious eastern region of his rising, may be heightened to us by setting beside it a ceremony of a darker faith, displaying the awe-struck horror of the western home of death. The antithesis to the eastward consecration by the orthodox Brahmans is the westward consecration by the Thugs, worshippers of Kali the death-goddess. In honour of Kali their victims were murdered, and to her the sacred pickaxe was consecrated, wherewith the graves of the slain were dug. At the time of the suppression of Thuggee, Englishmen had the consecration of the pickaxe performed in make-believe in their presence by those who well knew the dark ritual. On the dreadful implement no shadow of any living thing must fall, its consecrator sits facing the west to perform the fourfold washing and the sevenfold passing through the fire, and then, it being proved duly consecrated by the omen of the cocoa-nut divided at a single cut, it is placed on the ground, and the bystanders worship it, turning to the west.[2]

These two contrasted rites of east and west established

[1] Colebrooke, 'Essays,' vol. i., iv. and v.
[2] 'Illustrations of the History and Practices of the Thugs,' London, 1837, p. 46.

themselves and still remain established in modern European religion. In judging of the course of history that has brought about this state of things, it scarcely seems that Jewish influence was effective. The Jewish temple had the entrance in the east, and the sanctuary in the west. Sun-worship was an abomination to the Jews, and the orientation especially belonging to it appears as utterly opposed to Jewish usage, in Ezekiel's horror-stricken vision: "and, behold, at the door of the temple of Jehovah, between the porch and the altar, about five-and-twenty men, with their backs toward the temple of Jehovah, and their faces toward the east, and they worshipped the sun toward the east."[1] Nor is there reason to suppose that in later ages such orientation gained ground in Jewish ceremony. The solar rites of other nations whose ideas were prominent in the early development of Christianity, are sufficient to account for the rise of Christian orientation. On the one hand there was the Asiatic sun-worship, perhaps specially related to the veneration of the rising sun in old Persian religion, and which has left relics in the east of the Turkish empire into modern years; Christian sects praying toward the sun, and Yezidis turning to the east as their kibleh and burying their dead looking thither.[2] On the other hand, orientation was recognized in classic Greek religion, not indeed in slavish obedience to a uniform law, but as a principle to be worked out in converse ways. Thus it was an Athenian practice for the temple to have its entrance east, looking out through which the divine image stood to behold the rising sun. This rule it is that Lucian refers to, when he talks of the delight of gazing toward the loveliest and most longed-for of the day, of welcoming the sun as he peeps forth, of taking one's fill of light through the wide-open doors, even as the

[1] Ezek. viii. 16. Mishna, 'Sukkoth,' v. See Fergusson in Smith's 'Dictionary of the Bible,' s. v. 'Temple.'
[2] Hyde, 'Veterum Persarum Religionis Historia,' ch. iv. Niebuhr, 'Reisebeschreibung nach Arabien,' vol. i. p. 396. Layard, 'Nineveh,' vol. i. ch. ix.

ancients built their temples looking forth. Nor was the contrary rule as stated by Vitruvius less plain in meaning; the sacred houses of the immortal gods shall be so arranged, that if no reason prevents and choice is free, the temple and the statue erected in the cell shall look toward the west, so that they who approach the altar to sacrifice and vow and pray may look at once toward the statue and the eastern sky, the divine figures thus seeming to arise and look upon them. Altars of the gods were to stand toward the east.[1]

Unknown in primitive Christianity, the ceremony of orientation was developed within its first four centuries. It became an accepted custom to turn in prayer toward the east, the mystic region of the Light of the World, the Sun of Righteousness. Augustine says, "When we stand at prayer, we turn to the east, where the heaven arises, not as though God were only there, and had forsaken all other parts of the world, but to admonish our mind to turn to a more excellent nature, that is, to the Lord." No wonder that the early Christians were thought to practise in substance the rite of sun-worship which they practised in form. Thus Tertullian writes: "Others indeed with greater truth and verisimilitude believe the sun to be our God. . . . . the suspicion arising from its being known that we pray toward the region of the east." Though some of the most ancient and honoured churches of Christendom stand to show that orientation was no original law of ecclesiastical architecture, yet it became dominant in early centuries. That the author of the 'Apostolical Constitutions' should be able to give directions for building churches toward the east (ὁ οἶκος ἔστω ἐπιμήκης, κατ' ἀνατολὰς τετραμμένος), just as Vitruvius had laid down the rule as to the temples of the gods, is only a part of that assimilation of the church to the temple which took effect so largely in the scheme of worship. Of all Christian ceremony, however, it was in the rite of baptism that orientation took its fullest and most picturesque

[1] Lucian. De Domo, vi. Vitruv. de Architectura, iv. 5. See Welcker, vol. i. p. 403.

form. The catechumen was placed with face toward the west, and then commanded to renounce Satan with gestures of abhorrence, stretching out his hands against him, or smiting them together, and blowing or spitting against him thrice. Cyril of Jerusalem, in his 'Mystagogic Catechism,' thus depicts the scene: "Ye first came into the ante-room of the baptistery, and standing toward the west (πρὸς τὰς δυσμὰς) ye were commanded to put away Satan, stretching out your hands as though he were present. . . . . And why did ye stand toward the west? It was needful, for sunset is the type of darkness, and he is darkness and has his strength in darkness; therefore symbolically looking toward the west ye renounce that dark and gloomy ruler." Then turning round to the east, the catechumen took up his allegiance to his new master, Christ. The ceremony and its significance are clearly set forth by Jerome, thus: "In the mysteries [meaning baptism] we first renounce him who is in the west, and dies to us with our sins; and so, turning to the east, we make a covenant with the Sun of righteousness, promising to be his servants."[1] This perfect double rite of east and west, retained in the baptismal ceremony of the Greek Church, may be seen in Russia to this day. The orientation of churches and the practice of turning to the east as an act of worship, are common to both Greek and Latin ritual. In our own country they declined from the Reformation, till at the beginning of the present century they seemed falling out of use; since then, however, they have been restored to a certain prominence by the revived mediævalism of our own day. To the student of history, it is a striking example of the connexion of thought and rite through the religions of the lower and higher culture, to see surviving in our midst, with meaning dwindled into

---

[1] Augustin. de Serm. Dom. in Monte, ii. 5. Tertullian. Contra Valentin. iii.; Apolog. xvi. Constitutiones Apostolicæ, ii. 57. Cyril. Catech. Myst. i. 2. Hieronym. in Amos. vi. 14; Bingham, 'Antiquities of Chr. Church,' book viii. ch. 3, book xi. ch. 7, book xiii. ch. 8. J. M. Neale, 'Eastern Church,' part i. p. 956; Romanoff, 'Greco-Russian Church,' p. 67.

symbolism, this ancient solar rite. The influence of the divine Sun upon his rude and ancient worshippers still subsists before our eyes as a mechanical force, acting diamagnetically to adjust the axis of the church and turn the body of the worshipper.

The last group of rites whose course through religious history is to be outlined here, takes in the varied dramatic acts of ceremonial purification or Lustration. With all the obscurity and intricacy due to age-long modification, the primitive thought which underlies these ceremonies is still open to view. It is the transition from practical to symbolic cleansing, from removal of bodily impurity to deliverance from invisible, spiritual, and at last moral evil. Our language follows this ideal movement to its utmost stretch, where such words as cleansing and purification have passed from their first material meaning, to signify removal of ceremonial contamination, legal guilt, and moral sin. What we thus express in metaphor, the men of the lower culture began early to act in ceremony, purifying persons and objects by various prescribed rites, especially by dipping them in and sprinkling them with water, or fumigating them with and passing them through fire. It is the plainest proof of the original practicality of proceedings now passed into formalism, to point out how far the ceremonial lustrations still keep their connexion with times of life when real purification is necessary, how far they still consist in formal cleansing of the new-born child and the mother, of the manslayer who has shed blood, or the mourner who has touched a corpse. In studying the distribution of the forms of lustration among the races of the world, while allowing for the large effect of their transmission from religion to religion, and from nation to nation, we may judge that their diversity of detail and purpose scarcely favours a theory of their being all historically derived from one or even several special religions of the ancient world. They seem more largely to exemplify independent working out, in different directions, of an idea common to mankind at large. This view may be justified by surveying lustration through a series of typical instances, which show its appearance and character in savage and barbaric culture, as being an act belonging to certain well-marked events of human life.

The purification of the new-born child appears among the lower races in various forms, but perhaps in some particular instances borrowed from the higher. It should be noticed that though the naming of the child is often associated with its ceremonial cleansing, there is no real connexion between the two rites, beyond their coming due at the same early time of life. To those who look for the matter-of-fact origin of such ceremonies, one of the most suggestive of the accounts available is a simple mention of the two necessary acts of washing and name-giving, as done together in mere practical purpose, but not as yet passed into formal ceremony—the Kichtak Islanders, it is remarked, at birth wash the child, and give it a name.[1] Among the Yumanas of Brazil, as soon as the child can sit up, it is sprinkled with a decoction of certain herbs, and receives a name which has belonged to an ancestor.[2] Among some Jakun tribes of the Malay Peninsula, as soon as the child is born it is carried to the nearest stream and washed; it is then brought back to the house, the fire is kindled, and fragrant wood thrown on, over which it is passed several times.[3] The New Zealanders' infant baptism is no new practice, and is considered by them an old traditional rite, but nothing very similar is observed among other branches of the Polynesian race. Whether independently invented or not, it was thoroughly worked into the native religious scheme. The baptism was performed on the eighth day or earlier, at the side of a stream or elsewhere, by a native priest who sprinkled water on the child with a branch or twig; sometimes the child was immersed. With this lustration it received its name, the priest repeating a list of

[1] Billings, 'N. Russia,' p. 175.
[2] Martius, 'Ethnog. Amer.' vol. ⁚. p. 485.
[3] 'Journ. Ind. Archip.' vol. ii. p. 264.

ancestral names till the child chose one for itself by sneezing at it. The ceremony is of the nature of a dedication, and was accompanied by rhythmical formulas of exhortation. The future warrior was bidden to flame with anger, to leap nimbly and ward off the spears, to be angry and bold and industrious, to work before the dew is off the ground; the future housewife was bidden to get food and go for firewood and weave garments with panting of breath. In after years, a second sacred sprinkling was performed to admit a lad into the rank of warriors. It has to be noticed with reference to the reason of this ceremonial washing, that a new-born child is in the highest degree tapu, and may only be touched by a few special persons till the restriction is removed.[1] In Madagascar, a fire is kept up in the room for several days, then the child in its best clothes is in due form carried out of the house and back to its mother, both times being carefully lifted over the fire, which is made near the door.[2] In Africa, some of the most noticeable ceremonies of the class are these. The people of Sarac wash the child three days after birth with holy water.[3] When a Mandingo child was about a week old its hair was cut, and the priest, invoking blessings, took it in his arms, whispered in its ear, spat thrice in its face, and pronounced its name aloud before the assembled company.[4] In Guinea, when a child is born, the event is publicly proclaimed, the new-born babe is brought into the streets, and the headman of the town or family sprinkles it with water from a basin, giving it a name and invoking blessings of health and wealth upon it; other friends follow the example, till the child is thoroughly drenched.[5] In these various examples

[1] Taylor, 'New Zealand,' p. 184; Yate, p. 82; Polack, vol. i. p. 51; A. S. Thomson, vol. i. p. 118; Klemm, 'Cultur-Gesch.' vol. iv. p. 304. See Schirren, 'Wandersagen der Neuseeländer,' pp. 58, 183; Shortland, p. 145.
[2] Ellis, 'Madagascar,' vol. i. p. 152.
[3] Munzinger, 'Ost-Afrika,' p. 387.
[4] Park, 'Travels,' ch. vi.
[5] J. L. Wilson, 'Western Africa,' p. 399. See also Bastian, 'Mensch,'

of lustration of infants, the purifications by fire have the most importance ethnologically, not because this proceeding is more natural to the savage mind than that of bathing or sprinkling with water, but because this latter ceremony may have been imitated from Christian baptism. Thus, while there is nothing to prevent our supposing some rites of savage baptism to be of native origin, it seems unsafe to assert this in any individual case.

The purification of women at childbirth, etc., is ceremonially practised by the lower races under circumstances which do not suggest adoption from more civilized nations. The seclusion and lustration among North American Indian tribes have been compared with those of the Levitical law, but the resemblance is not remarkably close, and belongs rather to a stage of civilization than to the ordinance of a particular nation. It is a good case of independent development in such customs, that the rite of putting out the fires and kindling "new fire" on the woman's return is common to the Iroquois and Sioux in North America,[1] and the Basutos in South Africa. These latter have a well-marked rite of lustration by sprinkling, performed on girls at womanhood.[2] The Hottentots considered mother and child unclean till they had been washed and smeared after the uncleanly native fashion.[3] Lustrations with water were usual in West Africa.[4] Tatar tribes in Mongolia used bathing, while in Siberia the custom of leaping over a fire answered the purpose of purification.[5] The Mantras of the Malay Peninsula have made the bathing of the mother after

vol. ii. p. 279 (Watje); 'Anthropological Review,' Nov. 1864, p. 243 (Mpongwe); Barker-Webb and Berthelot, vol. ii. p. 163 (Tenerife).
[1] Schoolcraft, 'Indian Tribes,' part i. p. 261; part iii. p. 243, etc. Charlevoix, 'Nouvelle France,' vol. v. p. 425. Wilson in 'Tr. Eth. Soc.' vol. iv. p. 294.
[2] Casalis, 'Basutos,' p. 267.
[3] Kolben, vol. i. pp. 273, 283.
[4] Bosman, in 'Pinkerton,' vol. xvi. pp. 423, 527; Meiners, vol. ii. pp. 107, 463.
[5] Pallas, 'Mongolische Völkerschaften,' vol. i. p. 166, etc.; Strahlenberg, 'Siberia,' p. 97.

childbirth into a ceremonial ordinance.[1] It is so among the indigenes of India, where both in northern and southern districts the naming of the child comes into connexion with the purification of the mother, both ceremonies being performed on the same day.[2] Without extending further this list of instances, it is sufficiently plain that we have before us the record of a practical custom becoming consecrated by traditional habit, and making its way into the range of religious ceremony.

Much the same may be said of the purification of savage and barbaric races on occasion of contamination by bloodshed or funeral. In North America, the Dacotas use the vapour-bath not only as a remedy, but also for the removal of ceremonial uncleanness, such as is caused by killing a person, or touching a dead body.[3] So among the Navajos, the man who has been deputed to carry a dead body to burial, holds himself unclean until he has thoroughly washed himself in water prepared for the purpose by certain ceremonies.[4] In Madagascar, no one who has attended a funeral may enter the palace courtyard till he has bathed, and in all cases there must be an ablution of the mourner's garments on returning from the grave.[5] Among the Basutos of South Africa, warriors returning from battle must rid themselves of the blood they have shed, or the shades of their victims would pursue them and disturb their sleep. Therefore they go in procession in full armour to the nearest stream to wash, and their weapons are washed also. It is usual in this ceremony for a sorcerer higher up the stream to put in some magical ingredient, such as he also uses in the preparation of the holy water which is sprinkled over the people with a beast's tail at the frequent public purifications. These Basutos, moreover, use fumigation with burning wood to purify growing corn, and cattle taken from the

[1] Bourien in 'Tr. Eth. Soc.' vol. iii. p. 81.
[2] Dalton in 'Tr. Eth. Soc.' vol. vi. p. 22; Shortt, ibid. vol. iii. p. 375.
[3] Schoolcraft, 'Indian Tribes,' part i. p. 255.
[4] Brinton, 'Myths of New World,' p. 127.
[5] Ellis, 'Madagascar,' vol. i. p. 241; see 407, 419.

434   RITES AND CEREMONIES.

enemy. Fire serves for purification in cases too trifling to require sacrifice; thus when a mother sees her child walk over a grave, she hastens to call it, makes it stand before her, and lights a small fire at its feet.[1] The Zulus, whose horror of a dead body will induce them to cast out and leave in the woods their sick people, at least strangers, purify themselves by an ablution after a funeral. It is to be noticed that these ceremonial practices have come to mean something distinct from mere cleanliness. Kafirs who will purify themselves from ceremonial uncleanness by washing, are not in the habit of washing themselves or their vessels for ordinary purposes, and the dogs and the cockroaches divide between them the duty of cleaning out the milk-baskets.[2] Mediæval Tatar tribes, some of whom had conscientious scruples against bathing, have found passing through fire or between two fires a sufficient purification, and the household stuff of the dead was lustrated in this latter way.[3]

In the organized nations of the semi-civilized and civilized world, where religion shapes itself into elaborate and systematic schemes, the practices of lustration familiar to the lower culture now become part of stringent ceremonial systems. It seems to be at this stage of their existence that they often take up in addition to their earlier ceremonial significance an ethical meaning, absent or all but absent from them at their first appearance above the religious horizon. This will be made evident by glancing over the ordinances of lustration in the great national religions of history. It will be well to notice first the usages of two semi-civilized nations of America, which, though they have scarcely produced practical effect on civilization at large, give valuable illustration of a transition period in culture, leaving apart the obscure question of their special civiliza-

[1] Casalis, 'Basutos,' p. 258.
[2] Grout, 'Zulu-land,' p. 147; Backhouse, 'Mauritius and S. Africa,' pp. 213, 225.
[3] Bastian, 'Mensch,' vol. iii. p. 75; Rubruquis, in Pinkerton, vol. vii. p. 82; Plano Carpini in Hakluyt, vol. i. p. 37.

tion having been influenced in early or late times from the Old World.

In the religion of Peru, lustration is well-marked and characteristic. On the day of birth, the water in which the child had been washed was poured into a hole in the ground, charms being repeated by a wizard or priest; an excellent instance of the ceremonial washing away of evil influences. The naming of the child was also more or less generally accompanied with ceremonial washing, as in districts where at two years old it was weaned, baptized, had its hair ceremonially cut with a stone knife, and received its child-name; Peruvian Indians still cut off a lock of the child's hair at its baptism. Moreover, the significance of lustration as removing guilt is plainly recorded in ancient Peru; after confession of guilt, an Inca bathed in a neighbouring river and repeated this formula, "O thou River, receive the sins I have this day confessed unto the Sun, carry them down to the sea, and let them never more appear."[1] In old Mexico, the first act of ceremonial lustration took place at birth. The nurse washed the infant in the name of the water-goddess, to remove the impurity of its birth, to cleanse its heart and give it a good and perfect life; then blowing on water in her right hand she washed it again, warning it of forthcoming trials and miseries and labours, and praying the invisible Deity to descend upon the water, to cleanse the child from sin and foulness, and to deliver it from misfortune. The second act took place some four days later, unless the astrologers postponed it. At a festive gathering, amid fires kept alight from the first ceremony, the nurse undressed the child sent by the gods into this sad and doleful world, bade it receive the life-giving water, and washed it, driving out evil from each limb and offering to the deities appointed prayers for virtue and blessing. It

---

[1] Rivero and Tschudi, 'Peruvian Antiquities,' p. 180; J. G. Müller, 'Amer. Urrelig.' p. 389; Acosta, 'Ind. Occ.' v. c. 25; Brinton, p. 126. See account of the rite of driving out sicknesses and evils into the rivers, 'Rites and Laws of Incas,' tr. and ed. by C. R. Markham, p. 22.

was then that the toy instruments of war or craft or household labour were placed in the boy's or girl's hand (a custom singularly corresponding with one usual in China), and the other children, instructed by their parents, gave the newcomer its child-name, here again to be replaced by another at manhood or womanhood. There is nothing unlikely in the statement that the child was also passed four times through the fire, but the authority this is given on is not sufficient. The religious character of ablution is well shown in Mexico by its forming part of the daily service of the priests. Aztec life ended as it had begun, with ceremonial lustration; it was one of the funeral ceremonies to sprinkle the head of the corpse with the lustral water of this life.[1]

Among the nations of East Asia, and across the more civilized Turanian districts of Central Asia, ceremonial lustration comes frequently into notice; but it would often bring in difficult points of ethnography to attempt a general judgment how far these may be native local rites, and how far ceremonies adopted from foreign religious systems. As examples may be mentioned in Japan the sprinkling and naming of the child at a month old, and other lustrations connected with worship;[2] in China the religious ceremony at the first washing of the three days' old infant, the lifting of the bride over burning coals, the sprinkling of holy-water over sacrifices and rooms and on the mourners after a funeral;[3] in Birma the purification of the mother by fire, and the annual sprinkling-festival.[4] Within the range of Buddhism in its Lamaist form, we find such instances as the Tibetan and

---

[1] Sahagun, 'Nueva España,' lib. vi.; Torquemada, 'Monarquia Indiana,' lib. xii.; Clavigero, vol. ii. pp. 39, 86, etc.; Humboldt, 'Vues des Cordillères,' Mendoza Cod.; J. G. Müller, p. 652.

[2] Siebold, 'Nippon,' v. p. 22; Kempfer, 'Japan,' ch. xiii. in Pinkerton, vol. vii.

[3] Doolittle, 'Chinese,' vol. i. p. 120, vol. ii. p. 273. Davis, vol. i. p. 269.

[4] Bastian, 'Oestl. Asien,' vol. ii. p. 247; Meiners, vol. ii. p. 106; Symes in Pinkerton, vol. ix. p. 435.

Mongol lustration of the child a few days after birth, the lama blessing the water and immersing the child thrice, and giving its name; the Buraet consecration by threefold washing; the Tibetan ceremony where the mourners returning from the funeral stand before the fire, wash their hands with warm water over the hot coals, and fumigate themselves thrice with proper formulas.[1] With this infant baptism of Tibetans and Mongols may be compared the rite of their ethnological kinsfolk in Europe. The Lapps in their semi-Christianized state had a private form of baptism, in which a new name was given with a three-fold sprinkling and washing with warm water where mystic alder-twigs were put; this ceremony they called by the name of "laugo" or bathing, a word not of Lapp but Scandinavian origin; it might be repeatedly performed, and was considered a thoroughly native Lapp proceeding, utterly distinct from the Christian baptism to which the Lapps also conformed.[2] It is, however, the easiest ethnographic explanation of these two baptismal ceremonies in Central Asia and Northern Europe, to suppose imitation of Christianity either entirely bringing in a new rite, or adapting a previous native one.

Other Asiatic districts show lustration in more compact and characteristic religious developments. The Brahman leads a life marked by recurring ceremonial purification, from the time when his first appearance in the world brings uncleanness on the household, requiring ablution and clean garments to remove it, and thenceforth through his years from youth to old age, where bathing is a main part of the long minute ceremonial of daily worship, and further washings and aspersions enter into more solemn religious acts, till at last the day comes when his kinsfolk, on their way home from his funeral, cleanse themselves by a final bath

[1] Köppen, 'Religion des Buddha,' vol. ii. p. 320; Bastian, 'Psychologie,' pp. 151, 211, 'Mensch,' vol. ii. p. 499.
[2] Leems, 'Lapland,' in Pinkerton, vol. i. p. 483; Klemm, 'Cultur-Gesch.' vol. iii. p. 77.

from their contamination by his remains. For the means of some of his multifarious lustrations the Hindu has recourse to the sacred cow, but his more frequent medium of removing uncleanness of body and soul is water, the divine waters to which he prays, "Take away, O Waters, whatsoever is wicked in me, what I have done by violence or curse, and untruth!"[1] The Parsi religion prescribes a system of lustrations which well shows its common origin with that of Hinduism by its similar use of cow's urine and of water. Bathing or sprinkling with water, or applications of "nirang" washed off with water, form part of the daily religious rites, as well as of such special ceremonies as the naming of the new-born child, the putting on of the sacred cord, the purification of the mother after childbirth, the purification of him who has touched a corpse, when the unclean demon, driven by sprinkling of the good water from the top of the head and from limb to limb, comes forth at the left toe and departs like a fly to the evil region of the north. It is, perhaps, the influence of this ancestral religion, even more than the actual laws of Islam, that makes the modern Persian so striking an example of the way in which ceremony may override reality. It is rather in form than in fact that his cleanliness is next to godliness. He carries the principle of removing legal uncleanness by ablution so far, that a holy man will wash his eyes when they have been polluted by the sight of an infidel. He will carry about a water-pot with a long spout for his ablutions, yet he depopulates the land by his neglect of the simplest sanitary rules, and he may be seen by the side of the little tank where scores of people have been in before him, obliged to clear with his hand a space in the foul scum on the water, before he plunges in to obtain ceremonial purity.[2]

[1] Ward, 'Hindoos,' vol. ii. pp. 96, 246, 337; Colebrooke, 'Essays,' vol. ii. Wuttke, 'Gesch. des Heidenthums,' vol. ii. p. 378. 'Rig-Veda,' i. 22, 23.

[2] Avesta, Vendidad, v.-xii.; Lord, in Pinkerton, vol. viii. p. 570; Naoroji, 'Parsee Religion;' Polak, 'Persien,' vol. i. p. 355, etc., vol. ii. p. 271. Meiners, vol. ii. p. 125.

Over against the Aryan rites of lustration in the religions of Asia, may be set the well-known types in the religions of classic Europe. At the Greek amphidromia, when the child was about a week old, the women who had assisted at the birth washed their hands, and afterwards the child was carried round the fire by the nurse, and received its name ; the Roman child received its prænomen with a lustration at about the same age, and the custom is recorded of the nurse touching its lips and forehead with spittle. To wash before an act of worship was a ceremony handed down by Greek and Roman ritual through the classic ages; καθαραῖς δὲ δρόσοις, ἀφυδρανάμενοι στείχετε ναούς — eo lavatum, ut sacrificem. The holy-water mingled with salt, the holy-water vessel at the temple entrance, the brush to sprinkle the worshippers, all belong to classic antiquity. Romans, their flocks and herds and their fields, were purified from disease and other ill by lustrations which show perfectly the equivalent nature of water and fire as means of purification ; the passing of flocks and shepherds through fires, the sprinkling water with laurel branches, the fumigating with fragrant boughs and herbs and sulphur, formed part of the rustic rites of the Palilia. Bloodshed demanded the lustral ceremony. Hektor fears to pour with unwashen hands the libation of dark wine, nor may he pray bespattered with gore to cloud-wrapped Zeus ; Æneas may not touch the household gods till cleansed from slaughter by the living stream. It was with far changed thought that Ovid wrote his famous reproof of his too-easy countrymen, who fancied that water could indeed wash off the crime of blood :—

"Ah nimium faciles, qui tristia crimina cædis
Fluminea tolli posse putetis aqua."

Thus, too, the mourner must be cleansed by lustration from the contaminating presence of death. At the door of the Greek house of mourning was set the water-vessel, that those who had been within might sprinkle themselves and be clean; while the mourners returning from a Roman funeral, aspersed with water and stepping over fire, were this double process made pure.[1]

The ordinances of purification in the Levitical law relat especially to the removal of legal uncleanness connected with childbirth, death, and other pollutions. Washing was prescribed for such purposes, and also sprinkling with water of separation, water mingled with the ashes of the red heifer. Ablution formed part of the consecration of priests, and without it they might not serve at the altar nor enter the tabernacle. In the later ages of Jewish national history, perhaps through intercourse with nations whose lustrations entered more into the daily routine of life, ceremonial washings were multiplied. It seems also that in this period must be dated the ceremony which in after ages has held so great a place in the religion of the world, their rite of baptism of proselytes.[2] The Moslem lustrations are ablutions with water, or in default with dust or sand, performed partially before prayer, and totally on special days or to remove special uncleanness. They are strictly religious acts, belonging in principle to prevalent usage of Oriental religion; and their details, whether invented or adopted as they stand in Islam, are not carried down from Judaism or Christianity.[3] The rites of lustration which have held and hold their places within the pale of Christianity are in well-marked historical connexion with Jewish and Gentile ritual. Purification by fire has only appeared as an actual ceremony

---

[1] Details in Smith's 'Dic. of Gr. and Rom. Ant.' and Pauly, 'Real Encyclopedie,' s. v. 'amphidromia,' 'lustratio,' 'sacrificium,' 'funus;' Meiners, 'Gesch. der Religionen,' book vii.; Lomeyer, 'De Veterum Gentilium Lustrationibus;' Montfaucon, 'L'Antiquité Expliquée,' etc. Special passages: Homer. Il. vi. 266; Eurip. Ion. 96; Theocrit. xxiv. 95; Virg. Æn. ii. 719; Plaut. Aulular. iii. 6; Pers. Sat. ii. 31; Ovid. Fast. i. 669, ii. 45, iv. 727; Festus, s. v. 'aqua et ignis,' etc. The obscure subject of lustration in the mysteries is here left untouched.

[2] Ex. xxix. 4, xxx. 18, xl. 12; Lev. viii. 6, xiv. 8, xv. 5, xxii. 6; Numb. xix., etc.; Lightfoot in 'Works,' vol. xi.; Browne in Smith's 'Dic. of the Bible,' s. v. 'baptism;' Calmet, 'Dic.' etc.

[3] Reland, 'De Religione Mohammedanica;' Lane, 'Modern Eg.' vol. i. p. 98, etc.

among some little-known Christian sects, and in the European folk-lore custom of passing children through or over fire, if indeed we can be sure that this rite is lustral and not sacrificial.[1] The usual medium of purification is water. Holy water is in full use through the Greek and Roman churches. It blesses the worshipper as he enters the temple, it cures disease, it averts sorcery from man and beast, it drives demons from the possessed, it stops the spirit-writer's pen, it drives the spirit-moved table it is sprinkled upon to dash itself frantically against the wall; at least these are among the powers attributed to it, and some of the most striking of them have been lately vouched for by papal sanction. This lustration with holy water so exactly continues the ancient classic rite, that its apologists are apt to explain the correspondence by arguing that Satan stole it for his own wicked ends.[2] Catholic ritual follows ancient sacrificial usage in the priest's ceremonial washing of hands before mass. The priest's touching with his spittle the ears and nostrils of the infant or catechumen, saying, "Ephphetha," is obviously connected with passages in the Gospels; its adoption as a baptismal ceremony has been compared, perhaps justly, with the classical lustration by spittle.[3] Finally, it has but to be said that ceremonial purification as a Christian act centres in baptism by water, that symbol of initiation of the convert which history traces from the Jewish rite to that of John the Baptist, and thence to the Christian ordinance. Through later ages adult baptism carries on the Jewish ceremony of the admission of the proselyte, while infant baptism combines this with the lustration of the new-born infant. Passing through a range of meaning such as separates the sacrament of the Roman

[1] Bingham, 'Antiquities of Christian Church,' book xi. ch. 2. Grimm, 'Deutsche Mythologie,' p. 592; Leslie, 'Early Races of Scotland,' vol. i. p. 113; Pennant, in Pinkerton, vol. iii. p. 383.
[2] Rituale Romanum; Gaume, 'L'Eau Bénite;' Middleton, 'Letter from Rome,' etc.
[3] Rituale Romanum. Bingham, book x. ch. 2, book xv. ch. 8. See Mark vii. 34, viii. 23; John ix. 6.

centurion from the sacrament of the Roman cardinal, becoming to some a solemn symbol of new life and faith, to some an act in itself of supernatural efficacy, the rite of baptism has remained almost throughout the Christian world the outward sign of the Christian profession.

In considering the present group of religious ceremonies, their manifestations in the religions of the higher nations have been but scantily outlined in comparison with their rudimentary forms in the lower culture. Yet this reversal of the proportions due to practical importance in no way invalidates, but rather aids, the ethnographic lessons to be drawn by tracing their course in history. Through their varied phases of survival, modification, and succession, they have each in its own way brought to view the threads of continuity which connect the faiths of the lower with the faiths of the higher world; they have shown how hardly the civilized man can understand the religious rites even of his own land without knowledge of the meaning, often the widely unlike meaning, which they bore to men of distant ages and countries, representatives of grades of culture far different from his.

CHAPTER XIX.

CONCLUSION.

Practical results of the study of Primitive Culture—Its bearing least upon Positive Science, greatest upon Intellectual, Moral, Social, and Political Philosophy—Language—Mythology—Ethics and Law—Religion—Action of the Science of Culture, as a means of furthering progress and removing hindrance, effective in the course of Civilization.

IT now remains, in bringing to a close these investigations on the relation of primitive to modern civilization, to urge the practical import of the considerations raised in their course. Granted that archæology, leading the student's mind back to remotest known conditions of human life, shows such life to have been of unequivocally savage type; granted that the rough-hewn flint hatchet, dug out from amidst the bones of mammoths in a drift gravel-bed to lie on an ethnologist's writing-table, is to him a very type of primitive culture, simple yet crafty, clumsy yet purposeful, low in artistic level yet fairly started on the ascent toward highest development—what then? Of course the history and pre-history of man take their proper places in the general scheme of knowledge. Of course the doctrine of the world-long evolution of civilization is one which philosophic minds will take up with eager interest, as a theme of abstract science. But beyond this, such research has its practical side, as a source of power destined to influence the course of modern ideas and actions. To establish a connexion between what uncultured ancient men thought and did, and what cultured modern men think and do, is not a matter of inapplicable theoretic knowledge, for it raises the issue, how far are modern opinion and conduct based on the strong ground of soundest modern knowledge, or how far only on such knowledge as was available in the earlier and ruder stages of culture where their types were shaped. It has to be maintained that the early history of man has its bearing, almost ignored as that bearing has been by those whom it ought most stringently to affect, on some of the deepest and most vital points of our intellectual, industrial, and social state.

Even in advanced sciences, such as relate to measure and force and structure in the inorganic and organic world, it is at once a common and a serious error to adopt the principle of letting bygones be bygones. Were scientific systems the oracular revelations they sometimes all but pretend to be, it might be justifiable to take no note of the condition of mere opinion or fancy that preceded them. But the investigator who turns from his modern text-books to the antiquated dissertations of the great thinkers of the past, gains from the history of his own craft a truer view of the relation of theory to fact, learns from the course of growth in each current hypothesis to appreciate its raison d'être and full significance, and even finds that a return to older starting-points may enable him to find new paths, where the modern track seems stopped by impassable barriers. It is true that rudimentary conditions of arts and sciences are often rather curious than practically instructive, especially because the modern practitioner has kept up, as mere elementary processes, the results of the ancient or savage man's most strenuous efforts. Perhaps our toolmakers may not gain more than a few suggestive hints from a museum of savage implements, our physicians may only be interested in savage recipes so far as they involve the use of local drugs, our mathematicians may leave to the infant-school the highest flights of savage arithmetic, our astronomers may only find in the star-craft of the lower races an uninstructive combination of myth and commonplace. But there are departments of knowledge, of not less consequence than mechanics and medicine, arithmetic and astronomy, in which the study of the lowest stages, as influencing the practical acceptance of the higher, cannot be thus carelessly set aside.

If we survey the state of educated opinion, not within the limits of some special school, but in the civilized world at large, on such subjects especially as relate to Man, his intellectual and moral nature, his place and function among his fellow men and in the universe at large, we see existing side by side, as if of equal right, opinions most diverse in real authority. Some, vouched for by direct and positive evidence, hold their ground as solid truths. Others, though founded on crudest theories of the lower culture, have been so modified under the influence of advancing knowledge, as to afford a satisfactory framework for recognized facts; and positive science, mindful of the origin of its own philosophic schemes, must admit the validity of such a title. Others, lastly, are opinions belonging properly to lower intellectual levels, which have held their place into the higher by mere force of ancestral tradition; these are survivals. Now it is the practical office of ethnography to make known to all whom it may concern the tenure of opinions in the public mind, to show what is received on its own direct evidence, what is ruder ancient doctrine reshaped to answer modern ends, and what is but timehonoured superstition in the garb of modern knowledge.

Topic after topic shows at a glimpse the way in which ethnography bears on modern intellectual conditions. Language, appearing as an art in full vigour among rude tribes, already displays the adaptation of childlike devices in self-expressive sound and pictorial metaphor, to utter thoughts as complex and abstruse as savage minds demand speech for. When we consider how far the development of knowledge depends on full and exact means of expressing thought, is it not a pregnant consideration that the language of civilized men is but the language of savages, more or less improved in structure, a good deal extended in vocabulary, made more precise in the dictionary definition of words?

The development of language between its savage and cultured stages has been made in its details, scarcely in its principle. It is not too much to say that half the vast defect of language as a method of utterance, and half the vast defect of thought as determined by the influence of language, are due to the fact that speech is a scheme worked out by the rough and ready application of material metaphor and imperfect analogy, in ways fitting rather the barbaric education of those who formed it, than our own. Language is one of those intellectual departments in which we have gone too little beyond the savage stage, but are still as it were hacking with stone celts and twirling laborious friction-fire. Metaphysical speculation, again, has been one of the potent influences on human conduct, and although its rise, and one may almost say also its decline and fall, belong to comparatively civilized ages, yet its connexion with lower stages of intellectual history may to some extent be discerned. For example, attention may be recalled to a special point brought forward in this work, that one of the greatest of metaphysical doctrines is a transfer to the field of philosophy from the field of religion, made when philosophers familiar with the conception of object-phantoms used this to provide a doctrine of thought, thus giving rise to the theory of ideas. Far more fully and distinctly, the study of the savage and barbaric intellect opens to us the study of Mythology. The evidence here brought together as to the relation of the savage to the cultured mind in the matter of mythology has, I think, at any rate demonstrated this. With a consistency of action so general as to amount to mental law, it is proved that among the lower races all over the world the operation of outward events on the inward mind leads not only to statement of fact, but to formation of myth. It gives no unimportant clues to the student of mental history, to see by what regular processes myths are generated, and how, growing by wear and increasing in value at secondhand, they pass into pseudo-historic legend. Poetry is full of myth, and he who will understand it analytically will do well to study it ethnographically. In so far as myth, seriously or sportively meant, is the subject of poetry, and in so far as it is couched in language whose characteristic is that wild and rambling metaphor which represents the habitual expression of savage thought, the mental condition of the lower races is the key to poetry—nor is it a small portion of the poetic realm which these definitions cover. History, again, is an agent powerful and becoming more powerful, in shaping men's minds, and through their minds their actions in the world; now one of the most prominent faults of historians is that, through want of familiarity with the principles of myth-development, they cannot apply systematically to ancient legend the appropriate tests for separating chronicle from myth, but with few exceptions are apt to treat the mingled mass of tradition partly with undiscriminating credulity and partly with undiscriminating scepticism. Even more injurious is the effect of such want of testing on that part of traditional or documentary record which, among any section of mankind, stands as sacred history. It is not merely that in turning to the index of some book on savage tribes, one comes on such a suggestive heading as this, "Religion—*see* Mythology." It is that within the upper half of the scale of civilization, among the great historic religions of the world, we all know that between religion and religion, and even to no small extent between sect and sect, the narratives which to one side are sacred history, may seem to the other mythic legend. Among the reasons which retard the progress of religious history in the modern world, one of the most conspicuous is this, that so many of its approved historians demand from the study of mythology always weapons to destroy their adversaries' structures, but never tools to clear and trim their own. It is an indispensable qualification of the true historian that he shall be able to look dispassionately on myth as a natural and regular product of the human mind, acting on appropriate facts in a manner suited to the intellectual state of the people producing it, and that he shall treat it as an accretion to be deducted from professed history, whenever it is recognized by the tests of being decidedly against evidence as fact, and at the same time clearly explicable as myth. It is from the ethnographic study of savage and barbaric races that the knowledge of the general laws of myth-development, required for the carrying out of this critical process, may be best or must necessarily be gained.

The two vast united provinces of Morals and Law have been as yet too imperfectly treated on a general ethnographic scheme, to warrant distinct statement of results. Yet thus much may be confidently said, that where the ground has been even superficially explored, every glimpse reveals treasures of knowledge. It is already evident that inquirers who systematically trace each department of moral and legal institutions from the savage through the barbaric and into the civilized condition of mankind, thereby introduce into the scientific investigations of these subjects an indispensable element which merely theoretical writers are apt unscrupulously to dispense with. The law or maxim which a people at some particular stage of its history might have made fresh, according to the information and circumstance of the period, is one thing. The law or maxim which did in fact become current among them by inheritance from an earlier stage, only more or less modified to make it compatible with the new conditions, is another and far different thing. Ethnography is required to bridge over the gap between the two, a very chasm where the arguments of moralists and legists are continually falling in, to crawl out maimed and helpless. Within modern grades of civilization this historical method is now becoming more and more accepted. It will not be denied that English law has acquired, by modified inheritance from past ages, a theory of primogeniture and a theory of real estate which are so far from being products of our own times that we must go back to the middle ages for anything like a satisfactory explanation of them; and as for more absolute survival, did not Jewish disabilities stand practically, and the wager of battle nominally, in our law of not many years back ? But the point to be pressed here is, that the development and survival of law are processes that did not first come into action within the range of written codes of comparatively cultured nations. Admitted that civilized law requires its key from barbaric law; it must be borne in mind that the barbarian lawgiver too was guided in judgment not so much by first principles, as by a reverent and often stupidly reverent adherence to the tradition of earlier and yet ruder ages.

Nor can these principles be set aside in the scientific study of moral sentiment and usage. When the ethical systems of mankind, from the lowest savagery upward, have been analyzed and arranged in their stages of evolution, then ethical science, no longer vitiated by too exclusive application to particular phases of morality taken unreasonably as representing morality in general, will put its methods to fair trial on the long and intricate world-history of right and wrong.

In concluding a work of which full half is occupied by evidence bearing on the philosophy of religion, it may well be asked, how does all this array of facts stand toward the theologian's special province ? That the world sorely needs new evidence and method in theology, the state of religion in our own land bears witness. Take English Protestantism as a central district of opinion, draw an ideal line through its centre, and English thought is seen to be divided as by a polarizing force extending to the utmost limits of repulsion. On one side of the dividing line stand such as keep firm hold on the results of the 16th century reformation, or seek yet more original canons from the first Christian ages; on the other side stand those who, refusing to be bound by the doctrinal judgments of past centuries, but introducing modern science and modern criticism as new factors in theological opinion, are eagerly pressing toward a new reformation. Outside these narrower limits, extremer partizans occupy more distant ground on either side. On the one hand the Anglican blends gradually into the Roman scheme, a system so interesting to the ethnologist for its maintenance of rites more naturally belonging to barbaric culture; a system so hateful to the man of science for its suppression of knowledge, and for that usurpation of intellectual authority by a sacerdotal caste which has at last reached its climax, now that an aged bishop can judge, by infallible inspiration, the results of researches whose evidence and methods are alike beyond his knowledge and his mental grasp. On the other hand, intellect, here trampled under foot of dogma, takes full revenge elsewhere, even within the domain of religion, in those theological districts where reason takes more and more the command over hereditary belief, like a mayor of the palace superseding a nominal king. In yet farther ranges of opinion, religious authority is simply deposed and banished, and the throne of absolute reason is set up without a rival even in name; in secularism the feeling and imagination which in the religious world are bound to theological belief, have to attach themselves to a positive natural philosophy, and to a positive morality which shall of its own force control the acts of men. Such, then, is the boundless divergence of opinion among educated citizens of an enlightened country, in an age scarcely approached by any former age in the possession of actual knowledge and the strenuous pursuit of truth as the guiding principle of life. Of the causes which have brought to pass so perplexed a condition of public thought, in so momentous a matter as theology, there is one, and that a weighty one, which demands mention here. It is the partial and one-sided application of the historical method of enquiry into theological doctrines, and the utter neglect of the ethnographical method which carries back the historical into remoter and more primitive regions of thought. Looking at each doctrine by itself and for itself, as in the abstract true or untrue, theologians close their eyes to the instances which history is ever holding up before them, that one phase of a religious belief is the outcome of another, that in all times religion has included within its limits a system of philosophy, expressing its more or less transcendental conceptions in doctrines which form in any age their fittest representatives, but which doctrines are liable to modification in the general course of intellectual change, whether the ancient formulas still hold their authority with altered meaning, or are themselves reformed or replaced. Christendom furnishes evidence to establish this principle, if for example we will but candidly compare, the educated opinion of Rome in the 5th with that of London in the 19th century, on such subjects as the nature and functions of soul, spirit, deity, and judge by the comparison in what important respects the philosophy of religion has come to differ even among men who represent in different ages the same great principles of faith. The general study of the ethnography of religion, through all its immensity of range, seems to countenance the theory of evolution in its highest and widest sense. In the treatment of some of its topics here, I have propounded special hypotheses as to the order in which various stages of doctrine and rite have succeeded one another in the history of religion. Yet how far these particular theories may hold good, seems even to myself a minor matter. The essential part of the ethnographic method in theology lies in admitting as relevant the compared evidence of religion in all stages of culture. The action of such evidence on theology proper is in this wise, that a vast proportion of doctrines and rites known among mankind are not to be judged as direct products of the particular religious systems which give them sanction, for they are in fact more or less modified results adopted from previous systems. The theologian, as he comes to deal with each element of belief and worship, ought to ascertain its place in the general scheme of religion. Should the doctrine or rite in question appear to have been transmitted from an earlier to a later stage of religious thought, then it should be tested, like any other point of culture, as to its place in development.

The question has to be raised, to which of these three categories it belongs :—is it a product of the earlier theology, yet sound enough to maintain a rightful place in the later ;— is it derived from a cruder original, yet so modified as to become a proper representative of more advanced views ?—is it a survival from a lower stage of thought, imposing on the credit of the higher by virtue not of inherent truth but of ancestral belief? These are queries the very asking of which starts trains of thought which candid minds should be encouraged to pursue, leading as they do toward the attainment of such measure of truth as the intellectual condition of our age fits us to assimilate. In the scientific study of religion, which now shows signs of becoming for many a year an engrossing subject of the world's thought, the decision must not rest with a council in which the theologian, the metaphysician, the biologist, the physicist, exclusively take part. The historian and the ethnographer must be called upon to show the hereditary standing of each opinion and practice, and their enquiry must go back as far as antiquity or savagery can show a vestige, for there seems no human thought so primitive as to have lost its bearing on our own thought, nor so ancient as to have broken its connexion with our own life.

It is our happiness to live in one of those eventful periods of intellectual and moral history, when the oft-closed gates of discovery and reform stand open at their widest. How long these good days may last, we cannot tell. It may be that the increasing power and range of the scientific method, with its stringency of argument and constant check of fact, may start the world on a more steady and continuous course of progress than it has moved on heretofore. But if history is to repeat itself according to precedent, we must look forward to stiffer duller ages of traditionalists and commentators, when the great thinkers of our time will be appealed to as authorities by men who slavishly accept their tenets, yet cannot or dare not follow their methods through better evidence to higher ends. In either case, it is for those among us whose minds are set on the advancement of civilization, to make the most of present opportunities, that even when in future years progress is arrested, it may be arrested at the higher level. To the promoters of what is sound and reformers of what is faulty in modern culture, ethnography has double help to give. To impress men's minds with a doctrine of development, will lead them in all honour to their ancestors to continue the progressive work of past ages, to continue it the more vigorously because light has increased in the world, and where barbaric hordes groped blindly, cultured men can often move onward with clear view. It is a harsher, and at times even painful, office of ethnography to expose the remains of crude old culture which have passed into harmful superstition, and to mark these out for destruction. Yet this work, if less genial, is not less urgently needful for the good of mankind. Thus, active at once in aiding progress and in removing hindrance, **the science of culture is essentially a reformer's science.**

INDEX.

Abacus, i. 270.
Accent, i. 173.
Acephali, i. 390.
Achilles:—vulnerable spot, i. 358; dream, i. 444; in Hades, ii. 81.
Acosta, on American archetypal deities, ii. 244.
Adam, ii. 312, 315.
Ælian, i. 372, ii. 423; on Kynokephali, i. 389.
Æolus, i. 361, ii. 269.
Æsculapius:—incubation in temple, ii. 121; serpents of, ii. 241.
Affirmative and negative particles, i. 192.
Afghans, race-genealogy of, i. 403.
Agni, ii. 281, 386.
Agreement in custom and opinion no proof of soundness, i. 13.
Agriculture, god of, ii. 305.
Ahriman, ii. 328.
Ahura-Mazda, ii. 283, 328, 355.
Alexander the Great, i. 395, ii. 138.
Alfonso di Liguori, St., bilocation of, i. 447.
Alger, Rev. W. R., i. 471, 484, ii. 83.
Algonquin languages, animate and inanimate genders, i. 302.
Ali as Thunder-god, ii. 264.
All Souls', feast of dead, ii. 37.
Allegory, i. 277, 408.
Aloysius Gonzaga, St., letters to, ii. 122.
Alphabet, i. 171; by raps, i. 145; as numeral series, i. 258.
Amatongo, i. 443, ii. 115, 131, 313, 367, 387.
Amenti, Egyptian dead-land, ii. 67, 81, 96, 295, 311.
Amphidromia, ii. 439.
Analogy, myth product of, i. 297.
Ancestors, eponymic myths of, i. 398, ii. 234; worship of divine, ii. 113, 311; see Manes-worship, Totem-worship.
Ancestral names indicate re-birth of souls, ii. 5.
Ancestral tablet, Chinese, ii. 118, 152.

Andaman Islanders, mythic origin of, i. 369, 389.
Angang, omen from meeting animal, i. 120.
Angel, see Spirit; of death, i. 295, ii. 196, 322.
Angelo, St., legend of, i. 295.
Anima, animus, i. 433, 470.
Animals:—omens from, i. 120; calls to and cries of, 177; imitative names from cries, etc., 206; treated as human, i. 467, ii. 230; souls of, i. 469; future life and funeral sacrifice of, i. 469, ii. 75, etc.; entry and transmigration of souls into and possession by spirits, ii. 7, 152, 161, 175, 231, 241, 378, etc.; diseases transferred to, ii. 147; see spirits invisible to men, ii. 196.
Animals, sacred, incarnations or representatives of deities, ii. 231; receive and consume sacrifices, 378.
Animal-worship, i. 467, ii. 229, 378.
Animism:—defined, i. 424; is the philosophy of religion, i. 426, ii. 356; is a primitive scientific system of man and nature based on the conception of the human soul, i. 428, 499, ii. 108, 184, 356; its stages of development, survival, and decline, i. 499, ii. 181, 356. See Soul, Spirit, etc., etc.
Anra-Mainyu, ii. 328.
Antar, tumulus of, ii. 29.
Anthropomorphic conceptions of spirit and deity, ii. 110, 184, 247, 335.
Antipodes, i. 392.
Ape-men, i. 379; apes degenerate men, 376; can but will not talk, 379.
Apollo, ii. 294.
Apophis-serpent, ii. 241.
Apotheosis, ii. 120.
Apparitional soul, i. 428; its likeness to body, 450.
Apparitions, i. 143, 440, 445, 478, ii. 24, 187, 410, etc.
Archetypal deities and ideas, ii. 243.

Ares, ii. 308.
Argos Panoptes, i. 320.
Argyll, Duke of, on primæval man, i. 60.
Arithmetic, see Counting.
Arriero, i. 191.
Arrows, magic, i. 345.
Artemidorus, on dream-omens, i. 122.
Artemis, ii. 302.
Aryan race :—no savage tribe among, i. 49; antiquity of culture, i. 54.
Ascendant in horoscope, i. 129.
Ashera, worship of, ii. 166.
Ashes strewn for spirit-footprints, i. 455, ii. 197.
Asmodeus, ii. 254.
Association of ideas, foundation of magic, i. 116.
Astrology, i. 128, 291.
Atahentsic, ii. 299, 309, 323.
Atahocan, ii. 324, 339.
Atavism, explained by transmigration, ii. 3.
Atheist, use of word, i. 420.
Augury, etc., i. 119. See ii. 179, 231.
Augustine, St., i. 199, 441, ii. 54, 427; on dreams, i. 441; on incubi, ii. 190.
Augustus, genius of, ii. 202.
Avatars, ii. 239.
Avernus, Lake, ii. 45.
Ayenbite of Inwyt, i. 456.

Baal-Shemesh, ii. 295.
Bacon, Lord, on allegory, i. 277.
Bætyls, animated stones, ii. 166.
Baku, burning wells of, ii. 282.
Baldr, i. 464.
Bale, Bishop, i. 384; on witchcraft, i. 142.
Bands, clerical, i. 18.
Baptism, ii. 440; orientation in, 427.
Baring-Gould, Rev. S., on werewolves, i. 314.
Bastian, Prof. Adolf, Mensch in der Geschichte, i. vi.; ii. 209, 222, 242, 280, etc.
Baudet, etymology of, i. 413.
Beal, ii. 252, 408.
Bear, Great, i. 359.
Beast-fables, i. 381, 409.
Bees, telling, i. 287.
Bel, ii. 293, 380, 384.
Berkeley, Bishop, on ideas, i. 499; on force and matter, ii. 160.
Bewitching by objects, i. 116.
Bible and key, ordeal by, i. 128.
Bilocation, i. 447.
Bird, of thunder, i. 362; bird conveys spirit, ii. 161, 175.

Blemmyæ, headless men, i. 390.
Blood :—related to soul, i. 431; revives ghosts, ii. 48; offered to deities, 381; substitute for life, 402.
Blood-red stain, myths to account for, i. 406.
Bloodsuckers, ii. 191.
Blow-tube, i. 67.
Bo tree, ii. 218.
Boar's head, ii. 408.
Boats without iron, myth on, i. 374.
Bochica, i. 353, ii. 290.
Boehme, Jacob, on man's primitive knowledge, ii. 185.
Bolotu, ii. 22, 62, 310.
Boni Homines, i. 77.
Book of Dead, Egyptian, ii. 13, 96.
Boomerang, i. 67.
Boreas, i. 362, ii. 268.
Bosjesman, etymology of word, i. 381.
Bow and arrow, i. 7, 15, 64, 73.
Brahma, ii. 354, 425.
Brahmanism:—funeral rites, i. 465, etc.; transmigration, ii. 9, 20, 97; manes-worship, 119 ; stone-worship, 164 ; idolatry, 178 ; animal-worship, 238 ; sun-worship, 292; orientation, 425 ; lustration, 437.
Breath, its relation to soul, i. 432.
Bride-capture, game of, i. 73.
Bridge, first crossing, i. 106; of dead, i. 495, ii. 50, 94, 100, etc.
Brinton, Dr. D. G., i. 53, 361, ii. 90, 342; on dualistic myths, ii. 320.
Britain, eponymic kings of, i. 400; voyage of souls to, ii. 64.
Brosses, C. de, on degeneration and development, i. 36 ; origin of language, 161; fetishism, ii. 144; species-deities, 246.
Browne, Sir Thos., on magnetic mountain, i. 375.
Brutus, evil genius of, ii. 203.
Brynhild, i. 465.
Buck, buck, game of, i. 74.
Buddha, transmigrations of, i. 414, ii. 11.
Buddhism :—culture-tradition, i. 41; saints rise in air, i. 149 ; transmigration, ii. 11, 20, 97; nirvana, ii. 79 ; tree-worship, i. 476, ii. 21, serpent-worship, 240 ; religious formulas, 372.
Buildings, victim immured in foundation, i. 104, etc.; mythic founders of, i. 394.

INDEX. 457

Bull. Bishop, on guardian angels, ii. 203.
Bura Pennu, ii. 327, 350, 368, 404.
Burial, ghost wanders till. ii. 27; corpse laid east and west, 423.
Burning oats from straw, i. 44.
Burton, Capt. R. F., continuance-theory of future life, ii. 75; disease-spirits, 150.
Burton, Anatomy of Melancholy, incubi, etc., ii. 191.
Buschmann, Prof., on nature-sound, i. 223.
Butler, Bishop, on natural religion, ii. 356.

Cacodæmon, ii. 138.
Cæsar, on German deities, ii. 294.
Cagots, i. 115, 384.
Calderwood, Prof., i. vii., ii. 96.
Calls to animals, i. 177.
Calmet on spirits, ii. 188, etc.
Calumet, i. 210.
Candles against demons, ii. 194.
Cant, myth on word, i. 397.
Cardinal numbers, i. 257.
Cards, Playing, i. 82, 126.
Cassava, i. 63.
Castrén, ii. 80, 155, 177, 245, 351.
Cave-men, condition of, i. 59.
Ceremonies, religious, ii. 362, etc.
Ceres, ii. 306.
Chances, games of, their relation to arts of divination, i. 78.
Chanticleer, i. 413.
Charivari at eclipse, i. 329.
Charms:—objects, i. 118, ii. 148; formulas, their relation to prayers, ii. 373.
Charon, i. 490, ii. 93.
Chesterfield, Lord, on customs, i. 95; on omens, i. 118.
Chic, myth on word, i. 397.
Childbirth-goddess, ii. 305.
Children, numerical series of names for, i. 254; suckled by wild beasts, i. 281; receive ancestors' souls and names, ii. 4; sacrifice of, ii. 398, 403.
Children's language, i. 223.
China, religion of:—funeral rites, i. 464, 493; manes-worship, ii. 118; cultus of heaven and earth, 257, 272, 352; divine hierarchy, 352; prayer, 370; sacrifices, 385, 405.
Chinese culture-tradition, i. 40; remains in Borneo, i. 57.
Chiromancy or palmistry, i. 125.
Chirp or twitter of ghosts, etc., i. 453.

Christmas, origin of, ii. 297.
Chronology, limits of ancient, i. 54.
Cicero, on dreams, i. 444; sun-gods, ii. 294.
Civilization, see Culture.
Civilized men adopt savage life, i. 45.
Civilization-myths, i. 39, 353.
Clairvoyance, by objects, i. 116.
Clashing rocks, myth of, i. 348.
Clicks, i. 171, 192.
Cocoa-nut, divination by, i. 80.
Coin placed with dead, i. 490, 494.
Columba, St., legend of, i. 104.
Columbus, his quest of Earthly Paradise, ii. 61.
Comparative theology, ii. 251.
Comte, Auguste, i. 19: fetishism, i. 478, ii. 144, 354; species-deities, 242.
Confucius, i. 157; funeral sacrifice, i. 464, ii. 42; spirits, 206; name of supreme deity, 352.
Consonants, i. 169.
Constellations, myths of, i. 290, 356.
Continuance-theory of future life, ii. 75.
Convulsions:—by demoniacal possession, ii. 130; artificially produced, 416.
Convulsionnaires, ii. 420.
Copal incense, ii. 384.
Cord, magical connexion by, i. 117.
Corpse taken out by special opening in house, ii. 26, soul remains near, ii. 29, 150.
Cortes, i. 319.
Costume, i. 18.
Counting, art of, i. 21, 240, etc.; on fingers and toes, 244; by letters of alphabet, etc. 258; derivation of numeral words, 247; evidence of independent development of low tribes, 271.
Counting-games, i. 75, 87.
Couvade, in South India, i. 84.
Cow, name of, i. 208; purification by nirang, etc. ii. 438.
Cox, Mr. G. W., i. 341, 346, 362.
Creator, doctrine of, ii. 249, 312, 321, etc.
Credibility of tradition, i. 275, 370.
Crete, earth of, fatal to serpents, i. 372.
Cromlechs and menhirs objects of worship, ii. 164.
Culture:—definition of, i. 1; scale of, i. 26; primitive, represented by modern savages, i. 21, 68, ii. 443, etc.; development of, i. 21 etc., 62, etc., 237, 270, 417, etc, ii. 356, 445; evidence of independent progress

from low stages, i. 56, etc.; survival in culture, 70, etc.: evidence of early culture from language, 236; art of counting, 270: myth, 284; religion, i. 500, ii. 102, 184, 356, etc.; practical import of study of culture, 443.
Curtius, Marcus, leap of. ii. 378.
Curupa, cohoba, narcotic used in W. Ind. and S. Amer., ii. 416.
Customs, permanence of, i. 70, 156; rational origin of, 94.
Customs of Dahome, i. 462.
Cycle of Necessity, ii. 13.
Cyclops, i. 391.
Cyrus, i. 281, 286.

Dancing for religious excitement, ii. 133, 420.
Danse Macabre, myth on name, i. 397.
Dante, Divina Commedia, ii. 55, 220.
Daphne, ii. 220.
Dark, evil spirits in, ii. 194.
Darwin, Mr., i. vii..ii. 152, 223.
Dasent, Dr., i. 19.
Davenport Brothers, i. 152, 311.
Dawn, i. 338, 344, etc.
Day, sun as eye of, i. 350.
Day and Night, myths of, i. 322, 337, etc., ii. 48, 323.
Dead, use objects sacrificed for them, i. 485; feasts of, ii. 29; region of future life of, ii. 59, 74, 244; god and judge of, ii. 75, etc., 308.
Deaf and Dumb, counting, i. 244, 262; their mythic ideas, i. 298, 413.
Death:—ascribed to sorcery, i. 138; omens of, i. 145, 449: angel of, i. 295, ii. 196, 321; personification and myths of, i. 295, 349, 355, ii. 46, etc., 309; death and sunset, myths of, i. 335, ii. 48; exit of soul at death, i. 448, ii. 1, etc.; death of soul, ii. 22.
Death-watch, i. 146.
Decimal notation, i. 261.
Degeneration in culture, i. 35, etc.; is a secondary action, i. 38, 69; examples of, in Africa, North America, etc. i. 47.
Delphi, oracle of, i. 94, ii. 137.
Demeter, i. 328, ii. 273, 306.
Democritus, theory of ideas, i. 497.
Demons:— souls become, ii. 27, 111, etc.; iron, charm against, i. 140; pervade world, ii. 111, 137, 185, etc.; disease-demons, 126, etc., 177, 192, 215; water-demons, i. 109, ii. 209;
tree and forest demons, ii. 215, 222; possession and obsession by demons, i. 98, 152, 309, ii. 111, 123, etc., 179, 404; expulsion of, i. 103, ii. 125, 438; answer in own name through patient or medium, ii. 124. etc., 182, 366.
Dendid, creation-poem of, ii. 21.
Deodand, origin of, i. 287.
Destruction of objects sacrificed to dead, i. 483; to deities, ii. 376, etc.
Development of culture, see Culture.
Development myths, men from apes, etc. i. 376.
Devil:—as satyr, i. 307; devils' tree, ii. 148; devil-dancers, ii. 133; devil-worshippers, ii. 329.
Dice, for divination and gambling, i. 82.
Dies Natalis, ii. 202, 297.
Different al words, phonetic expression of distance and sex, i. 220.
Dirge, Lyke-wake, i. 495; of Hos, ii. 32.
Disease:—personification and myths of, i. 295; by exit of soul, i. 436; by demoniacal possession, etc., i. 127, ii. 114, 123, 404; disease-spirits, ii. 125, etc, 178, 215, 408; embodied in objects or animals, 146, 178, etc.; see Demons, Vampires.
Distance expressed by phonetic modification, i. 220.
Divination: lots, i. 78; symbolic processes, 81, 117; augury, etc., 119; dreams, 121; haruspication, 124; swinging ring, etc., 126; astrology 128; possessed objects, i. 125, ii. 155.
Divining rod and pendulum, i. 127.
Doctrines borrowed by low from high races:—on future life, ii. 91; dualism, 316; supremacy, 333.
Dodona, oak of, ii. 219.
Dog-headed men, i. 389.
Dolmens, etc., myths suggested by, i. 387.
Domina Abundia, ii. 389.
Dook, ghost, i. 433.
D'Orbigny, on religion of low tribes, i. 419; on sun-worship, ii. 286.
Dravidian languages, high and low gender, i. 302.
Dreams:—omens by, i. 121; by contraries, 122; caused by exit of soul, i. 440; by spiritual visit to soul, i. 442, 478; evidence of future life, ii. 24, 49, 75; oracular fasting for, 410; narcotizing for, 416.

Drift, stone implements from, i. 58.
Drivers' and Drovers' words, i. 180.
Drowning, superstition against rescuing from, i. 107; caused by spirits, 109,
Drugs used to produce morbid excitement, dreams, visions, etc., ii. 416.
Dual and plural numbers in primitive culture, i. 265.
Dualism :—good and evil spirits, ii. 186; good and evil genius, 202; good and evil deity, 316.
Dusii, ii. 190.
Dwarfs, myths of, i. 385.
Dyu, ii. 258.

Earth, myths of, i. 322, etc., 364, ii. 270, 320.
Earth-bearer, i. 364.
Earth-goddess and earth-worship, i. 322, etc., ii. 270, 306, 342.
Earth-mother, i. 326, etc., 365.
Earthquake, myths of, i. 364.
Earthly Paradise, ii. 57, etc.
Earthly resurrection, ii. 5.
East and West, burial of dead, turning to in worship, adjusting temples toward, ii. 383, 422.
Easter fires and festivals, ii. 297.
Eclipse, myths of, i. 288, 329, 356; driving off eclipse-monster, i. 328.
Ecstasy, swoon, etc. :—by exit of soul, i. 439; by demoniacal possession, ii. 130; induced by fasting, drugs, excitement, ii. 410, etc.
Edda, i. 84, ii. 77, etc.
Egypt, antiquity of culture, i. 54; religion of, transmigration, ii. 13; future life, 96; animal-worship, 238 ; sun-worship, 295, 311 ; dualism, 327; polytheism and supremacy, 355.
El, ii. 355.
Elagabal, Elagabalus, Heliogabalus, ii. 295, 398.
Elements, worship of the four, ii. 303.
Elf-furrows, myth of, i. 393.
Elijah as thunder-god, ii. 264.
Elysium, ii. 97.
Embodiment of souls and spirits, ii. 3, 123, etc.
Emotional tone, i. 166, etc.
Emphasis, i. 173.
Endor, witch of, i. 446.
Energumens or demoniacs, ii. 139.
Englishman, Peruvian myth of, i. 354.
Enigmas, Greek, i. 93.
Enoch, Book of, i. 408.

Enthusiasm, changed signification of, ii. 183.
Epicurean theory of development of culture, i. 37, 60 ; of soul, 456; of ideas, 497.
Epileptic fits by demoniacal possession, ii. 130, 137; induced, 419.
Eponymic ancestors, etc., myths of, i. 387, 398, etc., ii. 235.
Essence of food consumed by souls, ii. 39 ; by deities, 381.
Ethereal substance of soul, i. 454 ; of spirit, ii. 198.
Ethnological evidence from myths of monstrous tribes, i. 379, etc.; from eponymic race-genealogies, 401.
Etiquette, significance of, i. 95.
Etymological myths : — names of places, i. 395; of persons, 396 ; nations, cities, etc., traced to eponymic ancestors or founders, 398, etc.
Euhemerism, i. 279.
Evans, Mr. John, on stone implements, i., 65; Mr. Sebastian, i. 106, 453.
Evil deity, ii. 316, etc. ; worshipped only, 320.
Excitement of convulsions, etc., for religious purposes, ii. 133, 419.
Exeter, myth on name of, i. 396.
Exorcism and expulsion of souls and spirits, i. 102, 454, ii. 26, 40, 125, etc., 146, 179, 199, 433.
Expression of feature causes corresponding tone, i. 165, 183.
Expressive sound modifies words, i. 215.
Ex-voto offerings, ii. 406, 409.
Eye of day, of Odin, of Graiæ, i. 350.

Fables of animals, i. 381, 409.
Familiar spirits, ii. 199.
Fancy, in mythology, i. 315, 405.
Farrar, Rev. F. W., i. 161, ii. 83.
Fasting for dreams and visions, i. 306, 445, ii. 410.
Fauns and satyrs, ii. 227.
Feasts, of the dead, ii. 30 ; sacrificial banquets, 395.
Feralia, ii. 42.
Fergusson, Mr., on tree-worship, ii. 218 ; serpent-worship, 240.
Fetch or wraith, i. 448, 452.
Fetish, etymology of, ii. 143.
Fetishism :—defined, ii. 143 ; doctrine of, i. 477, ii. 157, etc., 175, 205, 215, 270, etc. ; survival of, ii. 160; its relation to philosophical theory of force, 160; to nature-worship, 205 ; to animal-worship, 231 ; tran-

sition to polytheism, 243 ; to supremacy, 335; to pantheism, 354.
Fiji and S. Africa, moon-myth common to, i. 355.
Finger-joints cut off as sacrifice, ii. 400.
Fingers and toes, counting on, i. 242.
Finns, as sorcerers, i. 84, 115.
Fire, passing through or over, i. 85, ii. 281, 429, etc. ; lighted on grave, i. 484; drives off spirits, ii. 194 ; new fire, ii. 278, 290, 297, 432; perpetual fire, 278 ; sacrifice by fire, 383, etc.
Fire-drill, i. 15, 50 ; antiquity of, ii. 280: ceremonial and sportive survival of, i. 75.
Fire-god and fire-worship, ii. 277, 376, etc., 403.
Firmament, belief in existence of, i. 299, ii. 70.
First Cause, doctrine of, ii. 335.
Food offered to dead, i. 485, ii. 30, etc.; to deities, ii. 397 ; how consumed, ii. 39, 376.
Footprints of souls and spirits, ii. 197.
Forest-spirits, ii. 215, etc.
Formalism, ii. 363, 371.
Formulas :—prayers, ii. 371 ; charms, 373.
Fortunate Isles, ii. 63.
Four winds, cardinal points, i. 361.
Frances, St., her guardian angels, ii. 203.
French numeral series in English, i. 268.
Fumigation, see Lustration.
Funeral procession :—horse led in, i. 463, 474 ; kill persons meeting, 464.
Funeral sacrifice :—attendants and wives killed for service of dead, i. 458 ; animals, 472; objects deposited or destroyed, 481 ; motives of, 458, 472, 483; survival of, 463, 474, 492 ; see Feasts of Dead.
Future Life, i. 419, 469, 480, ii. 1, etc., 100; transmigration of soul, ii. 2 ; remaining on earth or departure to spirit-world, 22 ; whether races without belief in, 20; connexion with evidence of senses in dreams and visions, 24, 49 ; locality of region of departed souls, 74 ; visionary visits to, 46; connexion of solar ideas with, 48, 74, 311, 422; character of future life, 74 ; continuance-theory, 75 ; retribution-theory, 83 ; introduction of moral element, 10, 83 ; stages of doctrine of future life, 100 ; its practical effect on mankind, 104 ; god of the dead, 308.

Gambling-numerals, i. 268.
Games :—children's games related to serious occupations, i. 72; counting-games, 74 ; games of chance related to arts of divination, 78.
Gataker, on lots, i. 79.
Gates of Hades, Night, Death, i. 347.
Gayatri, daily sun-prayer of Brahmans, ii. 292.
Genders, distinguished as male and female, animate and inanimate, etc., i. 301.
Genghis-Khan, worshipped, ii. 117.
Genius, patron or natal, ii. 199, 216; good and evil, 203 ; changed signification of word, 181.
German and Scandinavian mythology and religion :—funeral sacrifice, i. 464, 491 ; Walhalla, ii. 79, 88 ; Hel, i. 347, ii. 88 ; Odin, Woden, i. 351; 362, ii. 269 ; Loki, i. 83, 365 ; Thor, Thunder, ii. 266 ; Sun and Moon, i. 289, ii. 294.
Gesture-language, and gesture accompanying language, i. 163 ; effect of gesture on vocal tone, 165 ; gesture-counting original method, i. 246.
Ghebers or Gours, fire-worshippers, ii. 282.
Gheel, treatment of lunatics at, ii. 143.
Ghost :—ghost-soul, i. 142, 428, 433, 445, 488; seen in dreams and visions, 440, etc. ; voice of, 452; substance and weight of, 453; of men, animals, and objects, 429, 469, 479 ; popular theory inconsistent and brokendown from primitive, 479 ; ghosts as harmful and vengeful demons, ii. 27 ; ghosts of unburied wander, ii. 28 ; ghosts remain near corpse or dwelling, ii. 29, etc.; laying ghosts, ii. 153, 194.
Giants, myths of, i. 386.
Gibbon, on development of culture, i. 33.
Glanvil, Saducismus Triumphatus, ii. 140.
Glass-mountain, Anafielas, i. 492.
Godless month, ii. 350.
Gods :—seen in vision, i. 306; of waters, ii. 209 ; of trees, groves, and forests, 215 ; embodied in or represented by animals, 231 ; gods of species, 212; higher gods of polytheism, 247, etc. ; of dualism,

316; gods of different religions compared. 250; classified by common attributes, 254.
Gog and Magog, i. 386, etc.
Goguet, on degeneration and development, i. 32.
Gold, worshipped, ii. 154.
Good and evil, rudimentary distinction of, ii. 89, 318; good and evil spirits and dualistic deities, 317.
Goodman's croft, ii. 408.
Graiæ, eye of, i. 352.
Great-eared tribes, i. 388.
Greek mythology and religion:— nature-myths, i. 320, 328, 349; funeral rites, 464, 490; future life, ii. 53, 63, etc.; nature-spirits and polytheism, 206, etc.; Zeus, 258, etc., 355; Demeter, 273, 306; Nereus, Poseidon, 277; Hephaistos, Hestia, 284; Apollo, 294; Hekate, Artemis, 302; stone-worship, 165; sacrifice, 386, 396; orientation, 426; lustration, 439.
Grey, Sir George, i. 322.
Grote, Mr., on mythology, i. 276, 400.
Grove-spirits, ii. 215.
Guarani, name of, i. 401.
Guardian spirits and angels, ii. 199.
Gulf of dead, ii. 62.
Gunthram, dream of, i. 442.
Gypsies, i. 49, 115.

Hades, under-world of departed souls, i. 335, 340, ii. 65, etc., 81, 97, 309; descent into, i. 340, 345, ii. 45, 54, 83; personification of, i. 340, ii. 55, 309, 311.
Haetsh, Kamchadal, ii. 46, 313.
Hagiology, ii. 120, 261; rising in air, i. 151; miracles, i. 157, 371; second-sight, i. 449; hagiolatry, ii. 120.
Hair, lock of, as offering, ii. 401.
Half-blood, succession of forbidden, i. 20.
Half-men, tribes of, i. 391.
Haliburton, Mr., on sneezing-rite, i. 103.
Hamadryad, ii. 215.
Hand-numerals, from counting on fingers, etc., i. 246.
Hanuman, monkey-god, i. 378.
Hara kari, i. 463.
Harmodius and Aristogiton, ii. 63.
Harpies, ii. 269.
Harpocrates, ii. 295.
Haruspication, i. 123, ii. 179.

Harvest-deity, ii. 305, 364, 368.
Hashish, ii. 379.
Head-hunting, Dayak, i. 459.
Headless tribes myths of, i. 390.
Healths, drinking, i. 96.
Heart, related to soul, i. 431, ii. 152.
Heaven, region of departed souls, ii. 70.
Heaven and earth, universal father and mother, i. 322, ii. 272, 345.
Heaven-god, and heaven-worship, i. 306, 322, ii. 255, etc., 337, etc., 367, 395.
Hebrides, low culture in, i. 45.
Hekate, i. 150, ii. 302, 418.
Hel, death-goddess, i. 301, 347, ii. 88, 311.
Hell, ii. 56, 68, 97; related to Hades, ii. 74, etc.; as place of torment, not conception of savage religion, 103.
Hellenic race-genealogy, i. 402.
Hellshoon, i. 491.
Hephaistos, ii. 212, 280.
Hera, ii. 305.
Herakles, ii. 294; and Hesione, i. 339.
Hermes Trismegistus, ii. 178.
Hermotimus, i. 439, ii. 13.
Hero-children suckled by beasts, i. 281.
Hesiod, Isles of Blest, ii. 63.
Hestia, ii. 284.
Hiawatha, poem of, i. 345, 361.
Hide-boiling, i. 44.
Hierarchy, polytheistic, ii. 248, 337, 349, etc.
Hissing, for silence, contempt, respect, i. 197.
History, relation of myth to, i. 278, 416, ii. 447; criticism of, i. 280; similarity of nature-myth to, 320.
Hole to let out soul, i, 453.
Holocaust, ii. 385, 396.
Holyoake, Holywood, etc., ii. 229.
Holy Sepulchre, Easter fire at, ii. 297.
Holy water, ii. 188, 439.
Holy wells, ii. 214.
Horne Tooke on interjections, i. 175.
Horse, sacrificed or led at funeral, i. 463, 473.
Horseshoes, against witches and demons, i. 140.
House abandoned to ghost, ii. 25.
Hucklebones, i. 82.
Huitzilopochtli, ii. 254, 307.
Human sacrifice:—funerals, i. 458; to deities, ii. 271, 385, 389, 398, 403.
Humboldt, W. v., on continuity, i.

19; on language, 236; on numerals, 258.
Hume, Natural History of Religion, i. 477.
Huns. as giants. i. 386.
Hunting-calls, i. 181.
Hurricane, i. 363.
Hyades, i. 358.
Hysteria, etc., by possession, ii. 131, etc.; induced, 419.

amblichus, i. 150, ii. 187.
deas:—Epicurean related to object-souls. i. 497; Platonic related to species-deities, ii. 244.
Idiots, inspired, ii. 117.
Idol, see Image.
Idolatry as related to fetishism, ii. 168.
Images:—fallen from heaven, i. 157; as substitutes in sacrifice, i. 463, ii. 405; fed and treated as alive, ii. 170; moving, weeping, sweating, etc., 171; animated by spirits or deities, 172.
Imagination, based on experience, i. 273, 298, 304.
Imitative words, i. 200; verbs. etc., of blowing, swelling, mumbling, spitting, sneezing, eating, etc., 203, etc.; names of animals, 206; names of musical instruments, 208; verbs, etc., of striking, cracking, clapping, falling, etc., 211; prevalence of imitative words in savage language, 212; imitative adaptation of words, 214.
Immateriality of soul, not conception of lower culture, i. 455, ii. 198.
Immortality of soul, not conception of lower culture, ii. 22.
Implements, inventions of, i. 64, etc.
Incas, myth of ancestry and civilization, i. 288, 354, ii. 290, 301.
Incense, ii. 383.
Incubi and succubi, ii. 189.
Indigenes of low culture, i. 50, etc.; considered as sorcerers, 113; myths of, as monsters, 376, etc.
Indo-Chinese languages, musical pitch of vowels, i. 169.
Indra, i. 320, ii. 265.
Infant, lustration of, ii. 430, etc.
Infernus, ii. 81.
Innocent VIII., bull against witchcraft, i. 139, ii. 190.
Inspired idiot, ii. 128.
Interjectional words:—verbs, etc., of wailing, laughing, insulting, complaining, fearing, driving, etc., i. 187; hushing, hissing, loathing, hating, etc., 197.
Interjections, i. 175; sense-words used as, 176; directly expressive sounds, 183.
Intoxicating liquor, absence of, i. 63.
Intoxication as a rite, ii. 417.
Inventions. development of, i. 14, 62; myths of, 39, 392.
Iosco, Ioskeha and Tawiscara, myth of, i. 288, 348, ii. 323.
Ireland, low culture in, i. 44.
Iron, charm against witches, elves, etc., i. 140.
Islands. earth of, fatal to serpents, i. 372; of Blest, ii. 57.
Italian numeral series in English, i. 268.

Jameson, Mrs., on parables. i. 414.
Januarius, St., blood of, i. 157.
Jawbone, mythic, i. 344.
Jerome, St., ii. 428.
Jew's harp, vowels sounded with, i. 168.
John, St., Midsummer festival of, ii. 298.
Johnson, Dr., i. 6, ii. 24.
Jonah, i. 329.
Jones, Sir W., on nature-deities. ii. 253, 286.
Joss-sticks, ii. 384.
Journey to spirit-world, region of dead, i. 431, ii. 44, etc.
Judge of dead, ii. 92, 314.
Julius Cæsar, i. 320.
Jupiter, i. 350, ii. 258, etc.

Kaaba, black stone of, ii. 166.
Kalewala, Finnish epic, ii. 46, 80, 93, 261.
Kali, ii. 425.
Kami-religion of Japan, ii. 117, 301, 350.
Kang-hi on magnetic needle, i. 375.
Kathenotheism, ii. 354.
Keltic counting by scores continued in French and English, i. 263.
Kepler on world-soul, ii. 354.
Kimmerian darkness, ii. 48.
Kissing, i. 63.
Kitchi Manitu and Matchi Manitu, Great and Evil Spirit, ii. 324.
Klemm, Dr., on development of implements, i. 64.
Kobong or totem, ii. 235.
Koran, i. 407, ii. 77, 296.
Kottabos, game of, i. 82.

Kronos swallowing children, i. 341.
Kynokephali, i. 389.

Lake-dwellers, i. 61.
Language: i. 17, 236, ii. 445:—directly expressive element in, i. 160; correspondence of this in different languages, 163; interjectional forms, 175; imitative forms, 200; differential forms, 220: children's language, 223; origin and development of language, 229; relation of language to mythology, 299; gender, 301; language attributed to birds, etc., 19, 469; place of language in development of culture, ii. 445.
Languedoc, etc., i. 193.
Last breath, inhaling, i. 433.
Laying ghosts, ii. 25, 153.
Legge, Dr., on Confucius, ii. 352.
Leibnitz, i. 2.
Lewes, Mr. G. H., i. 497.
Liebrecht, Prof. F., i. vii., 108, 177, 348-9, ii. 24, 164, 195, etc.
Life caused by soul, i. 436.
Light and darkness, analogy of good and evil, ii. 324.
Likeness of relatives accounted for by re-birth of soul, ii. 3.
Limbus Patrum, ii. 83.
Linnæus, name of, ii. 229.
Little Red Riding-hood, i. 341.
Loki, i. 83, 365.
Lots, divination and gambling by, i. 78.
Lubbock, Sir J.:—evidence of metallurgy and pottery, against degeneration-theory, i. 57; on low tribes described as without religious ideas, i. 421; on water-worship, ii. 210; on totem-worship, 236.
Lucian, i. 149, ii. 13, 52, 67, 302, 426.
Lucina, ii. 302, 305.
Lucretius, i. 40, 60.
Lunatics, demoniacal possession of, ii. 124, etc.
Lustration, by water and fire, ii. 429, etc.; of new-born children, 430; of women, 432; of those polluted by blood or corpse, 433; general, 434, etc.
Luther, on witches, i. 137; on guardian angels, ii. 203.
Lyell, Sir C., on degeneration-theory, i. 57.
Lying in state, of King of France, ii. 35.
Lyke-wake dirge, i. 495.

M'Lennan, Mr., theory of totemism, ii. 236.
Macrocosm, i. 350, ii. 354.
Madness and idiocy by possession, ii. 128, etc., 179.
Magic:—origin and development, i. 112, 132; belongs to low level of culture, 112; attributed to low tribes, 113; based on association of ideas, 116; processes of divination, 78, 118; relation to Stone Age, 127; see Fetishism.
Magnetic Mountain, philosophical myth of, i. 374.
Maine, Sir H. S., i. 20.
Maistre, Count de, on degeneration in culture, i. 35; astrology, 128; animation of stars, 291.
Makrokephali, i. 391.
Malleus Maleficarum, ii. 140, 191.
Man, primitive condition of, i. 21, ii. 443; see Savage.
Man of the woods, bushman, orang-utan, i. 381.
Man swallowed by monster, nature-myth of, i. 335, etc.
Manco Ccapac, i. 354.
Manes and manes-worship, i. 98, 143, 434, ii. 8, 111, etc., 129, 162, 307, 364; theory of, ii. 113, etc.; divine ancestor or first man as great deity, 311, 347.
Manichæism, ii. 14, 330.
Manitu, ii. 249, 324, 341.
Manoa, golden city of, ii. 249.
Manu, laws of:—ordeal by water, i. 141; pitris, ii. 119.
Marcus Curtius, leap of, ii. 378.
Margaret, St., i. 340.
Markham, Mr. C. R., i. vii., ii. 263, 337, 366, 392, etc.
Marriages in May, i. 70.
Mars, ii. 308.
Martius, Dr., on dualism, ii. 325.
Maruts, Vedic, i. 362, ii. 268.
Mass, ii. 410.
Master of life or breath, ii. 60, 339, etc., 365.
Materiality of soul, i. 453; of spirit, ii. 198.
Maui, i. 335, 343, 360, ii. 253, 267, 279.
Maundevile, Sir John, i. 375, ii. 45.
Medicine, of N. A. Indians, ii. 154, 200, 233, 372, etc., 411.
Meiners, History of Religions, ii. 27, 48, etc.
Melissa, i. 491.

Men descended from apes, myths of, i. 376; men with tails, 383.
Menander, guardian genius, ii. 201.
Merit and demerit, Buddhist, ii. 12, 98.
Messalians, i. 103.
Metaphor, i. 234, 297; myths from, 405.
Metaphysics, relation of animism to, i. 497, ii. 242, 311.
Metempsychosis. i. 379, 409, 469, 476, ii. 2; origin of, ii. 16.
Micare digitis, i. 75.
Middleton, Dr., i. 157, ii. 121.
Midgard-snake, ii. 241.
Midsummer festival, ii. 298.
Milk and blood, sacrifices of, ii. 48; see Blood.
Milky Way, myths of, i. 359, ii. 72.
Mill, Mr. J. S., on ideas of number, i. 240.
Milton, on eponymic kings of Britain, i. 400.
Minne, drinking, i. 96.
Minucius Felix, on spirits, etc., ii. 179.
Miracles, i. 276, 371, ii. 121.
Mithra, i. 351, ii. 293, 297.
Moa, legend of, ii. 50.
Mohammed, legend of, i. 407.
Moloch, ii. 281, 403.
Money borrowed to be repaid in next life, i. 491.
Monkeys, preserved as dwarfs, i. 388; see Apes.
Monotheism, ii. 331.
Monster, driven off at eclipse, i. 328; hero or maiden devoured by, 335.
Monstrous mythic human tribes, ape-like, tailed, gigantic and dwarfish, noseless, great-eared, dog-headed, etc., i. 376, etc.; their ethnological significance, 379, etc.
Month's mind, i. 88.
Moon :— omens and influence by changes, i. 130; myths of, 288, 354; inconstant, 354; changes typical of death and new life, i. 354, ii. 300; moon-myths common to S. Africa and Fiji, i. 354, and to Bengal and Malay Peninsula, 356; moon abode of departed souls, ii. 70.
Moon-god and moon-worship, i. 289, ii. 299, etc., 323.
Moral and social condition of low tribes, i. 29, etc.
Moral element in culture, i. 28; absent or scanty in lower religions, i. 427, ii. 361; divides lower from higher religions, ii. 361; introduced in funeral sacrifice, i. 495; in trans-migration, ii. 12; in future life. 85, etc.; in dualism, 316, etc.; in prayer, 373; in sacrifice, 386, etc.; in lustration, 429.
Morals and Law, ii. 448.
Morbid imagination related to myth, i. 305.
Morbid excitement for religious purposes, ii. 416, etc.
Morning and evening stars, myths of, i. 344, 350.
Morra, game of, in Europe and China, i. 75.
Morzine, demoniacal possessions at, i. 152, ii. 141.
Mound-builders, i. 56.
Mountain, abode of departed souls on, ii. 60, ascending for rain, 260.
Mouth of Night and Death, myths of, i. 347.
Müller, Prof. J. G., on future life, ii. 90, etc.
Müller, Prof. Max :—on language and myth, i. 299; funeral rites of Brahmans, 466; heaven-god, ii. 258; 353; sun-myth of Yama, 314; Chinese religion, 352; kathenotheism, 354.
Mummies, ii. 19, 34, 151.
Musical instruments named from sound, i. 208.
Musical tone used in language, i. 168, 174.
Mutilation of soul with body, i. 451.
Mythology :—i. 23, 273, etc.; formation and laws of, 273, etc.; allegorical interpretation, 277; mixture with history, 278; rationalization, euhemerism, etc., 278; classification and interpretation, 281, 317, etc.; nature-myths, 284, 316, etc.; personification and animation of nature, 285; grammatical gender as related to, 301; personal names of objects as related to, 303; morbid delusion, 305; similarity of nature-myths to real history, 319; historical import of mythology, i. 416, ii. 446; its place in culture, ii. 446; philosophical myths, i. 366; explanatory legends, 392; etymological myths, 395; eponymic myths, 399; legends from fancy and metaphor, 405; realized or pragmatic legends, 407; allegory and parables, 408.
Myths :—myth-riddles, i. 93; origin of sneezing-rite, 101; foundation-sacrifice, 104; heroes suckled by beasts, 281; sun, moon, and stars, 288,

etc.; eclipse, 288; water-spout, 292; sand-pillar, 293; rainbow, 293, 297; waterfalls, rocks, etc., 295; disease, death, pestilence, 295; phenomena of nature, 297, 320; heaven and earth, i. 322, ii. 345; sunrise and sunset, day and night, death and life, i. 335, ii. 48, 62, 322; moon, inconstant, typical of death, i. 353; civilization-legends, 39, 353; winds, i. 361, ii. 266; thunder, i. 362, ii. 264; men and apes, development and degeneration, i. 378; ape-men, 379; men with tails, 382; giants and dwarfs, 385; monstrous men, 389; personal names introduced, 394; race-genealogies of nations, 402; beast-fables, 409; visits to spirit-world, ii. 46, etc.; giant with soul in egg, 153; transformation into trees, 219; dualistic myth of two brothers, 320.

Nagas, serpent-worshippers, ii. 213, 240.
Names:—of children in numerical series, i. 254; of objects as related to myth, 303; of personal heroes introduced into myths, 394; of places, tribes, countries, etc., myths formed from, 396; ancestral names given to children, ii. 4; name-giving ceremonials, ii. 429.
Natural religion, i. 427, ii. 103, 356.
Nature, conceived of as personal and animated, i. 285, 478, ii. 184.
Nature-deities, polytheistic, ii. 255, 376.
Nature-myths, i. 284, 316, etc., 326.
Nature-spirits, elves, nymphs, etc., ii. 184, 204, etc.
Necromancy, i. 143, 312, 446; see Manes.
Negative and affirmative particles, i. 192.
Negroes re born as whites, ii. 5.
Neo or Hawaneu, ii. 333.
Neptune, ii. 276.
Nereus, ii. 274.
Neuri, i. 313.
New birth of soul, ii. 3.
Newton, Sir Isaac, on sensible species, i. 498.
Nicene Council, spirit-writing at, i. 148.
Nicodemus, Gospel of, ii. 54.
Niebuhr, on origin of culture, i. 41.
Night, myths of, i. 334, ii. 48, 61.
Nightmare-demon, ii. 189, 193.
Nilsson, Prof., on development of culture, i. 61, 64.

Nirvana, ii. 12, 79.
Nix, water-demon, i. 110, ii. 213.
Norns or Fates, i. 352.
Noseless tribes, i. 388.
Notation, arithmetical, quinary, decimal, vigesimal, i. 261.
Numerals:—low tribes only to 3 or 5, i. 242; derivation of numerals from counting fingers and toes, 246; from other significant objects, 251; series of number-names of children, 254; new formation of numerals, 255; etymology of, 259, 270; numerals borrowed from foreign languages, 266; initials of numerals, used as figures, 269; see Notation.
Nympholepsy, ii. 137.
Nymphs:—water-nymphs, ii. 212; tree-nymphs, 219, 227.

Objectivity of dreams and visions, i. 442, 479; abandoned, 500.
Objects treated as personal, i. 286, 477, ii. 205; souls or phantoms of objects, i. 478, 497, ii. 9; dispatched to dead by funeral sacrifice, i. 481.
Occult sciences, see Magic.
Odin, or Woden, as heaven-god, i. 351, 362, ii. 269; one-eyed, i. 351.
Odysseus, unbinding of, i. 153; descent to Hades, i. 346, ii. 48, 65.
Ohio, Ontario, i. 190.
Ojibwa, myth of, i. 345, ii. 46.
Oki, demon, ii. 208, 255, 342.
Old man of sea, ii. 277.
Omens, i. 97, 118, etc., 145, 449.
Omophore, Manichæan, i. 365.
One-eyed tribes, i. 391.
Oneiromancy, i. 121.
Opening to let out soul, i. 453.
Ophiolatry, see Serpent-worship.
Ophites, ii. 242.
Oracles, i. 94, ii. 411; by inspiration or possession, ii. 124, etc., 179.
Orang-utan, i. 381.
Orcus, ii. 67, 80.
Ordeal by fire, i. 85; by sieve and shears, 128; by water, 140; by bear's head, ii. 231.
Ordinal numbers, i. 257.
Oregon, Orejones, i. 389.
Origin of language, i. 231; numerals, 247.
Orion, i. 358, ii. 81.
Orientation, solar rite or symbolism, ii. 422.
Ormuzd, ii. 283, 328.
Orpheus and Eurydike, i. 346, ii. 48.
Osiris, ii. 67, 295; and Isis, i. 289.
Otiose supreme deity, ii. 320, 336, etc.

Outcasts, distinct from savages, i. 43, 49.
Owain, Sir, visit to Purgatory, ii. 56.
Pachacamac, ii. 337, 366.
Pandora, myth of, i. 408.
Panotii, i. 389.
Pantheism, ii. 332, 341, 354.
Papa, mamma, etc., i. 223.
Paper figures substitutes in sacrifice, i. 464, 493, ii. 405.
Parables, i. 411.
Pars pro toto in sacrifice, ii. 399.
Parthenogenesis, ii. 190, 307.
Particles, affirmative and negative, i. 192; of distance, 220.
Passage de l'Enfer, ii 65.
Patrick, St., i. 372; his Purgatory, ii. 45, 55.
Patroklos, i. 444, 464.
Patron saints, ii. 120; patron spirits, 199.
Pattern and matter, ii. 246.
Pennycomequick, i. 396.
Periander, i. 491.
Perkun, Perun, ii. 266.
Persian race-genealogy, i. 403.
Persephone, myth of, i. 321.
Perseus and Andromeda, i. 339.
Personal names, in mythology, i. 303, 394, 396.
Personification:—natural phenomena, i. 285, etc., 320, 477, ii. 205, 254; disease, death, etc., i. 295; ideas, 300; tribes, cities, countries, etc., 339; Hades, i. 339, ii. 55.
Pestilence, personification and myths of, i. 295.
Peter and Paul, Acts of, i. 372.
Petit bonhomme, game of, i. 77.
Petronius Arbiter, i. 75, ii. 261.
Philology, Generative, i. 198, 230.
Philosophical myths, i. 368.
Phrase-melody, i. 174.
Pillars of Hercules, i. 395.
Pipe, i. 208.
Pithecusæ, i. 377.
Places, myths from names of, i. 395.
Planchette, i. 147.
Plants, souls of, i. 474.
Plath, Dr. on Chinese religion, ii. 352, etc.
Plato, on transmigration, ii. 13; Platonic ideas, 244.
Pleiades, i. 291, 358.
Pliny on magic, i. 133; on eclipses, 334.
Plurality of souls, i. 433.
Plutarch, visits to spirit-world, ii. 53.
Pneuma, psyche, i. 433, 437.
Pointer-facts, i. 62.

Polytheism, ii. 247, etc.; based on analogy of human society, ii. 248, 337, 349, 352; classification of deities by attributes, 255; heaven-god, 255, 334, etc.; rain-god, 259; thunder-god, 262; wind-god, 266; earth-god, 270; water-god, 274; sea-god, 275; fire-god, 277; sun-god, 286, 335, etc.; moon-god, 299; gods of childbirth agriculture, war, etc, 304; god and judge of dead, 308; first man, divine ancestor, 311; evil deity, 316; supreme deity, 332; relation of polytheism to monotheism, 333.
Popular rhymes, etc., i. 86; sayings, i. 19, 83, 122, 313, ii. 268, 353.
Poseidon, i. 365, ii. 277, 378.
Possession and obsession, see Demons, Embodiment.
Pott, Prof., on reduplication, i. 219; on numerals, 261.
Pottery, evidence from remains, i. 56; absence of potter's wheel, 45, 63.
Pozzuoli, myth of subsidence of, i. 372.
Pragmatic or realized myths, i. 407.
Prayer:—doctrine of, ii. 364, etc.; relation to nationality, 371; introduction of moral element, 373; prayers, i. 98, ii. 136, 208, 261, 280, 292, 329, 338, 364, etc., 435; rosary, ii. 372; prayer-mill and prayer-wheel, 372.
Prehistoric archæology, i. 55, etc.; ii. 443.
Priests consume sacrifices, ii. 379.
Prithivi, i. 327, ii. 258, 272.
Procopius, voyage of souls to Britain, ii. 64.
Progression in culture, i. 14, 32; inventions, 62, etc.; language, 236; arithmetic, 270; philosophy of religion, see Animism.
Prometheus, i. 365, ii. 400.
Proverbs, i. 84, etc.; see Popular Sayings.
Psychology, i. 428.
Pupil of eye, related to soul, i. 431.
Purgatory, ii. 68, 92; St. Patrick's, 55.
Purification, see Lustration.
Puss, i. 178.
Pygmies, myths of, i. 385; connected with dolmens, 387; monkeys as, 388.
Pythagoras, ii. 13, 137, 187.

Quaternary period, i. 58.
Quetelet, M., on social laws, i. 11.
Quinary numeration and notation, i. 261; in Roman numeral letters, 263.

INDEX.   467

Races:—distribution of culture among, i. 49; culture of mixed races, Gauchos, etc., 46, 52; ethnology in eponymic genealogies, 401; moral condition of low races, 26; considered as magicians, 113; as monsters, 380.
Rahu and Ketu, eclipse-monsters, i. 379.
Rain-god, ii. 254, 259.
Rainbow, myths of, i. vii. 293, ii. 239.
Ralston, Mr., i. vii., 342, ii. 245, etc.
Rangi and Papa, i. 322, ii. 345.
Rapping, omens and communications by, i. 144, ii. 221.
Rationalization of myths, i. 278.
Red Swan, myth of, i. 345.
Reduplication, i. 219.
Reid, Dr., on ideas, i. 499.
Relics, ii. 150.
Religion, i. 22, ii. 357, 449; whether any tribes without, i. 417; accounts misleading among low tribes, 419; rudimentary definition of, 424; adoption from foreign religions, future life, ii. 91; ideas and names of deities, 254, 309, 331, 344; dualism, 316, 322; supreme deity, 333; natural religion, i. 427, ii. 103, 356.
Resurrection, ii. 5, 18.
Retribution-theory of future life, ii. 83; not conception of lower culture, 83.
Return and restoration of soul, i. 436.
Revival, in culture, i. 136, 141.
Revivals, morbid symptoms in religious, ii. 421.
Reynard the Fox, i. 412.
Riddles, i. 90.
Ring, divination by swinging, i. 126.
Rising in air, supernatural, i. 149, ii. 415.
Rites, religious, ii. 362, etc.
River of death, i. 473, 480, ii. 23, 29, 51, 94.
River-gods and river-worship, ii. 209.
River-spirits, i. 109, ii. 209, 407.
Rock, spirit of, ii. 207.
Roman mythology and religion:— funeral rites, ii. 42; future life, 45, 67, 81; nature-spirits, 220, 227; polytheism, 251; Jupiter, 258, 265; Neptune, 277; Vesta, 285; Lucina, 302, 305, etc.
Roman numeral letters, i. 263.
Romulus, patron deity of children, ii. 121; and Remus, i. 281.
Rosary, ii. 372.

Sabæism, ii. 296.

Sacred springs, streams, etc., ii., 209; trees and groves, 222; animals, 234, 378.
Sacrifice :—origin and theory of, ii. 375, etc., 207, 269; manner of consumption or reception by deity, 216, 376, etc., see 39; motive of sacrificer, 393, etc.; substitution, 399; survival, i. 76, ii. 214, 228, 406.
Saint-Foix, i. 474, ii. 35.
Saints, worship of, ii. 120.
Samson's riddle, i. 93.
Sanchoniathon, ii. 221.
Sand-pillar, myths of, i. 293.
Sanskrit roots, i. 197, 224.
Savage, man of woods, i. 382.
Savage culture as representative of primitive culture :—i. 21, ii. 443; magic, witchcraft, and spiritualism, i. 112, etc.; language, i. 236, ii. 445; numerals, i. 242; myth, 284, 324; doctrine of souls, 499; future life, ii. 102; animistic theory of nature, i. 285, ii. 180, 356; polytheism, 248; dualism, 317; supremacy, 334; rites and ceremonies, 363, 375, 411, 421, 429.
Savitar, ii. 292.
Scalp, i. 460.
Scores, counting by, i. 263.
Sea, myths of, ii. 275.
Sea-god and sea-worship, ii. 275, 377.
Second-death, ii. 22.
Second sight, i. 143, 447.
Semitic race, no savage tribe among, i. 49; antiquity of culture, 54; race-genealogy, 404.
Sennaar, i. 395.
Serpent, emblem of immortality and eternity, ii. 241.
Serpent-worship, ii. 8, 239, 310, 347.
Sex distinguished by phonetic modification, i. 222.
Shadow related to soul, i. 430, 435; shadowless men, 85, 430.
Shell-mounds, i. 61.
Sheol, ii. 68, 81; gates of, i. 347.
Shingles, disease, i. 307.
Shoulder-blade, divination by, i. 124.
Sieve and shears, oracle by, i. 128.
Silver at new moon, ii. 302.
Sing-bonga, ii. 291, 350.
Skylla and Charybdis, ii. 208.
Slaves sacrificed to serve dead, i. 458.
Sling, i. 73.
Snakes, destroyed in Ireland, etc., i. 372.
Sneezing, salutation on, i. 97; connected with spiritual influence, 97.
Social rank retained in future life, ii. 22, 84.

Sokrates, ii. 137, 294; demon of, 202; prayer of, 373.
Soma, Haoma, ii, 418.
Soul, doctrine of, definition and general course in history. i. 428, 499; cause of life, 428; qualities as conceived by lower races, 428; conception of, related to dreams and visions, i. 429, ii. 24, 410; related to shadow, heart, blood, pupil of eye, breath, i. 430; plurality or division of, 434; exit of, i. 309, 438, etc., 448, ii. 50; restoration of, i. 436, 475; trance, ecstasy, 439; dreams, 440; visions, 445: soul not visible to all, 446; likeness to body, i. 450; mutilated with body, 451; voice, a whisper, chirp, etc., 452; material substance of soul, i. 453, ii. 198; ethereality not immateriality of, in lower culture, i. 456; human souls transmitted by funeral sacrifice to future life, i. 458, ii. 31; souls of animals, i. 467, ii. 41; their future life and transmission by funeral sacrifice, i. 469; souls of plants, trees, etc., i. 474, ii. 10; souls of objects, i. 476, ii. 9, 75, 153, etc.; transmission by funeral sacrifice, i. 481; conveyed or consumed in sacrifice to deities, ii. 216, 389; object-souls related to ideas, i. 497; existence of soul after death of body, i, 428, etc., ii. 1, etc.; transmigration or metempsychosis, ii. 2; new birth in human body, 3; in animal body, plant, inert object, 9, etc.; souls remain on earth among survivors, near dwelling, corpse, or tomb, i. 148, 447, ii. 25, etc., 150; souls called up by necromancer or medium, i. 143, 312, 446, ii. 136, etc ; food set out for, ii. 30, etc.; region of departed souls, ii. 59, etc., 73, 244; future life of, i. 458, etc., ii. 74, etc.; relation of soul to spirit in general, ii. 109; souls pass into demons, patron-spirits, deities, 111, 124, 192, 200, 364, 375; manes-worship, 112, etc.; souls embodied in men, animals, plants, objects, 147, 153, 192, 232; mystic meaning of word soul, 359.
Soul of world, ii. 335, etc., 355. 366.
Soul-mass cake, ii. 43.
Sound-words, i. 231.
Speaking machine, i. 170.
Spear-thrower, i. 66.
Species-deities, ii. 242.
Spencer, Mr. Herbert, i. vii., ii. 236.
Sphinx, i. 90.
Spirit:—course of meaning of word. i. 433, ii. 181, 206, 359; animism, doctrine of spirits, i. 424, ii. 108, 356; doctrine of spirit founded on that of soul, ii. 109; spirits connected and confounded with souls, ii. 109, 363; spirits seen in dreams and visions, i. 306, 440, ii. 154, 189, 194, 411; action of spirits, i. 125, ii. 111, etc.; embodiment of spirits, ii. 123; disease by attack of, 126; oracular inspiration by, 130; whistling, etc., voice of, i. 453, ii. 135; act through fetishes, ii. 143, etc.; through idols, 167; spirits causes of nature, 185, 204. etc., 250; good and evil spirits, 186, 319; spirits swarm in dark, fire drives off, 194; seen by animals, 196; footprints of, i. 455, ii. 197; ethereal-material substance of, ii. 198; exclusion, expulsion, exorcism of, 125, 199; patron, guardian, and familiar spirits, 199; nature-spirits of volcanos, whirlpools, rocks, etc., 207; water-spirits and deities, 209, 407; tree-spirits and deities, 215; spirits subordinate to great polytheistic deities, 248, etc.; spirits receive prayer, 363; sacrifice, 375; see Animism, etc.
Spirit, Great, ii. 256, 325, 335, etc., 353, 365, 395.
Spirit-footprints, i. 455, ii. 197.
Spiritualism, modern:—its origin in savage culture, i. 141, 155, 426, ii. 25, 39; spirit-rapping, i. 144, ii. 193, 221, 407; spirit-writing, 147; rising in air, 149; supernatural unbinding, 153; moving objects, etc., i. 439, ii. 156, 319, 441; mediums, i. 146, 312, ii, 132, 410; oracular possession, i. 148, ii. 135, 141.
Spirit-world, journey or visit to, by soul, i. 439, 481, ii. 44, etc.
Spitting, i. 103; lustration with spittle, ii. 439, 441.
Standing-stones, objects of worship, ii. 164.
Stanley, Dean, ii. 387.
Stars, myths of, i. 288, 356; souls of, i. 291.
Staunton, William, his visit to Purgatory, ii. 58.
Stock-and-stone-worship, ii. 161, etc., 254, 388.
Stone, myths of men turned to, i. 353; stone-worship, ii. 160, etc., 254, 388.
Stone Age, i. 56, etc.; magic as belonging to, 140; myths of giants and dwarfs as belonging to, 385.

INDEX. 469

Storm, myths of, i. 322; storm-god, i. 323, ii. 266.
Strut, i. 62.
Substitutes in sacrifice, i. 106, 463, ii. 399, etc.
Succubi, see Incubi.
Sucking cure, ii. 146.
Suicide, body of, staked down, ii. 29, 193.
Sun, myths of, i. 288, 319, 335, etc., ii. 48, 66, 323; sunset, myths of, connected with death and future life, i. 335, 345, ii. 48, etc., 311; sun abode of departed souls, ii. 69.
Sun-god and sun-worship, i. 99, 288, 353, ii. 263, 285, 323, etc, 376, etc., 408, 422, etc.; sun and moon as good and evil deity, ii. 324, etc.
Superlative, triple, i. 265.
Superstition, case of survival, i. 16, 72, etc.
Supreme deity, ii. 332, 367; heaven-god, etc., as, 255, 337, etc.; sun-god as, 290, 337. etc. ; conception of, in manes worship, 334 ; as chief of divine hierarchy, 335, etc.; first cause, 335.
Survival in culture, i. 16, etc., 70, etc., ii. 403; children's games, i. 72; games of chance, etc., 78; proverbs, 89; riddles, 91; sneezing-salutation, 98; foundation-sacrifice, 104; not save drowning, 108; magic, witchcraft, etc., 112; spiritualism, 141; numeration, 262, 271; deodand, 287; werewolves, 313; eclipse-monster, 330; animism, i. 500, ii. 356; funeral sacrifice, i. 463, 474, 492; feasts of dead, ii. 35, 41 ; possession, 140 ; fetishism, 159 ; stone-worship, 168 ; water-worship, 213 ; fire-worship, 285; sun-worship, 297; moon-worship, 302 ; heaven-worship, 353 ; sacrifice, 406, etc.
Susurrus necromanticus, i. 453, ii. 135.
Suttee, i. 465.
Swedenborg, spiritualism of, i. 144, 450, ii. 18, 204.
Symbolic connexion in magic, etc., i. 116, etc., ii. 144; symbolism in religious ceremony, ii. 362, etc.
Symplegades, i. 350.

Tabor, i. 209.
Tacitus, i. 333, ii. 228, 273.
Tailed men, i. 383.
Tangaroa, Taaroa, ii. 345.
Tari Pennu, ii, 271, 349, 368, 404.
Taronhiawagon, ii. 256, 309.
Tarots. i. 82.
Tartarus, ii. 97.

Tatar race, culture of, i. 51; race-genealogy of, 404.
Tattooing, mythic origin of, i. 393.
Taylor, Jeremy, on lots, i. 79.
Teeth-defacing, mythic origin of, i. 393.
Temple, Jewish, ii. 426.
Tertullian, i. 456, ii. 188, 427.
Tezcatlipoca, ii 197, 344, 391.
Theodorus, St., church of, ii. 121.
Theophrastus, ii. 165.
Theresa, St., her visions, ii. 415.
Thor, ii. 266.
Thought, conveyance of, by vocal tone, i. 166; Epicurean theory of, 497; savage conception of, ii. 311.
Thousand and One Nights:—waterspout and sand-pillar, i. 292 ; Magnetic Mountain, 374; Abdallah of Sea and Abdallah of Land, ii. 106.
Thunder-bird, myths of, i. 363, ii. 262 ; thunder-bolt, ii. 262.
Thunder-god, ii. 262, 305, 312, 337, etc.
Tien and Tu, ii. 257, 272, 352.
Tlaloc, Tlalocan, ii. 61, 274, 309.
Tobacco smoked as sacrifice or incense. ii. 287, 343, 383 ; to cause morbid vision, etc., 417.
Torngarsuk, ii. 340.
Tortoise, World-, i. 364.
Totem-ancestors, i. 402, ii. 235; totem-worship, ii. 235.
Traditions, credibility of, i. 275, 280, 370 ; of early culture, i. 39, 52.
Transformation-myths, i. 308, 377, ii. 10, 220.
Transmigration of souls, i. 379, 409, 469, 476, ii. 2, etc. ; theory of, ii. 16.
Trapezus. i. 396.
Trees, objects suspended to. ii. 150, 223.
Tree-souls, i. 475, ii. 10, 215 ; tree-spirits, i. 476, ii. 148, 215.
Tribe-names, mythic ancestors, i. 398; tribe-deities, ii. 234.
Tribes without religion, i. 417.
Tuckett, Mr., i. 373.
Tumuli, remains of funeral sacrifice in, i. 486.
Tupan, ii. 263, 305, 333.
Turks, race-genealogy of, i. 403.
Turnskins, i. 308. etc.
Twin brethren, N. A. dualistic myth, ii. 320, etc.
Two paths, allegory of, i. 409.

Uiracocha, ii. 338, 366.
Ukko, ii. 257, 261, 265.
Ulster, mythic etymology of, ii. 65.
Unbinding, supernatural. i. 153.
Under-world, sun and souls of dead descend to, ii. 66 ; see Hades.
Unkulunkulu, ii. 116, 313, 347.

Vampires, ii. 191.
Vapour-bath, narcotic, of Scyths and N. A. Indians, ii. 417.
Vasilissa the Beautiful, i, 342.
Vatnsdæla Saga, i. 439.
Veda, i. 55, 351, 362, 465, ii. 72, 265, 281, 354, 371, 386.
Vegetal, sensitive, and rational souls, i. 435.
Ventriloquism, i. 453, ii. 132, 182.
Vergil, Polydore, ii. 409.
Versipelles, i. 308, etc.
Vesta, ii. 285.
Vigesimal notation, i. 261; survival in French and English, 263.
Visions:—mythic fancy in, i. 305; are apparitions of spirits, 143, 445, 478, ii. 194, 410; as evidence of future life, 24, 49; fasting for, 410; use of drugs to cause, 416.
Visits to spirit-world, i. 436, 481, ii. 46, etc.
Vitruvius, on orientation, ii. 427.
Vocal tone, i. 166, etc.
Voice of ghosts and other spirits, whisper, twitter, murmur, i. 452, ii. 134.
Volcano, mouth of underworld, i. 344, 364, ii. 69; caused by spirits, 207.
Vowels, i. 168.
Vulcan, ii. 280, 284.

Wainamoinen, ii. 46, 93.
Waitz, Prof., Anthropologie der Naturvölker, i. vi.; fetishism, ii. 157, 176.
Walhalla, i. 491, ii. 77, 88.
War-god, ii. 306.
Warriors, fate of souls of, ii. 87.
Wassail, i. 97, 101.
Water, spirits not cross, i. 442.
Waterfalls and waterspouts, myths of, i. 292, 294.
Water-gods and water-worship, ii. 209, 274, 376, 407.
Water-spirits and water-monsters, i. 110, ii. 208, etc.
Watling Street, Milky Way, i. 360.
Weapons, i. 64, etc.; personal names given to, 303.
Wedgwood, Mr. Hensleigh, on imitative language, i. 161.

Weight of soul, i. 455; of spirit, ii. 198.
Well-worship, ii. 209, etc.
Werewolves, etc., doctrine of, i. 113, 308, etc. 435, ii. 193.
West, mythic conceptions of, as region of night and death, i. 337, 343, ii. 48, 61, 66, 311, etc., 422, etc.; see East and West.
Whately, Archbishop, on origin of culture, i. 38, 42.
Wheatstone, Sir C., i. 170.
Wheel-lock, i. 15.
Whirlpool, spirit of, ii. 207.
Widow-sacrifice, i. 458.
Wild Hunt, i. 362, ii. 269.
Wilson, Dr. D., on dual and plural, i. 265.
Wind-gods, ii. 266.
Winds, myths of, i. 360.
Witchcraft, i. 116, etc.; origin in savage culture, 138; mediæval revival, 138; iron charm against, 140; ordeal by water, 140; rising in air, 152; doctrine of werewolves, 312; incubi and succubi, ii. 190; witch-ointment, 418.
Woden, see Odin.
Wolf of night, i. 341.
Wong, ii. 176, 205, 348.
World pervaded by spirits, ii. 137, 180, 185, 205, 250.
Worship as related to belief, i. 427, ii. 362.
Wraith or fetch, i. 448, 451.
Wright, Mr. T., ii. 56, 65.
Wuttke, Prof., i. 456, etc.

Xerxes, i. 286, ii. 378.

Yama, ii. 54, 314.
Yawning, possession, i. 102.
Yezidism, ii. 329.

Zend-Avesta, i. 116, 351, ii. 98, 293, 328, 438.
Zeus, i. 328, 350, ii. 258, etc., 353.
Zingani, myth of name, i. 409.
Zoroastrism, ii. 20, 98, 282, 319, 323, 354, 374, 400, 438.

THE END.